Marine Biology, Ecology and Zoology

Marine Biology, Ecology and Zoology

Editor: Agnes Wilson

Callisto Reference

www.callistoreference.com

Callisto Reference,
118-35 Queens Blvd., Suite 400,
Forest Hills, NY 11375, USA

Visit us on the World Wide Web at:
www.callistoreference.com

ISBN: 978-1-64116-059-9 (Hardback)

Cataloging-in-Publication Data

Marine biology, ecology and zoology / edited by Agnes Wilson.
p. cm.
Includes bibliographical references and index.
ISBN 978-1-64116-059-9
1. Marine biology. 2. Marine ecology. 3. Zoology. 4. Ecology. I. Wilson, Agnes.
QH91 .M37 2019
578.77--dc23

Table of Contents

Preface

In my initial years as a student, I used to run to the library at every possible instance to grab a book and learn something new. Books were my primary source of knowledge and I would not have come such a long way without all that I learnt from them. Thus, when I was approached to edit this book; I became understandably nostalgic. It was an absolute honor to be considered worthy of guiding the current generation as well as those to come. I put all my knowledge and hard work into making this book most beneficial for its readers.

Marine biology, zoology and ecology are sub-disciplines of biology. Marine biology is the study of marine life on Earth and strives to categorize organisms as per their functions. Marine zoology, on the other hand, is the science concerned with the study of all aspects of the marine animal kingdom and the interactions between the species within this kingdom. The study of the relations of marine organisms with their surroundings is called marine ecology. The aim of this book is to provide information about the theories and techniques central to these fields while also shedding light on the theoretical advancements achieved in the scientific research. It also delves into upcoming concepts and theories that are associated with these three disciplines in comprehensive detail. This book, with its in-depth analyses and data, will prove immensely beneficial to professionals and students involved in this area at various levels.

I wish to thank my publisher for supporting me at every step. I would also like to thank all the authors who have contributed their researches in this book. I hope this book will be a valuable contribution to the progress of the field.

Editor

1

Structure, function and cell dynamics during chaetogenesis of abdominal uncini in *Sabellaria alveolata* (Sabellariidae, Annelida)

Ekin Tilic* and Thomas Bartolomaeus

Abstract

Background: Dynamic apical microvilli of a single cell, called the chaetoblast, inside an ectodermal invagination form the template of annelid chaetae. Changes in the pattern of microvilli are frozen in time by release of chitin, such that the structure of the definitive chaeta reflects its formation. Cellular interactions during chaetogenesis also influence the structure of the chaeta. Analysing chaetogenesis allows for testing hypotheses on the homology of certain chaetal types. We used this approach to test whether the unusual uncini in *Sabellaria alveolata* are homologous to apparently similar uncini in other annelid taxa.

Results: Our study reveals unexpected details of sabellariid uncini, which mechanically reinforce the neuropodia enabling their use as paddles. The final structure of the chaeta is caused by pulses of microvilli formation and dynamic interaction between the chaetoblast and adjoining follicle cells. Cell dynamics during chaetogenesis of the uncini in *Sabellaria alveolata* exceeds by far that reported in previous studies on the formation of this type of chaetae.

Conclusion: Despite the superficial similarity of uncini in sabellariids and other annelids, differences in structure and details of formation do not support the homology of this type of chaetae. Chaetogenesis of sabellariid uncini involves unexpected microvilli and cell dynamics, and provides evidence that interactions between cells play a larger role in chaetogenesis than previously expected. In addition to their function as anchors, uncini in Sabellaridae stabilize the paddle-shaped notopodia, as each uncinus possesses a long, thin rod that extends deeply into the notopodium. The rods of all uncini in a single row form a bundle inside the notopodium that additionally serves as a muscle attachment site and thus have a similar function to the inner chaeta (acicula) of errant polychaetes (Aciculata).

Keywords: TEM, cLSM, Chaetogenesis, Hooked chaetae, Polychaetes, Sedentaria, Functional morphology

Background

Chaetae are chitinous extracellular structures that are important diagnostic characters in Annelida [1, 2]. Chaetae are formed within an ectodermal invagination, the chaetal follicle, which consists of a terminal chaetoblast and a few follicle cells [3–5]. Each chaetoblast has an array of apical microvilli that are modified in time and space while chitin polymerizes alongside the microvilli. Controlled modification of the microvilli pattern, thus, gives the chaeta its final shape [4, 6–8]. Alterations in spatiotemporal patterns allow the formation of a plethora of different chaetal types that can range from highly complex compound hooked chaetae to simple capillaries. Chaetogenesis involves such an elaborate interplay of cellular instruments that genetic programming and regulation is necessary to establish consistent chaetal arrangement and structure within annelid species and supraspecific taxa [9, 10]. Given that genetic information underlies chaetogenesis, we assume that any hypothesis on the homology of chaetae can be tested. Identical formation processes are expected for homologous and structurally similar chaetae.

* Correspondence: etilic@evolution.uni-bonn.de
Institute of Evolutionary Biology and Ecology, Rheinische Friedrich Wilhelms Universität Bonn, An der Immenburg 1, 53121 Bonn, Germany

Hooked chaetae and uncini possess several small apical teeth giving the chaetae a saw- or rasp-shaped appearance when viewed from above. These teeth may or may not surmount a single large tooth. Small apical teeth and the main tooth, if present, are curved relative to the shaft which represents the main axis of the chaeta. Uncini and hooked chaetae are discriminated by the length of the shaft, although its length is an imprecise character that varies intra- and supraspecifically [11]. Studies into the chaetogenesis of the hooked chaetae and uncini in certain sedentary polychaetes has revealed that the structure of these chaetae actually results from a uniform formation process (Sabellidae and Serpulidae [11–13]; Arenicolidae [14, 15]; Maldanidae [10]; Psammodrilida [16], Terebellida [12, 17]; Oweniidae [18]; Siboglinidae [19]). One of the major conclusions of these studies is that substructures and course of formation support the homology hypothesis for hooked chaetae and uncini, at least for the taxa studied [20]. Sabellariidae, which possess uncini that are aligned in a transverse row at the outer rim of the abdominal notopodia, however, have not been included in such comparative studies to date. At least on the level of light microscopy, sabellariid uncini do not seem to differ from uncini of the other taxa studied so far.

In the present study, we have investigated the ultrastructure and chaetogenesis of abdominal uncini in *Sabellaria alveolata* (Sabellariidae). Assuming all uncini are homologous, one would expect significant similarities in chaetal ultastructure and formation. Although for epistemological reasons it is impossible to prove non-homology, recognizable differences in mode of chaetogenesis would not support the homology of sabellaridan uncini to those of other hemisessile and sessile annelids with hooked chaetae, but would rather allow alternative hypotheses for the position of Sabellariidae.

Fig. 1 **a** Collection site of *Sabellaria alveolata*, showing the distinctive tubes under a rock crevice. **b** Habitus of *Sabellaria alveolata*, showing the different body regions

Material and methods

Animals

Sabellaria alveolata (Linnaeus, 1767) (Fig. 1) was collected in March 2013 in the rocky intertidal of Concarneau (Brittany, France). Here, *S. alveolata* occurs in dense colonies in sheltered rock crevices, building distinctive hard tubes from the sediment (Fig. 1a). The tubes were removed from the rocks with the help of a spatula and the animals were fixed in the field immediately after being removed from their tube.

Light microcopy (LM), histology and 3D reconstruction

The specimens of *Sabellaria alveolata* used for the serial semi-thin sections and the 3D reconstruction was fixed in 1.25 % glutaraldehyde buffered in 0.05 M phosphate buffer with 0.3 M NaCl for 1.5–2 hours. The fixed animals were stored in the same buffer until they were postfixed in 1 % OsO_4 for 45 min. The specimens were dehydrated in an acetone series right after the postfixation, transferred in propylene oxide and embedded in araldite. If necessary, specimens were sectioned into smaller pieces within the resin. Polymerization was started with BDMA (Benzyldimethylamine). A series of one micrometer sections were cut with a diamond knife (Diatome Histo Jumbo) on a Leica Ultracut S ultramicrotome, following the method described by Blumer et al. [21]. The sections were stained with toluidine blue (1 % toluidine, 1 % sodium-tetraborate and 20 % saccharose) and covered with a cover slip mounted with araldite. The semi-thin sections were analyzed with an Olympus microscope (BX-51) and photographed with an Olympus camera (Olympus cc12), equipped with the dot slide system (2.2 Olympus, Hamburg). Images were aligned using IMOD (Boulder Laboratories [22]) and IMOD-align (http://www.evolution.uni-bonn.de/mitarbeiter/bquast/software).

3D modelling of the chaetae was performed using the software 3ds max 13. Histological images were imported as surface materials (discreet) and the chaetae were modeled using standard cylindrical objects. When necessary, these were modified as NURBS

(Nonuniform rational B-Splines)-surfaces. The outline of the neuropodial torus was created using another NURBS surface.

Using the same method a second 3D model was constructed with the aligned TEM-images of the formative site. Here all of the studied developmental stages were modeled in order to visualize their topological position within the formative site.

Single chaetae analyzed using a confocal laser scanning microscope and Nomarsky differential interference contrast under an Olympus BX-51 microscope were isolated from pieces of PFA (1 h in 4 % paraformaldehyde) fixed specimens of *Sabellaria alveolata* by incubation in 5 % NaOH for 4–5 h. The chaetae were rinsed in distilled water, mounted on microscopical slides and examined.

Confocal laser scanning microscopy (CLSM)

The specimens used for confocal laser scanning microscopy were fixed in 4 % paraformaldehyde for 1 h and later stored in 0.1 M PBS (phosphate buffered with saline) containing 0.01 % NaN_3. The chaetigers were dissected to separate single parapodia. Isolated parapodia and segments were permeabilized in four 5-min changes of PBS with 0.1 % Triton X-100 (Fisher Scientific). The samples were then stained overnight in 4 °C with TRITC phalloidin at a dilution of 1:100. After staining, parapodia were rinsed in three quick changes and subsequently in two 10-min changes of PBS with 0.1 % Triton and one 10 min rinse in PBS without Triton. Stained and rinsed samples were dehydrated in isopropanol (2 min 70 %, 2 min 85 %, 2 min 95 %, 2 min 100 %, 2 min 100 %) and cleared in three 15-min changes of Murray Clear. The samples were placed in hollow-ground slides, mounted in Murray Clear, and sealed with nail polish.

The upper layers of musculature were partially removed from the confocal z stack, digitally, using Photoshop CS6 to allow viewing the chaetae within the torus. The entire CLSM image stack is available for download (link provided under data repository).

Electron microscopy (TEM, SEM)

Specimens used for transmission electron microscopy were fixed using the same fixation method described above for semi-thin sectioning (1.25 % glutaraldehyde buffered in 0.05 M phosphate buffer with 0.3 M NaCl for 1.5–2 h, postfixation with 1 % OsO_4 for 45 min). These specimens were also embedded in araldite and sectioned into a complete series of silver-interference coloured (70–75 nm) sections using a diamond knife (Diatome Histo Jumbo) on a Leica Ultracut S ultramicrotome. The serial section ribbons were placed on Formvar-covered, single-slot copper grids and stained with uranyl acetate and lead citrate in an automated TEM stainer (QG-3100, Boeckeler Instruments). The sections were examined using a Zeiss Libra 120 kV transmission electron microscope.

The chaetal formation was reconstructed using the information gathered from serial ultrathin sections and series of semi-thin sections of *S. alveolata*. The coverage of different stages of chaetogenesis was, with 14 consecutive developmental stages, dense enough to allow insights into the dynamics of the entire process that will be described in the following. The entire aligned stacks of ultra-thin and semi-thin sections are available for download (links provided under data repository).

For scanning electron microscopy (SEM) *Sabellaria alveolata was* fixed in Bouin's fluid, dehydrated in an alcohol series and dried with CO_2 in a critical point dryer (BALZERS). After dehydration the samples were sputtered with gold (BALZERS Sputter coater) and examined with a XL 30 SFEG (Philips Electron Optics) scanning electron microscope. During dehydration the animals were sonicated to remove debris and sand particles from the chaetae.

Results

Parapodial structure and chaetal arrangement

The body of *Sabellaria alveolata* is divided into four regions that are characteristic for Sabellariidae; the thorax, parathorax, abdomen, and the cauda (Fig. 1b). Chaetal elements in the thorax and parathorax comprise of opercular paleae, oar-shaped notochaetae and capillary chaetae. The abdomen of *S. alveolata* forms the largest part of the animal's body and bears segmental biramous parapodia with notopodial uncini and neuropodial capillaries. The cauda has the appearance of an unsegmented tube and is achaetous. Aciculae are absent in all segments (Fig. 1b).

The abdominal notopodia are paddle-like appendages on either side of the animal's body (Figs. 1b and 2). Those of the first few abdominal segments are broad and large, towards the posterior end they become narrower and elongate. Paired dorsal branchiae appear on the parathoracic segments and in the first 15–20 abdominal segments. They become gradually smaller along the antero-posterior axis and disappear completely in the posterior segments of the abdomen. The uncini are located at the apical margin, where they are aligned to form a single transverse row. Each chaeta arises from a chaetal follicle and all follicles are aligned within a single chaetal sac without being separated by an extracellular matrix.

Small, needle-shaped rods originate from the rostral and adrostal portion of each uncinus and extend into the notopodium. Apically these rods are aligned in a row, but as they reach deeper into the notopodium they form a bundle (Figs. 2, 3a). Each rod is surrounded by a follicle cell. The follicle cells of all rods comprise the

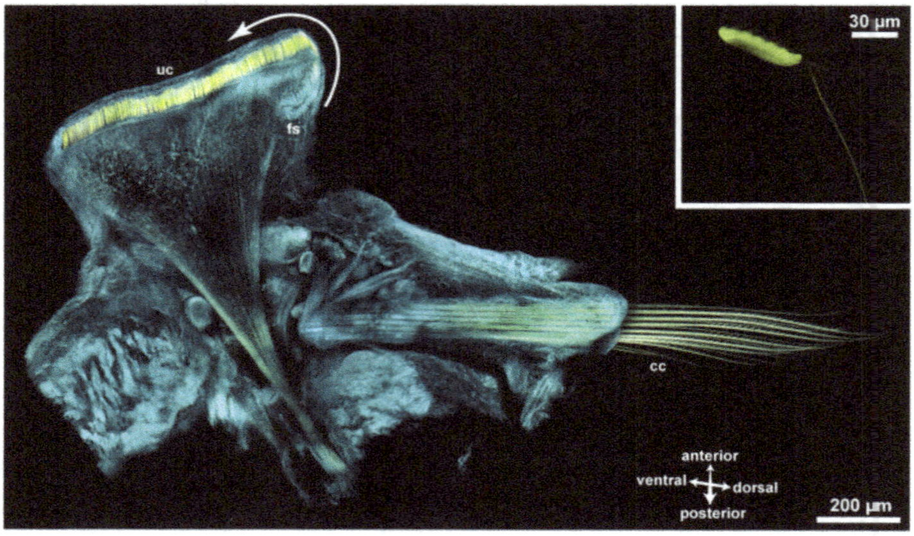

Fig. 2 Confocal *z*-projection of a phalloidin stained preparation of a single abdominal parapodium of *Sabellaria alveolata*. *cyan* phalloidin, *yellow* chaetal autofluorescence *uc* uncini, *fs* formative site, *arrow* marks the direction of chaetal development *cc* capillary chaetae, *inset* detail image of an isolated uncinus

inner end of the chaetal sac and rest on a common extracellular matrix (ECM). Follicle cells and ECM connect the bundle of rods to the parapodial musculature (see cLSM stack) in such a way that only the entire bundle can be moved, but not an individual chaeta. The formative site of the uncini is located at the ventral edge of the chaetal row and contains numerous developing chaetae (Figs. 2; 3a, b; 5a), so that chaetogenesis could be inferred in detail from an ultrastructural analysis of a series of different stages.

The neuropodia of the abdomen only possess capillary chaetae that are either simple or pinnate (Figs. 2, 3d). Neuropodial capillary chaetae are long and also reach deeply into the parapodium. Their overall position is right-angled to the bundle of rods of the notopodial hooks (Fig. 2). The basis of the neuropodial chaetal sac is connected with a network of radial chaetal muscles to the outer body wall, giving the chaetae the characteristic arrangement similar to an arrow pulled back in a bow (Fig. 2). Upon contraction, these muscles shorten and push the chaetae out of the body surface.

Structure of the uncini

Uncini in *Sabellaria alveolata* have a complex structure. The apical portion of a chaeta consists of a single median tooth followed by 5 to 6 pairs of teeth (Fig. 4d, e). The single small median tooth, called rostrum here, marks the rostral face of the uncinus; the size of the paired teeth decreases along a rostral-adrostral gradient, such that the adrostral pair of teeth is smaller than the rostral ones. All teeth and the rostrum originate from a blade-like shaft toward which they bend by 40°. Light microscopy shows that the shaft is composed of two different parts, separated by a fine rostro-adrostral refracting seam (Fig. 4d, e). The portion above this refracting seam directly underlies the teeth. It is small and dense, and we refer to this structure as the "base" hereunder. The portion of the shaft below the refracting line is keel-shaped and bright, and will be referred to as the "socket". This socket has a length of ± 45 µm from rostral to adrostral. Under Nomarsky contrast, small, densely-packed vertical lines that originate in the teeth proceed into the base to end at the refracting seam. In the socket several lines can be seen running longitudinally and almost parallel to the refracting line. The main axes of both, vertical and longitudinal lines, form an angle of ± 40° (Fig. 4d). The teeth, the small underlying base and a tiny portion of the rostral rod are the only externally visible structures in SEM preparations (Fig. 4a, b). As mentioned above, each uncinus possesses two vertical rods, a shorter adrostral one and a bipartite rostral one. Soon after its origin the rostral rod splits into a short anterior and a long posterior rod. While the shorter (anterior) rostral rod is almost as long as the adrostral rod, the longer (posterior) rostral rod extends up to 1.5 mm deep into the notopodium. The posterior rostral rod is nearly 80 times longer than the entire apical portion (shaft plus teeth; ±20 µm) (Figs. 2, 4e). All anterior rods of the notopodial uncini form the above described intranotopodial fiber bundle that serves as the attachment site of notopodial muscles. Both rostral rods have a similar diameter (±1.5 µm) to that of the adrostral rod.

Chaetogenesis

Chaetogenesis occurs continuously within the formative site and the TEM study of fixed material allows inferring the entire process of chaetal formation from different developmental stages within a single formative site of the chaetal sac (Figs. 5, 6, 7 and 8). Uncini are formed

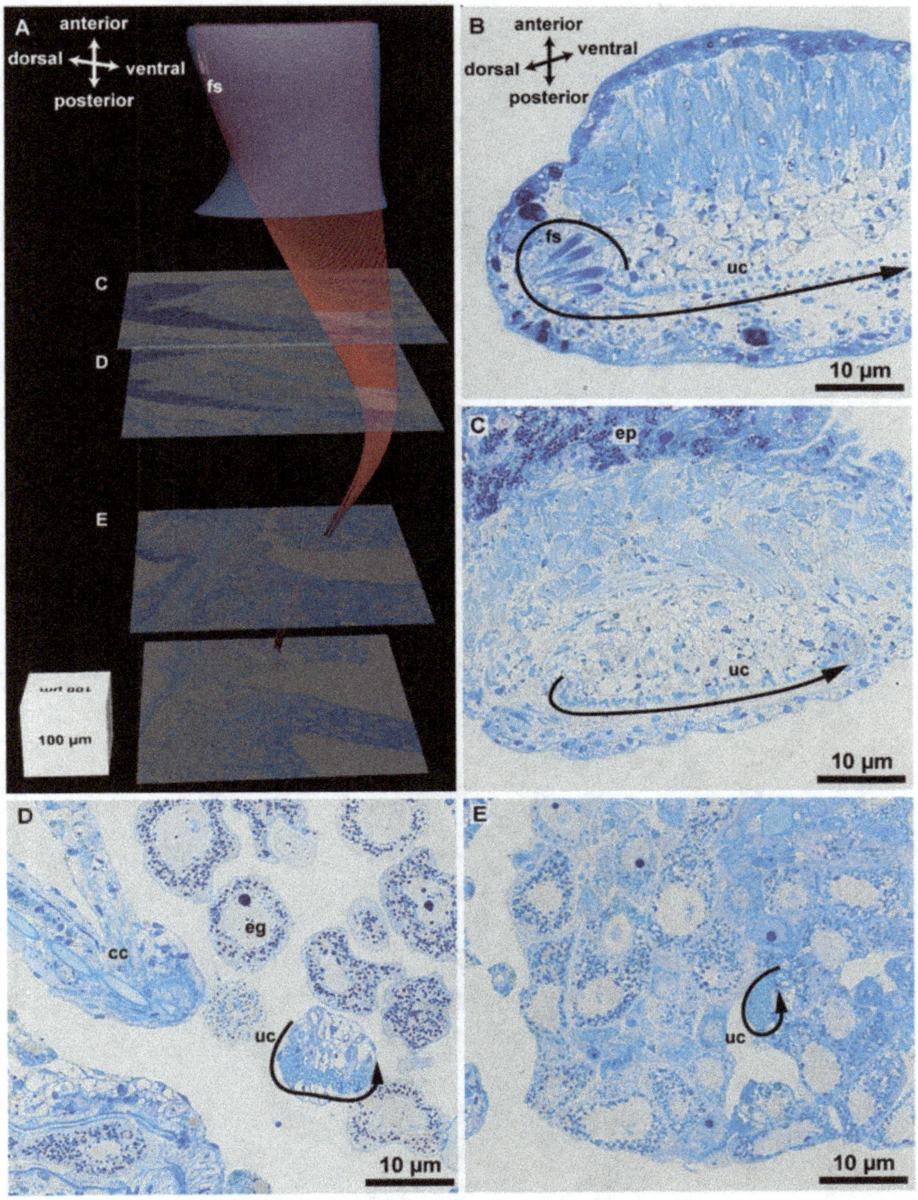

Fig. 3 a 3D model of the chaetal arrangement inside an abdominal torus. **b–e** Aligned semi-thin sections used to construct the 3D model. Corresponding section planes are marked in A. The *arrow* indicates the direction of chaetal formation. *fs* formative site, *uc* uncini, *cc* capillary chaetae

within an ectodermal invagination (chaetal follicle) consisting of the chaetoblast and at least five follicle cells. All cells are epithelial, interconnected by adluminal adherens junctions (belt desmosomes) and septate junctions, and rest on a common matrix that surrounds the chaetal sac. All cells surround a small compartment, the chaetal compartment, and bear several short microvilli that reach into the compartment. This compartment narrows to become a small canal that extends towards the epidermis where it opens to the exterior by a small pore. During chaetogenesis the chaeta is secreted into the chaetal compartment. The basalmost four cells are actively involved in chaetogenesis, i.e. the chaetoblast at the base of the chaetal follicle and three adjacent follicle cells. The fourth and fifth follicle cells form a ring that surrounds the chaetal compartment and the proximal section of the canal. Each of these cells possesses a subapically located diplosome (Fig. 6a, b). In young follicles one of the two diplosomes may contact the apical cell membrane, but this never induces a cilium (Fig. 6a). The microvilli of the chaetoblast and the first two follicle cells are set more densely and are longer than those of the remaining follicle cells; the microvilli of the chaetoblast are slightly larger in diameter than those of the

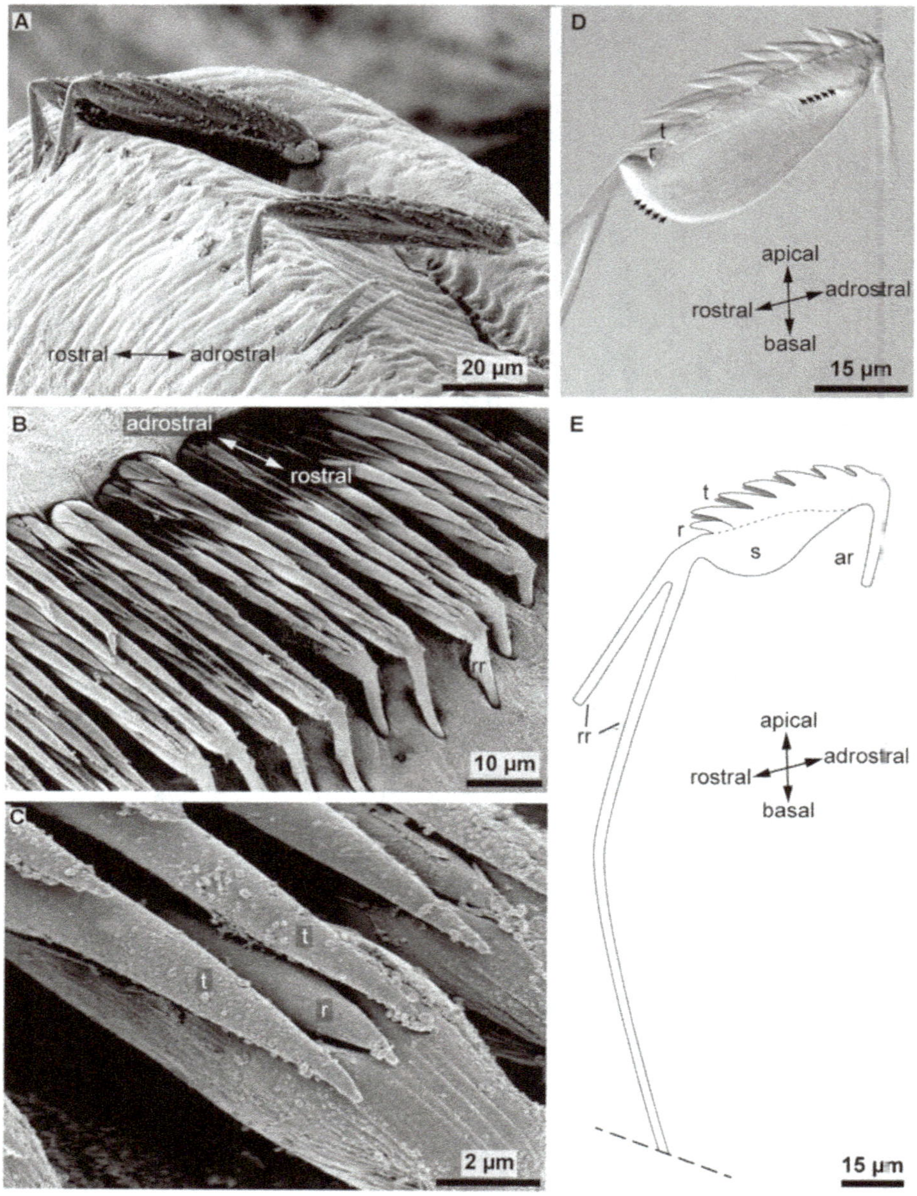

Fig. 4 a. SEM image of detached abdominal uncini. **b** SEM image of the row of uncini. **c** SEM image showing the rostral portion of an uncinus in detail. **d** Micrograph showing the apical portion of an abdominal hooked chaeta. *arrows* mark the direction of the internal canals. **e** Schematic drawing of an abdominal uncinus, in scale. *dotted line* illustrates the refracting seam at merger of the chaetal socket and base, *rr* rostral rod, *ar* adrostral rod, *r* rostrum, *t* tooth, *s* socket

follicle cells (Fig. 6b). The latter form the template of each substructure of the chaeta. Continuous polymerization of chitin between the bases of the microvilli enlarges the developing chaeta. Given that the microvilli have a constant length, sooner or later the developing chaeta will exceed the microvilli in length and electron-lucent canals will remain inside the chaeta where microvilli had once been. These canals may be filled up secondarily by electron-dense material.

In this study, 14 developmental stages of uncini were found in a single formative site that was cut into a series of ultrathin sections, analysed for ultrastructural details and reconstructed. Nine stages are shown in Fig. 9 and the topological position of these stages within the formative site can be seen in Fig. 5a. Chaetogenesis of uncini in *Sabellaria alveolata* can be divided into three steps: (1) formation of the rostrum, teeth and base (Fig. 9a–c), (2) formation of the socket (Fig. 9d–f), (3) formation of the rostral and adrostral rods (Fig. 9f–j).

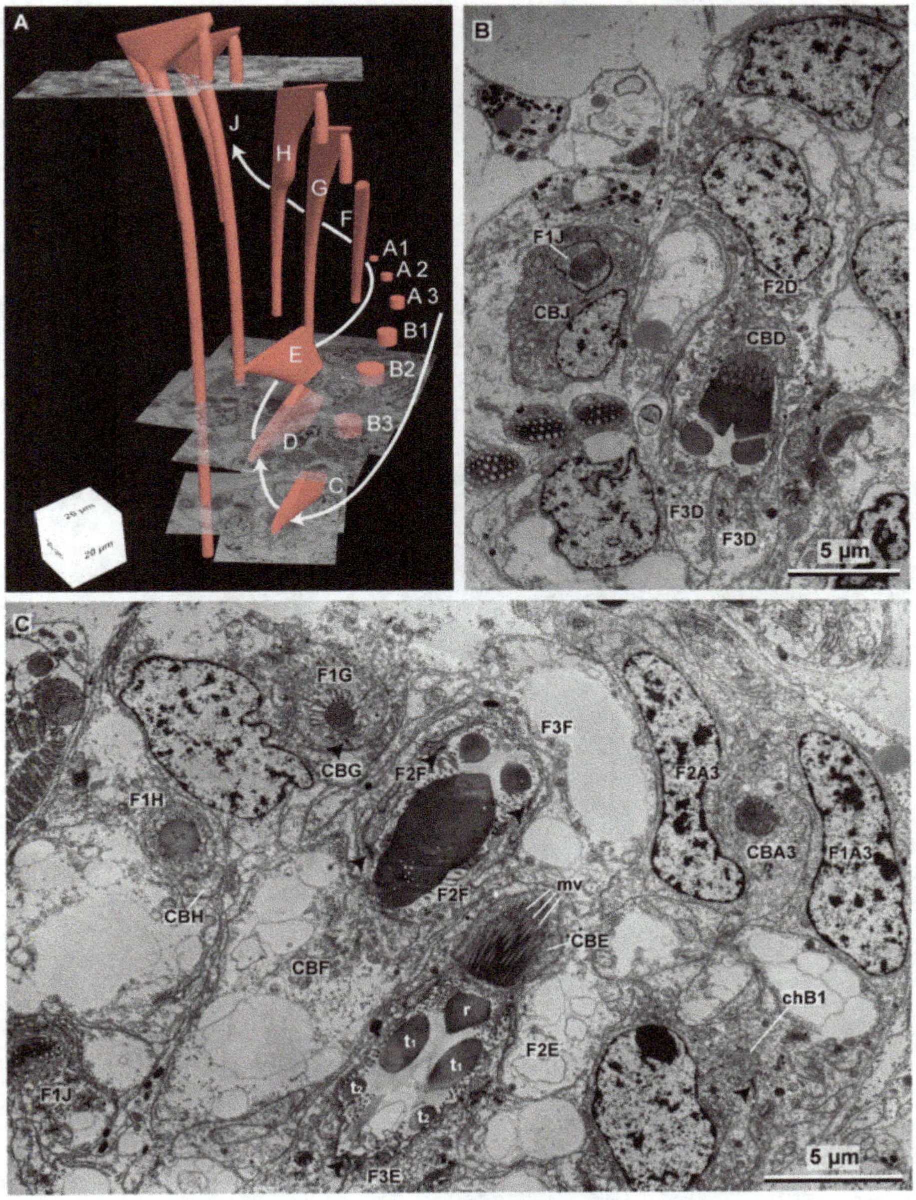

Fig. 5 **a** 3D model of the chaetal formative site, reconstructed using the aligned serial ultra-thin sections. Consequent developmental stages of uncini are labeled from A1–J. This numbering is employed all through the images when referring to these specific developmental stages. **b** TEM image of the formative site showing the formation of the long rostral rod in J and the formation of the socket in D. **c** TEM image of the formative site showing the formation of the adrostral rod in H and G, the formation of the socket in E and F, and the formation of the rostrum in A3. Note the chaetal canal of B1 (*chB1*) in the lower right. *F1–F3* follicle cells, *CB* chaetoblast, *r* rostrum, *t* tooth, *arrow heads* mark the adluminal adherens junctions

Formation of rostrum, teeth and base

Chaetogenesis begins when a small cluster of microvilli emerge on the surface of the chaetoblast (Figs. 6a–c, 9a). These microvilli form the template of the anteriormost tooth, the rostrum, and extend into the chaetal compartment. Chitin polymerizes between the bases of the microvilli and forms the tip of the rostrum. Additional microvilli that appear peripheral to the initial cluster broaden the rostrum. Subsequently, two additional clusters of microvilli are formed adrostrally on either side of the developing rostrum (Fig. 6d, e). These are the template of the first pair of teeth. In the same manner five additional pairs of teeth are subsequently added along a rostro-adrostral gradient, so that finally the sixth pair of teeth is situated adrostrally (Figs. 5c, 9b-c). All teeth

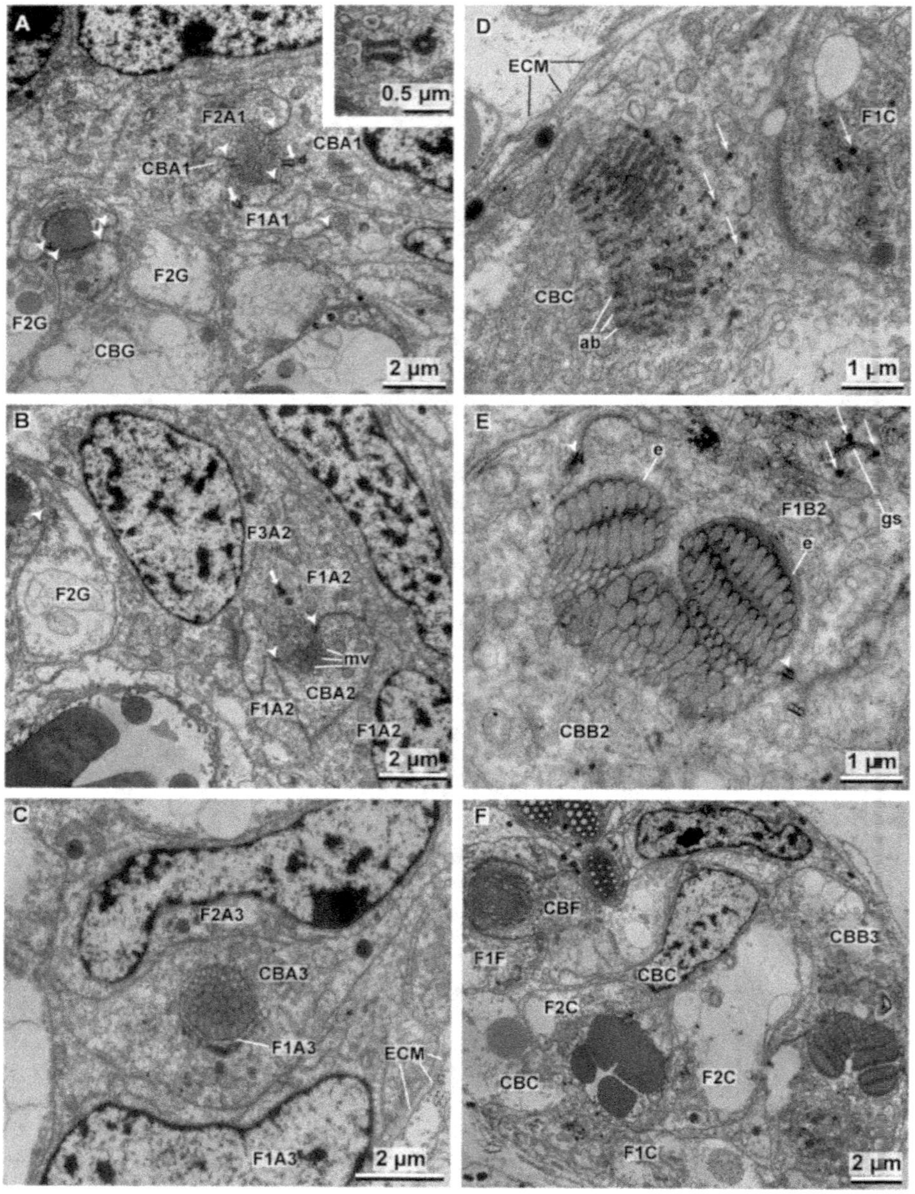

Fig. 6 a–c TEM images of A1–3 showing the initial stage of chaetogenesis and the formation of a rostrum. *Inset* high magnification of a diplosome **d** Production of chaetal material and the subsequent transportation to the chaetal *anlage* via vesicles. **e** Formation of the adrostral teeth in B2. **f** TEM image of the formative site showing the formation of teeth in B3 with multiple rows of microvilli, older teeth in C with almost completely filled canals and the adrostral portion of the chaeta in F. *F1–F3* follicle cells, *CB* chaetoblast, *arrow heads* mark the adluminal adherens junctions, *short arrows* mark centrioles, *long arrows* mark vesicles containing electron-dense chaetal material, *ECM* extra-cellular matrix, *ab* actin bundles, *mv* microvilli, *e* enamel, *gs* golgi stack

and the unpaired rostrum have nearly the same size. Their templates were all once formed by 2–3 rows, each consisting of 9–12 microvilli. Since the number of microvilli increased towards the base of the teeth, all microvilli finally form a broad and uniform field, which is the template for the base underlying the teeth. Electron-dense material released from vesicles of the first two follicle cells forms an enamel that covers and smoothens the irregular surface of the teeth (Fig. 6e). This material is produced inside Golgi stacks and transported to the chaetal surface in vesicles (Fig. 6d, e). While more rows of teeth are added and the developing chaeta enlarges, the canals left by the templating microvilli, become more or less completely filled with electron dense material (Fig. 6f). At the end of this first step of chaetogenesis the rostrum, four pairs of teeth,

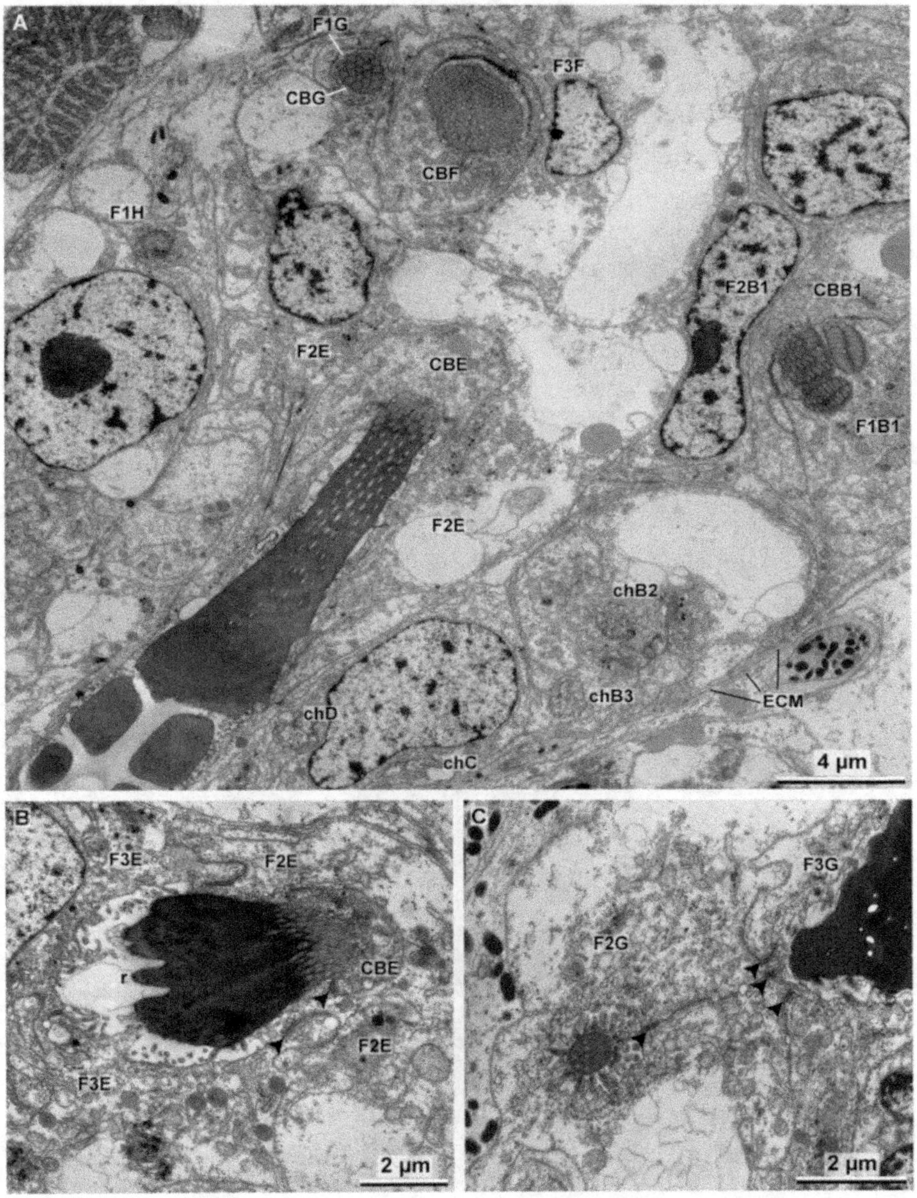

Fig. 7 **a** TEM image of the formative site showing the formation of the short rostral rod in G, the formation of the subrostral portion of the socket in F, the formation of the socket in E and the formation of teeth in B1. Note the canals (*chB2–D*) that connect inferior developmental stages to the outer surface. **b** Formation of the rostral part of the socket in E. **c** Adrostral rod of G surmounted by F2. *F1–F3* follicle cells, *CB* chaetoblast, *arrow heads* mark the adluminal adherens junctions, *ECM* extra-cellular matrix, *r* rostrum

and the anterior part of the base are formed. The microvilli are completely retracted from the rostral three quarters of the developing chaeta; the canals the microvilli left, are refilled with electron-dense deposits. The chaetoblast merely underlies the adrostral half of the developing chaeta, whereas the rostral half is underlain by the planar apical cell membrane of the second follicle cell. Chaetogenesis is interrupted in this region. The entire *anlage* is oriented vertically within the chaetal compartment.

Formation of the socket

Once all teeth and the base are formed, the chaetoblast grows towards the rostrum again to underlie the entire base and slightly exceed it rostrally. The chaetoblast then forms microvilli that form a homogeneous field These microvilli have a vertical orientation and thus are more or less longitudinal relative to the *anlage*. They form the template of the socket and chitin polymerizing between the microvilli is added to the base of the *anlage*. The longitudinal refracting seam visible under Nomarsky

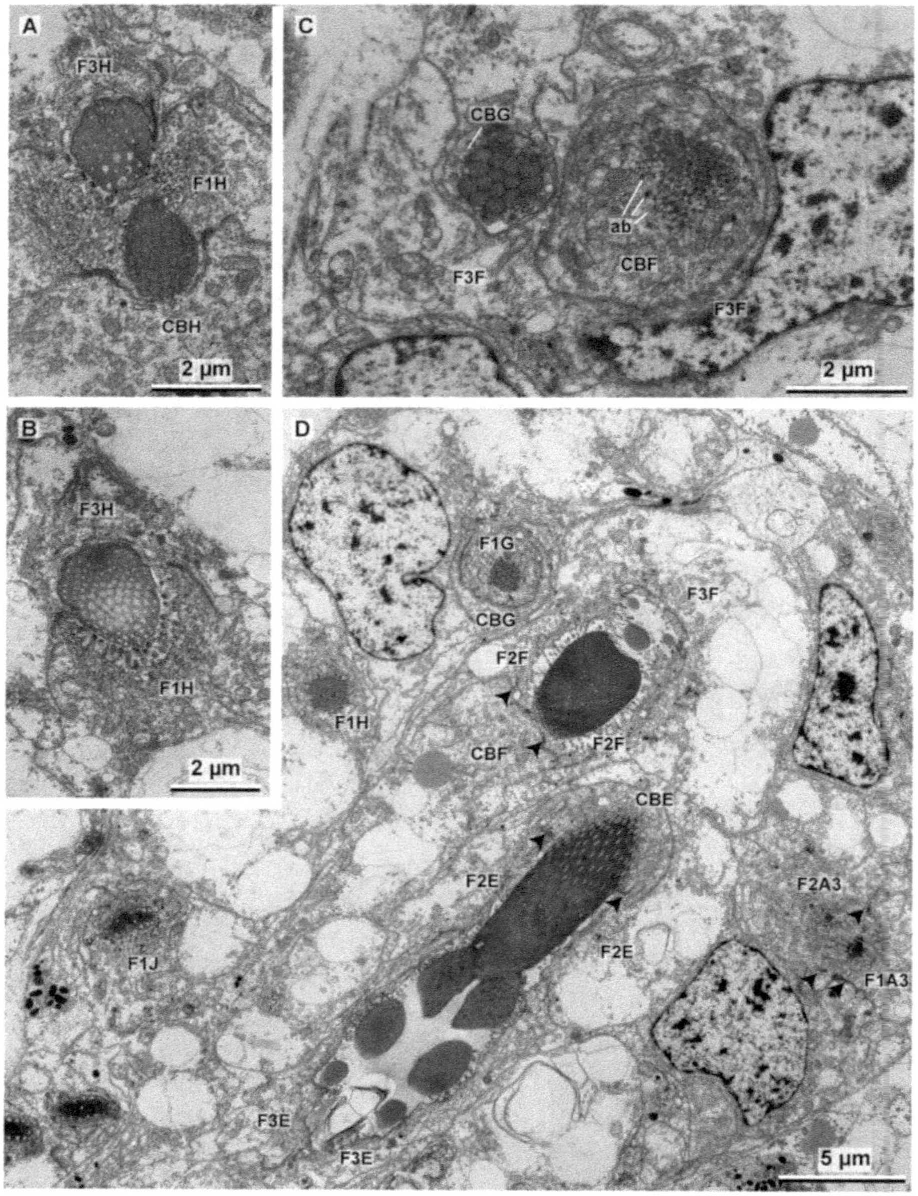

Fig. 8 a–b TEM images showing the merger of the long and short rostral rod in H. Note the newly developing long rod and the fully differentiated short rod in A. **c** Formation of the short rostral rod in G and the formation of the subrostral process in F, note the bundles of actin filaments that are located under the microvilli. **d** TEM image of the formative site showing the fully differentiated long rostral rod in _ developing long rod of H, formation of the short rostral rod in G, formation of the socket in F–E, and the tip of the rostrums in A3. *F1–F3* follicle cells, *CB* chaetoblast, *arrow heads* mark the adluminal adherens junctions, *ab* actin bundles

contrast between base and socket results from the break in chitin polymerization after teeth and base were formed (Figs. 5c, 7a,b, 8d, 9d-f). Re-orientation of the microvilli is also clearly visible under Nomarsky contrast in fully differentiated chaeta as longitudinally arranged lines inside the socket. These lines are actually canals left by microvilli inside the socket during formation (Fig. 4d, e). A group of microvilli remains at the apico-adrostral part of the chaeta, while those in the subapical part disappear. Bare cell membrane of the chaetoblast underlies this portion and no chitin is formed (Fig. 6f, 9e, f). At this time the entire *anlage* begins to alter its position within the chaetal sac again. Since the microvilli are always vertically oriented they also alter their position relative to the developing chaeta. The subapical group of microvilli forms an adrostral cap while the socket increases in size. Finally the microvilli that formed the template of

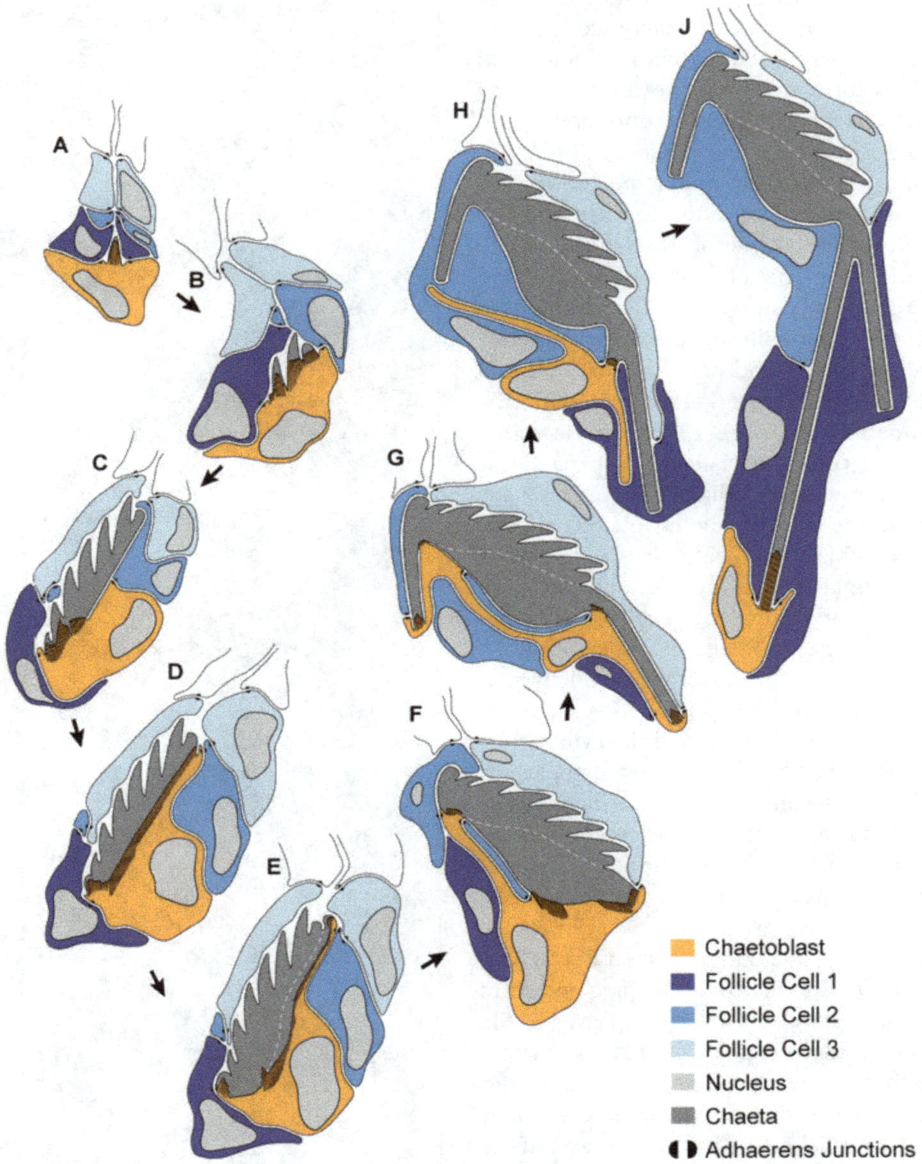

Fig. 9 Schematic illustration of chaetogenesis and the interaction between the chaetoblast and the follicle cells as a series of sagittal sections of subsequent representative stages of the chaetal formation. Topological position of corresponding development stages are marked in the 3D model in Fig. 5. **a** Earliest stage of chaetogenesis; formation of the rostrum. **b**–**c** formation of the teeth. **d**–**e** formation of the chaetal socket. **f**–**g** Formation of the adrostral rod and the short rostral rod. **h**–**j** Final step of chaetogenesis; formation of the long rostral rod

the adrostral portion of the socket retract and disappear, except for those microvilli that formed the adrostral cap. The same occurs rostrally, here leaving a large apico-rostral group of microvilli (Fig. 9f). Adrostrally, the second follicle cell expands into the gap between the developing chaeta and the chaetoblast, so that this part of the developing chaeta is now underlain by the apical cell membrane of the second follicle cell. The subrostral portion of the socket is underlain by the apical cell membrane of the chaetoblast. After the socket has been completed two groups of microvilli remain, a rostral and an adrostral one. The entire *anlage* now has a horizontal position within the chaetal compartment (Fig. 9f).

Formation of the rostral and adrostral rods

After the socket is completed the microvilli of the chaetoblast are almost completely reduced, except for a rostral and an adrostral group (Fig. 9f). The adrostral group of microvilli elongates and forms the template for the adrostral rod. Chitin polymerization happens rapidly and the adrostral rod elongates, parallel to the apico-basal axis of the uncinus (Fig. 9g). The rostral group of microvilli actually consists of two adjacent, but perpendicular patches of microvilli (Figs. 7a, 8c). They form the template for the rostral rod, which initially is rather massive and oblique to the rostro-adrostral axis of the chaeta. Later the microvilli split into an anterior and a posterior one. The microvilli of the anterior group elongate and become the template for the anterior rostral rod, while the posterior group consists of short microvilli and remains in its original position. Chitin polymerizes rapidly between the microvilli to form the anterior rostral rod. Anterior rostral rod and the adrostral rod are formed simultaneously; one keeps pace with the other during formation (Fig. 9g). During these initial steps of forming the rods the chaetoblast expands as the anterior rostral rod grows slightly oblique to the apico-basal axis of the developing chaeta. The perikaryon of the chaetoblast is located rostrally and a small adrostral cytoplasmic bridge underneath the socket connects the perikayon to the adrostral group of microvilli (Figs. 5, 9g). A small rostral cytoplasmic bridge connects the rostral group of microvilli that is the template of the anterior rostral rod. After both are completed, the microvilli are reduced and the cytoplasmic bridges are withdrawn (Fig. 9h). The adrostral cytoplasmic bridge is replaced by the second follicle cell, which already grew between the median portion of the socket and the chaetoblast earlier during chaetogenesis (Fig. 9f). The rostral cytoplasmic bridge is replaced by the first follicle cell. During withdrawal the last group of microvilli which remained posterior while the anterior rostral rod was formed, becomes active. Its microvilli elongate and form the template of the posterior rostral rod (Figs. 5b, 9j). While chitin polymerisation elongates the rod, the chaetoblast forms a cup that surrounds the developing posterior rostral rod. The posterior rostral rod increases very rapidly in length and grows parallel to the baso-apical axis of the chaeta (Fig. 10). No further modification of the microvilli pattern occurs in this last phase of chaetogenesis. During elongation the newly formed chaeta is pushed towards the surface, and finally becomes visible externally and aligns itself at the ventral edge of the chaetal row. The canals left by the microvilli during growth of the posterior rostral rod are not filled by any material and remain electron-translucent. The same is true for the anterior rostral rod and the adrostral rod. When the formation is complete, intermediate filaments appear inside the follicle cells and the chaetoblast. Hemidesmosomes connect them to the chaeta to mechanically link the chaeta to the perifollicular ecm (Fig. 10c). The chaetoblast remains cup-like at the chaetal base and the microvilli that formed the long rostral rod remain inside the basalmost part of the chaeta (Fig. 10b, c).

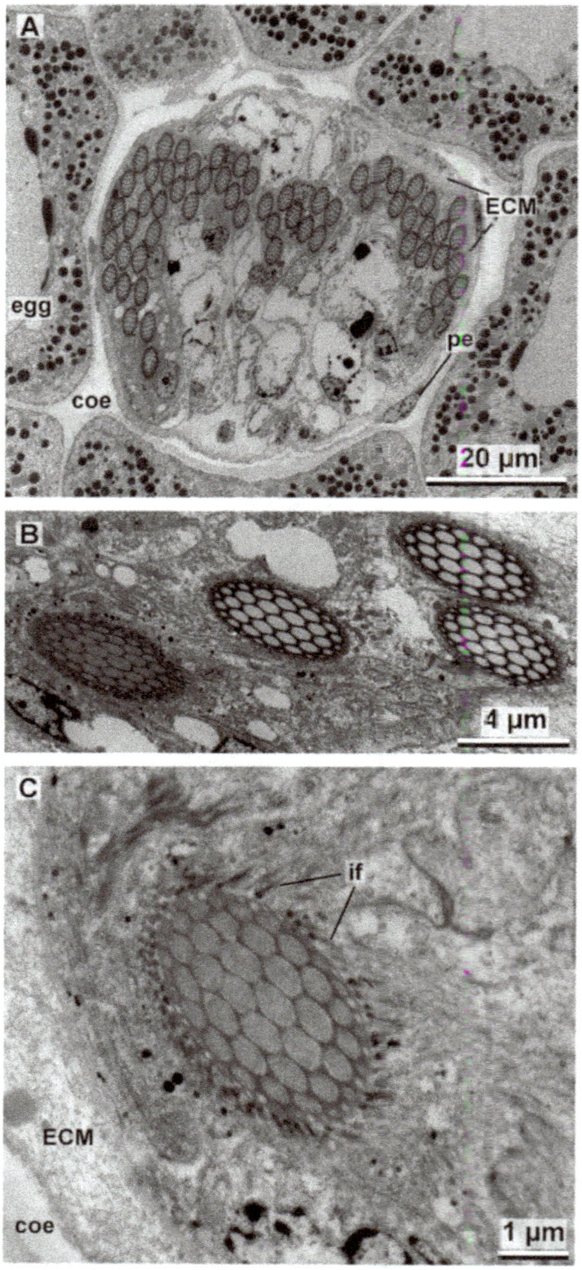

Fig. 10 a TEM image of the chaetal bundle showing the arrangement of fully differentiated chaetae. **b** Canals of the youngest chaetae stilled filled with microvilli in contrast to the hollow canals of older chaetae. **c** Detail image of the youngest chaetae, note the intermediary filaments (*if*) attached to the chaeta via hemidesmosomes. *coe* coelom, *pe* peritoneum, *ECM* extra-cellular matrix

Discussion

Uncini have repeatedly been described and illustrated for different sabellariid species [23–26], a practice that supports the adoption of *Sabellaria alveolata* as a representative for the entire group. Due to structural similarities of these chaetae across sabellariids we also assume that formation of them is largely identical in sabellariid species. However, the tremendous length of the posterior rostral rod has remained largely unnoticed. We suppose that this is caused by its extremely delicate structure, making it difficult to identify the actual extension of this rod without sectioning. Despite missing evidence from other species, we assume that a rostral rod extending deeply into the notopodium is characteristic for all sabellariid species, an assumption that has to be confirmed in subsequent studies. In the following we discuss our results in terms of function, homology, and phylogenetic significance, and highlight the cellular dynamics underlying chaetogenesis in *Sabellaria alveolata.*

Function

It has repeatedly been shown that hooked chaetae and uncini correlate with a tubiculous lifestyle and are used to withstand drag forces by interacting with the inner texture of the tube [27, 28]. Roy [29] reports that the sabellariid *Phragmatopoma californica* maintains its position in the center of the tube by extending the notopodia so that they contact the wall. Thereby, the notopodia must be of a certain length to maintain water currents inside the tube for oxygen supply and feces removal. We assume that the uncini serve as anchors to adhere to the visco-elastic wall of the tube [30]. In *Sabellaria alveolata* a second function is related to these chaetae, the structural correlate of which is the rods. The posterior rostral rod is 80 times longer than the shaft and extends deeply into the neuropodium. A rod consists of a few hollow chitin tubes wrapped in an enamel and serve as a rigid, but extremely flexible stick. Since the uncini of *S. alveolata* are aligned in a transverse row, these rods form a planar array in the tip of the notopodium that thus serves as a broad paddle and allows a maximum contact surface between the row of uncini and tube wall. Deeper inside the notopodium the rods converge to form a bundle that serves as the attachment site for parapodial muscles. The notopodium contains part of the body coelom, which functions as a hydroskeleton that guarantees the stiffness of the notopodium, but does not allow mobility. Since the bundle of rods is highly flexible and serves as an attachment site for transversal muscles, the notopodium can be moved backward and forward. Due to the mechanical properties of the chitinous rods inside, it will always return to its original structure after relocation. The rods thus serve in stability of the notopodium and allow moving it without influencing the shape of the notopodium. Roy [29] actually mentions that *Phragmatopoma californica* uses the notopodia to perform rear-to-front motions. These anteriorly directed strokes are used for backward moving when the animal rapidly withdraws into the tube. The structural prerequisite of such a notopodial performance is the internal bundle of rods. In this respect the bundle of rods in sabellariid notopodia has a similar function as the acicula of errant (aciculate) annelids. The aciculae also function as "skeletal" rods of parapodia to which the parapodial musculature is attached. In certain terebellids (i.e. *Terebella lapidaria*) similar long shafts/basal processes reach deep inside the parapodia (unpublished data). This indicates a convergent evolution of rod-like elements inside the parapodia, be it bundles of thin rods like in sabellarids, other chaetal protrusions like in terebellids or large and robust aciculae.

Homology

According to Holthe [31] hooked chaetae (= dentate hooks in Rouse & Plejel [32]) consist of a main tooth or rostrum, a capitium surmounting the rostrum, and smaller teeth and a manubrium or shaft. The rostrum and teeth of the capitium are curved and bend towards the shaft. Sometimes a subrostral process or distal expansion of the manubrium is found underneath the rostrum. Such chaetae are known from Sabellida, Terebellida, Oweniidae, adult Arenicolidae [11, 12, 15, 18]. If the shaft is shorter than the dentate distal section or virtually absent, the hooked chaetae will be called uncini (Sabellida, Terebellida, Chaetopteridae, Sabellariidae) [32]. In certain groups the dentate apex (rostrum plus capitium) is partly enveloped by hair-like protrusions of the subrostrum (juvenile Arenicolidae, Maldanidae, Psammodilidae) [10, 15, 16] or a hood (Capitellidae, Spionidae, certain Eunicida) [9, 33, 34]. It is thus unsurprising that testing for homology by studying chaetogenesis has revealed that hooked chaetae and uncini of certain taxa share identical steps during chaetogenesis (Sabellidae and Serpulidae: [11] Arenicolidae: [14, 15], Maldanidae: [10], Psammodrilida:[16], Pectinariidae: [12] , Terebellidae: [17], Oweniidae: [18]). In these taxa, the rostrum is always the very first structure that develops during genesis and is invariably preformed by a group of microvilli. Subsequently, each teeth of the capitium are formed by a large microvillus. Microvilli that served as template for the rostrum and the capitium later merge and form the shaft, which always is perpendicular to the rostrum. These characteristics are found in capitellid's hooded hooks, i.e., chaetae, in which a hood surrounds the distal section of the hooked chaeta [33]. The identity of the structure and formation patterns of hooked chaetae, uncini and hooded hooks across the above mentioned annelid taxa led to the hypothesis of their homology, which could be substantiated by several corresponding structural and developmental details [4, 17]. Other hooked chaetae with a hood differ from this pattern. In spionid and lumbrinerid species several microvilli and not a single microvillus form the template for the smaller spines that surmount the rostrum (for *Scolelepis squamata* [34]; for *Prionospio fallax* [35], for *Lumbrineris tetraura* [9]). In addition formation of the hood differs between spionid, capitellid and lumbrinerid species and does not support homology of the hood [9].

The structure and chaetogenesis of uncini in *Sabellaria alveolata* differ significantly from any hooked chaeta described so far, which poses problems in the application of terminology. Although the unpaired rostral tooth should be termed the rostrum, as it is the first structure formed during chaetogenesis and preformed by several microvilli, further groups of microvilli serve as the template for the following teeth, which thus should not be termed capitial teeth. The shaft consists of two sections, the base and socket; both are separated by a refracting line and are preformed by microvilli of different orientation. A shaft that is composed of two parts because of controlled spatial and temporal intermissions in the formation processes is thus far unknown. The subcuticular portion of sabellarid uncini consists of an adrostral and a bipartite rostral rod. Similar rod-like processes are known from terebellids (*Nicolea zostericola*; [17]) and chaetopterids (Tilic & Bartolomaeus, unpubl. data for *Chaetopterus variopedatus* and *Telepsaphus costarum*). However, the formation of these processes differs from the rods in *Sabellaria alveolata*. Two small groups of microvilli, one rostral and one adrostral remain in the mentioned terebellid and chaetopterid species after the microvilli were withdrawn from the shaft after its completion. Polymerization of chitin between the microvilli of both groups then gives rise to both processes, which thus are rather parts of the shaft than additional structures. These differences do not support a homology between the rod in *Sabellaria alveolata* and the rod-like processes of the manubrium in terebellid and chaetopterid species. Structure and chaetogenesis of uncini in *S. alveolata* thus differ in several aspects from that of other annelids with uncini and hooked chaeta. These differences either result from transformation or convergent evolution. A decision between both alternatives, however, depends on the phylogenetic position of Sabellariidae.

Phylogenetic implications

Sabellariidae were first described as a subgroup of Sabellida by Lamarck [36], and later moved to Terebellida by Savigny [37]. Levinsen [38] placed them as a separate suborder, using the name Hermelliformia, which was first coined by Malmgren [39]. Phylogenetic analyses based on morphological data [40–42] suggest a sister group relationship with Sabellidae. One decisive morphological character in favour of the close relationship to Sabellidae is the so-called "chaetal inversion" (for review [43]; [44–48].) Sabellaridae and Sabellidae show a unique chaetal arrangement with abdominal uncini in a notopodial position. This was considered to be an undisputed synapomorphy until Kieselbach and Hausen [43] provided evidence that the specific chaetal arrangement of Sabellidae and Sabellariidae arose independently (see also [49]). Kieselbach and Hausen [43] also emphasize that the homology of the uncini of sabellids and those of sabellarids is yet to be established. More recent molecular phylogenies of annelids [47, 43, 50–54] group them together with Spionida. A sabellariid-spionid sister group relationship had already been suggested by Wilson [55] and later Dales [56] and Rouse & Pleijel [32, 57]. Wilson [55] substantiated this hypothesis with characters of the larval organisation, since both possess long larval chaetae inserting posterior to the prototroch. Kieselbach [49] described a specialized ciliated sensory organ in the prostomium, of larval *Sabellaria alveolata* that was thus far only known from Spionida [58]. Except for being paired in Spionida, this organ shows an identical organization and the same substructures like Spionida, so that this sense organ supports the hypothesis of a sister group relationship of Spionida and Sabellariidae.

The differences of sabellid and sabellariid uncini in terms of substructures and chaetogenesis, however, do not provide evidence for a sister group relationship between both groups. Moreover, the fact that several microvilli and not a single big microvillus form the template for each adrostral teeth is identical in the spionids studied thus far [34, 35] and in *S. alveolata*. The better supported alternative hypothesis of a spionid-sabellariid sister group relationship presently argues against homology of sabellid and sabellariid uncini and for transformations that need to be analysed in subsequent studies.

Cell dynamics

In a recent essay Warren [59] compared the microvilli of the chaetoblast with the printing head of a 3D-printer, as they ensure assembly of a complex structure by selective addition of material in time and space. Chaetogenesis in *Sabellaria alveolata* illustrates the complexity of this process and provides empirical evidence that in addition to dynamic microvilli cell dynamics influences proper formation of the chaeta. Beside repeated formation of microvilli, the position of the chaetoblast within the formative site and the speed in which chitin polymerizes are important factors shaping the final structure of the sabellariid uncinus. Tilting the axis of the developing chaeta is a prerequisite to form the basal part of the shaft, the socket, as well as the proper orientation of the rods. The chaetoblast itself expands during chaetogenesis, relocates the perikaryon and finally remains as a cup-

Table 1 Structure and chaetogenesis of hooked chaetae and uncini in Annelids

Taxon	Species	Substructure								Formation		Reference
		Rostrum/main tooth	Adrostral teeth	Capitium	Shaft bipartite	Shaft length	Hood	Beard	Basal extensions	Rostrum	Adrostral teeth	
										Microvilli	Microvilli	
Sabellariidae	*Sabellaria alveolata*, abd	+	+	-	+	s	-	-	r	several	several	this study
Spionida	*Scolelepis squamata*	+	+	-	-	l	+	-	-	several	several	Hausen & Bartolomaeus 1998 [34]
	Malacoceros fuliginosus	+	+	-	-	l	+	.	-	several	several	Hausen & Bartolomaeus 1998 [34]
	Prionospio fallax	+	+	-	-	l	+	-	-	several	several	Hausen 2001 [35]
	Spirorbis spirorbis, abd	-	+	+	-	s	-	-	-	-	single	Bartolomaeus 1995 [12]
Sabellida	*Fabricia stellaris*, tho	+	+	+	-	l	-	-	-	several	single	Bartolomaeus 2002 [11]
	Fabricia stellaris, abd	-	+	+	-	s	-	-	-	-	single	Bartolomaeus 2002 [11]
	Branchiomma bombyx, abd	+	+	+	-	s	-	-	-	several	single	unpubl. data
	Pseudopotamilla reniformis, abd	+	+	+	-	s	-	-	-	several	single	Kolbasova et al. 2014 [13]
Terebellida	*Pectinaria koreni*	-	+	+	-	s	-	-	-	-	single	Bartolomaeus 1995 [12]
	Pectinaria auricoma	-	+	+	-	s	-	-	-	-	single	Bartolomaeus 1995 [12]
	Nicolea zostericola	+	+	+	-	s	-	-	p	several	single	Bartolomaeus 1998 [17]
Chaetopteridae	*Telepsaphus costarum*	-	+	+	-	s	-	-	p	-	single	unpubl. data
	Chaetopterus variopedatus	-	+	+	-	s	-	-	p	-	single	unpubl. data
Arenicolidae	*Arenicola marina*, juvenile	+	+	+	-	l	-	+	-	several	single	Bartolomaeus & Meyer 1997 [15]
	Arenicola marina	+	+	+	-	l	-	+	-	several	single	Bartolomaeus & Meyer 1997 [15]
Maldanidae	*Clymenura clypeata*	+	+	+	-	l	-	+	-	several	single	Tilic et al. 2015 [10]
	Johnstonia clymenoides	+	+	+	-	l	-	+	-	several	single	Tilic et al. 2015 [10]
Psammodrilidae	*Psammodrilus balanoglossoides*	+	+	+	-	l	-	+	-	several	single	Meyer & Bartolomaeus 1997 [16]
Capitellidae	*Capitella capitata*	+	+	+	-	l	+	-	-	several	single	Schweigkofler et al. 1998 [33]
Oweniidae	*Owenia fusiformis*	-	+	+	-	l	-	-	-	-	single	Meyer & Bartolomaeus 1996 [18]
Lumbrineridae	*Lumbrineris tetraura*	+	+	-	-	l	+	-	-	several	several	Tilic et al. 2014 [9]

abd. **abdomen,** ***tho.*** **thorax, + present, – absent,** ***s*** **short,** ***l*** **long,** ***r*** **rods,** ***p*** **processes**

like structure at the base of the rostral rod. During this final step of chaetogenesis the follicle cell expands tremendously as it surrounds the entire posterior rostral rod. Since the rod, when completed is 80 times longer than the shaft, the follicle cell expands to 80-fold of its initial length.

Fixation of a continuous developmental process causes that this process is divided into different stages. The formative site of *S. alveolata* studied in this paper, shows 13 of these stages (Fig. 5a). Provided that chaetogenesis is a continuous process, one would expect that the time passed between the stages is always identical, even though it is not exactly known. According to this consideration, the initial phase of chaetogenesis lasts rather long, since we found six subsequent stages showing increasing numbers of apical teeth. The remaining steps are rather rapid events, because six steps later the entire chaeta is complete, except for the posterior branch of the rostral rod. One step further this structure attained an enormous length. Chitin is produced by the chitin synthases that are located in the cell membrane and has been shown to appear at the bases of microvilli [60–62]. Provided that chitin synthase is also located in microvillar membrane, one would expect that the longer the microvilli are the higher is the rate of chitin synthesis. Although this remains to be shown experimentally, there is a remarkable correlation between the length of the microvilli and the speed of growth of chaeta in support of this anticipation: the longest microvilli can be found where chaetal elongation occurs rapidly.

Conclusions

Despite superficial similarities to the uncini of Sabellida, Terebellida and other smaller annelid groups, the uncini of *S. alveolata* differ in substructures and formation (Table 1). These differences concern (1) formation of adrostral teeth by groups of microvilli instead of one large microvillus, (2) bipartition of the shaft and its formation in temporally separated steps and (3) formation of rostral and adrostral manubrial extensions (4) followed by the formation of an adrostral and a bipartite rostral rod. These differences either result from transformations of an ancestral structure or from convergent evolution. Given that recent molecular and morphological data provide strong support for a sister group relationship between Spionida and Sabellariidae, the uncini in sabellariids on one hand and those of terebellids, sabellids and a few other smaller annelid taxa on the other hand appear to have evolved convergently. Since Spionidae possess hooded hooks consisting of apically dentate chaeta with a hood and since all apical teeth are pre-formed by groups of microvilli, it is likely that the sabellariid uncini evolved by transforming such dentate chaetae into uncini. Our study also shows that this transformation went along with changing functional demands. In contrast to spionid species, sabellariids live in a reinforced visco-elastic tube to which they are able to firmly adhere, using the uncini as anchors. The specific structure of the notopodium optimizes the contact surface towards the tube wall. In addition the notopodia are used for rapid withdrawal and must be movable. Since they are rather long structures they need some internal reinforcement that acts as attachment site for the transversal muscles. These attachment sites are provided by the rods originating from the uncini, since they form a central, flexible structure comparable to the acicula in aciculate annelids.

Data repository

To allow full transparency of the data presented in this study, all of the aligned serial semi-thin and ultra-thin sections and the confocal z-stack of the phalloidin stained parapodium are freely accessible in the morphological database, MorphDBase: www.morphdbase.de [63].

Complete series of aligned ultra-thin sections:

Direct link: www.morphdbase.de/?E_Tilic_20151015-M-27.1

Complete series of aligned semi-thin sections:

Part 1– Direct link: www.morphdbase.de/?E_Tilic_20151015-M-29.1

Part 2 – Direct link: www.morphdbase.de/?E_Tilic_20151015-M-28.1

Confocal z-stack of the phalloidin stained parapodium:

Direct link: www.morphdbase.de/?E_Tilic_20151015-M-30.1

Competing interests

The authors declare that they have no competing interests.

Authors' contributions

ET conducted the experimental work, analyzed the data, drafted the manuscript and prepared the figures. TB significantly contributed to the analysis and interpretation of the data and improved the final manuscript.

Acknowledgements

We would like to thank the staff of "Laboratoire de Biologie Marine" in Concarneau (France) for their hospitality during our collection trips. Furthermore we thank Claudia Müller and Tatjana Bartz for technical assistance.

References

1. Fauchald K. The polychaete worms: definitions and keys to the orders, families and genera. CA, USA: Natural History Museum of Los Angeles County Los Angeles; 1977.
2. Schroeder P. Chaetae. In: Bereiter-Hahn J, Matoltsy AG, Richards KS, editors. Biology of the Integument. Berlin Heidelberg: Springer; 198-. p. 297–309.
3. Bouligand Y. Les soies et les cellules associées chez deux Annelides Polychètes. Z Zellforsch Mikro Anat. 1967;79:332–363.
4. Hausen H. Chaetae and chaetogenesis in polychaetes (Annelida). Hydrobiologia. 2005;535–536:37–52.

5. Specht A, Westheide W. Intra-and interspecific ultrastructural character variation: The chaetation of the Microphthalmus listensis species group (Polychaeta, Hesionidae). Zoomorphology. 1988;107:371–376.
6. Ogawa Y, Kimura S, Wada M. Electron diffraction and high-resolution imaging on highly-crystalline β-chitin microfibril. J Struct Biol. 2011. doi:10.1016/j.jsb.2011.07.001.
7. Koide Y, Hasan I, Asanuma A, Fujii Y, Ogawa Y, Kobayashi H, et al. Expression of Various Types of Glycans in the Lugworm. Ann Mar Biol Res. 2015;2:1005.
8. Souza CP, Almeida BC, Colwell RR, Rivera ING. The importance of chitin in the marine environment. Mar Biotechnol. 2011;13:823–30.
9. Tilic E, Hausen H, Bartolomaeus T. Chaetal arrangement and chaetogenesis of hooded hooks in Lumbrineris (Scoletoma) fragilis and Lumbrineris tetraura (Eunicida, Annelida). Invertebr Biol. 2014;133:354–70. doi:10.1111/ivb.12066.
10. Tilic E, von Döhren J, Quast B, Beckers P, Bartolomaeus T. Phylogenetic significance of chaetal arrangement and chaetogenesis in Maldanidae (Annelida). Zoomorphology. 2015;134:383–401. doi:10.1007/s00435-015-0272-9.
11. Bartolomaeus T. Structure and Formation of Thoracic and Abdominal Uncini in *Fabricia stellaris* (Müller, 1774) - Implication for the Evolution of Sabellida (Annelida). Zool Anz - A J Comp Zool. 2002;241:1–17.
12. Bartolomaeus T. Structure and formation of the uncini in *Pectinaria koreni*, *Pectinaria auricoma* (Terebellida) and *Spirorbis spirorbis* (Sabellida): implications for annelic phylogeny and the position of the Pogonophora. Zoomorphology. 1995;115:161–77.
13. Kolbasova GD, Tzetlin AB, Kupriyanova EK. Chaetal loss and replacement in Pseudopotamilla reniformis (Sabellida, Annelida). Invertebr Biol. 2014;133:261–73. doi:10.1111/ivb.12061.
14. Bobin G. Morphogénèse des soies chez les annélides polychètes. Ann L'institut Océanographique. 1944;22:1–106.
15. Bartolomaeus T, Meyer K. Development and phylogenetic significance of hooked setae in Arenicolidae (Polychaeta, Annelida). Invertebr Biol. 1997;116:227–42.
16. Meyer R, Bartolomaeus T. Ultrastruktur und Morphogenese der Hakenborsten bei Psammocrilus balanpglossoides - Bedeutung für die Stellung der Psammodrilida (Annelida). Microfauna Marina. 1997;11:87–113.
17. Bartolomaeus T. Chaetogenesis in polychaetous Annelida: significance for annelid systematics and the position of the Pogonophora. Zoology. 1998;100:348–64.
18. Meyer K, Bartolomaeus T. Ultrastructure and formation of the hooked setae in *Owenia fusiformis* delle Chiaje, 1842: implications for annelid phylogeny. Can J Zool. 1996;74:2143–53. doi:10.1139/z96-243.
19. Schulze A. Ultrastructure of opisthosomal chaetae in Vestimentifera (Pogonophora, Obturata) and implications for phylogeny. Acta Zool. 2001;82:127–35.
20. Bartolomaeus T, Purschke G, Hausen H. Polychaete phylogeny based on morphological data - a comparison of current attempts. In: Bartolomaeus T, Purschke G, editors. Morphology, Molecules, Evolution and Phylogeny in Polychaeta and Related Taxa. Netherlands: Springer; 2005. p. 341–56.
21. Blumer MJF, Gahleitner P, Narzt T, Handl C, Ruthensteiner B. Ribbons of semithin sections: an advanced method with a new type of diamond knife. J Neurosci Methods. 2002;120:11–6.
22. Kremer JR, Mastronarde DN, McIntosh JR. Computer visualization of three-dimensional image data using IMOD. J Struct Biol. 1996;116:71–6.
23. Lana PC, Gruet Y. Sabellaria wilsoni sp. n. (Polychaeta, Sabellariidae) from the southeast coast of Brazil. Zool Scr. 1989;18:239–44.
24. Kirtley DW. A review and taxonomic revision of the family Sabellariidae Johnston, 1865 (Annelida: Polychaeta). Vero Beach, FL: Sabecon Press; 1994.
25. Capa M, Faroni-perez L, Hutchings P. Sabellariidae from Lizard Island, Great Barrier Reef, including a new species of *Lygdamis* and notes on external morphology of the median organ. Zootaxa. 2015;4019:184–206.
26. Souza dos Santos A, Dos Santos Brasil AC, Christoffersen ML. *Sabellaria* and *Lygdamis* (Polychaeta: Sabellariidae) from reefs off northeastern Brazil including a new species of Sabellaria. Zootaxa. 2014;3881(2):125–44.
27. Woodin SA, Merz RA. Holding on by their hooks: anchors for worms. Evolution. 1987;41:427–32.
28. Merz RA. Textures and traction: how tube-dwelling polychaetes get a leg up. Invertebr Biol. 2015;134:61–77. doi:10.1111/ivb.12079.
29. Roy PA. Tube dwelling behavior in the marine annelid Phragmatopoma californica (Fewkes)(Polychaeta: Sabellariidae). Bull South Calif Acad Sci. 1974;73:117–25.
30. Le Cam J-B, Fournier J, Etienne S, Couden J. The strength of biogenic sand reefs: Visco-elastic behaviour of cement secreted by the tube building polychaete Sabellaria alveolata, Linnaeus, 1767. Estuar Coast Shelf Sci. 2011;91:333–9.
31. Holthe T. Evolution, systematics and distribution of the Polychaeta Terebellomorpha, with a catalogue of the taxa and a bibliography. Gunneria. 1986;55:1–236.
32. Rouse GW, Pleijel F. Polychaetes. Oxford: Oxford University Press; 2001.
33. Schweigkofler M, Bartolomaeus T, von Salvini-Plawen L. Ultrastructure and formation of hooded hooks in *Capitella capitata* (Annelida, Capitellida). Zoomorphology. 1998;118:117–28.
34. Hausen H, Bartolomaeus T. Setal Structure and Chaetogenesis in *Scolelepis squamata* and *Malacoceros fuliginosus* (Spionidae, Annelida). Acta Zool. 1998;79:149–61. doi:10.1111/j.1463-6395.1998.tb01154.x.
35. Hausen H. Untersuchungen zur Phylogenie "spiomorpher" Polychaeten (Annelida). Berlin: Logos-Verl.; 2001.
36. de Lamarck JB. Histoire naturelle des Animaux sans Vertèbres, 5. Paris: Déterville and Verdère; 1818.
37. Savigny JC. Systèmes de diverses classes d'animaux sans vertèbres. Système des annélides, Princ. celles des côtes l'Égypte la Syrie, offrant les caractères tant distinctifs que Nat. des ordres, Fam. genres, avec la Descr. des espèces; 1822.
38. Levinsen GMR. Systematisk-geografisk oversigt over den nordiske Annulata, Gephyrea, Chætognathi og Balanoglossi. B. Lunos Kgl. Hof-bogtrykkeri; 1883.
39. Malmgren AJ. Annulata Polychaeta: Spetsbergiae, Groenlandiae, Islandiae et Scandinaviae. Hactenus Cognita. Ex Officina Frenckelliana; 1867.
40. Smith RS. Relationships within the order Sabellida (Polychaeta). Ophelia Suppl. 1991;5:249–60.
41. Rouse GW, Fauchald K. Cladistics and polychaetes. Zool Scr. 1997;26:139–204. doi:10.1111/j.1463-6409.1997.tb00412.x.
42. Schulze A. Phylogeny of Vestimentifera (Siboglinidae, Annelida) inferred from morphology. Zool Scr. 2003;32:321–342.
43. Kieselbach D, Hausen H. Chaetal arrangement provides no support for a close relationship of Sabellidae and Sabellariidae (Annelida). J Morphol. 2008;117:104–17. doi:10.1002/jmor.
44. Fauchald K, Rouse G. Polychaete systematics: Past and present. Zool Scr. 1997;26:71–138. doi:10.1111/j.1463-6409.1997.tb00411.x.
45. Rousset V, Rouse GW, Siddall ME, Tillier A, Pleijel F. The phylogenetic position of Siboglinidae (Annelida) inferred from 18S rRNA, 28S rRNA and morphological data. Cladistics. 2004;20:518–33.
46. Kupriyanova EK, Rouse GW. Yet another example of paraphyly in Annelida: molecular evidence that Sabellidae contains Serpulidae. Mol Phylogenet Evol. 2008;46:1174–81. doi:10.1016/j.ympev.2007.10.025.
47. Capa M, Hutchings P, Peart R. Systematic revision of Sabellariidae (Polychaeta) and their relationships with other polychaetes using morphological and DNA sequence data. Zool J Linn Soc. 2012;164:245–84. doi:10.1111/j.1096-3642.2011.00767.x.
48. Capa M, Hutchings P, Aguado TM, Bott NJ. Phylogeny of Sabellidae (Annelida) and relationships with other taxa inferred from morphology and multiple genes. Cladistics. 2011;27:449–69. doi:10.1111/j.1096-0031.2010.00341.x.
49. Kieselbach D. Untersuchungen zur Phylogenie sedentärer Polychaeten (Annelida). Bonn: Universitäts- und Landesbibliothek Bonn; 2012.
50. Struck TH, Golombek A, Weigert A, Franke FA, Westheide W, Purschke G, et al. The Evolution of Annelids Reveals Two Adaptive Routes to the Interstitial Realm. Curr Biol. 2015; 1–7. doi: 10.1016/j.cub.2015.06.007.
51. Weigert A, Helm C, Meyer M, Nickel B, Arendt D, Hausdorf B, et al. Illuminating the Base of the Annelid Tree Using Transcriptomics. Mol Biol Evol. 2014;31:1391–401. doi:10.1093/molbev/msu080.
52. Andrade SCS, Novo M, Kawauchi GY, Worsaae K, Pleijel F, Giribet G, et al. Articulating "archiannelids": Phylogenomics and annelid relationships, with emphasis on meiofaunal taxa. Mol Biol Evol. 2015. doi:10.1093/molbev/msv157.
53. Struck TH, Schult N, Kusen T, Hickman E, Bleidorn C, McHugh D, et al. Annelid phylogeny and the status of Sipuncula and Echiura. BMC Evol Biol. 2007;7:57. doi:10.1186/1471-2148-7-57.
54. Zrzavý J, Říha P, Piálek L, Janouskovec J. Phylogeny of Annelida (Lophotrochozoa): total-evidence analysis of morphology and six genes. BMC Evol Biol. 2009;9:189. doi:10.1186/1471-2148-9-189.
55. Wilson DP. The larvae of the British sabellarians. J Mar Biol. 1929;16:221–68.
56. Dales RP. The development and structure of the anterior region of the body in the Sabellariidae, with special reference to Phragmatopoma californica. Q J Microsc Sci. 1952;3:435–52.
57. Rouse GW, Pleijel F. Problems in polychaete systematics. Hydrobiologia. 2003;496:175–189.

58. Hausen H. Ultrastructure of presumptive light sensitive ciliary organs in larvae of Poecilochaetidae, Trochochaetidae, Spionidae, Magelonidae (Annelida) and its phylogenetic significance. Zoomorphology. 2007;126:185–201.
59. Warren G. In praise of other model organisms. J Cell Biol. 2015;208:387–9.
60. Peters W, Latka I. Electron microscopic localization of chitin using colloidal gold labelled with wheat germ agglutinin. Histochemistry. 1986;84:155–60.
61. Zimoch L, Merzendorfer H. Immunolocalization of chitin synthase in the tobacco hornworm. Cell Tissue Res. 2002;308:287–97.
62. Moussian B, Letizia A, Martínez-Corrales G, Rotstein B, Casali A, Llimargas M. Deciphering the genetic programme triggering timely and spatially-regulated chitin deposition. PLoS Genet. 2015;11:1.
63. Grobe P, Vogt L. Morph.D.Base 2.0. In: A public date base Morphol. data, metadata, phylogenetic matrices. http://www.morphdbase.de. 2009.

2

The life of the freshwater bryozoan *Stephanella hina* (Bryozoa, Phylactolaemata)—a crucial key to elucidating bryozoan evolution

Thomas Schwaha[1*], Masato Hirose[2] and Andreas Wanninger[1]

Abstract

Background: Phylactolaemata is the earliest branch and the sister group to all extant bryozoans. It is considered a small relict group that, perhaps due to the invasion of freshwater, has retained ancestral features. Reconstruction of the ground pattern of Phylactolaemata is thus essential for reconstructing the ground pattern of all Bryozoa, and for inferring phylogenetic relationships to possible sister taxa. It is well known that *Stephanella hina*, the sole member of the family Stephanelllidae, is probably one of the earliest offshoots among the Phylactolaemata and shows some morphological peculiarities. However, key aspects of its biology are largely unknown. The aim of the present study was to analyze live specimens of this species, in order to both document its behavior and describe its colony morphology.

Results: The colony morphology of *Stephanella hina* consists of zooidal arrangements with lateral budding sites reminiscent of other bryozoan taxa, i.e., Steno- and Gymnolaemata. Zooids protrude vertically from the substrate and are covered in a non-rigid jelly-like ectocyst. The latter is a transparent, sticky hull that for the most part shows no distinct connection to the endocyst. Interestingly, individual zooids can be readily separated from the rest of the colony. The loose tube-like ectocyst can be removed from the animals that produces individuals that are unable to retract their lophophore, but merely shorten their trunk by contraction of the retractor muscles.

Conclusions: These observations indicate that *S. hina* is unique among Phylactolaemata and support the notion that bryozoans evolved from worm-like ancestors. In addition, we raise several arguments for its placement into a separate family, Stephanellidae, rather than among the Plumatellidae, as previously suggested.

Keywords: Ectoprocta, Stephanellidae, Phylactolaemata, Ectocyst evolution, Solitary zooids

Background

Bryozoa or Ectoprocta is a group of sessile, colonial filter-feeders of approximately 6000 extant species. Its phylogenetic relationship to other lophotrochozoan taxa remains controversial (e.g. [1–4]). Due to shared morphological features in their general organization, including their feeding apparatus, the lophophore, this group was traditionally united with the Brachiopoda and Phoronida as 'Lophophorata' or 'Tentaculata' [5]. However, many molecular phylogenies have failed to support the 'Lophophorata' concept. Most have supported instead a close relationship between phoronids and brachiopods (e.g. [3, 6]), whereas only few recent studies have lent support to the 'lophophorate' grouping [7–9].

Bryozoans typically consist of several individuals, called zooids, which constitute the colony. Each zooid is typically divided into the cystid, which constitutes the body-wall, which is fortified with different extracellular secretions, and the polypide, which consists of the soft-body parts. Bryozoa can be divided into three class-level taxa: Phylactolaemata, Stenolaemata and Gymnolaemata. Morphological and molecular analyses agree that

* Correspondence: thomas.schwaha@univie.ac.at
[1]Department of Integrative Zoology, University of Vienna, Althanstraße 14, Vienna 1090, Austria
Full list of author information is available at the end of the article

Phylactolaemata is the basal-most branch and sister-group to the remaining taxa [10–12]. Phylactolaemata is a small group of approximately 80 species worldwide within six families that only occur in freshwater habitats [13]. Their members show different variations of the ectocyst, which may be either chitinous or gelatinous. Various analyses agree that Stephanellidae with the sole currently recognized species *Stephanella hina* Oka, 1908 is one of the earliest branches among the Phylactolaemata [14]. This species is almost exclusively reported from Japan and Southeast Asia [15, 16] and the east coast of North America [17], and it shows a unique colony morphology with stolon-like connections and lateral budding loci [18]. In addition, its statoblasts and in particular the formation of the sessoblasts seems to differ from all other phylactolaemate bryozoans [19]. This species is also unique in that it starts to grow in December and starts to decay in mid-April when other phylactolaemates start to thrive after overwintering [20]. Only few other species are known to survive cold winter temperatures as adults [13].

Surprisingly, almost no data are currently available for this species; even photographic documentation is almost entirely lacking [16, 18, 21]. In addition, whether *Stephanella* constitutes a separate family-level taxon or is part of the Plumatellidae, the largest of all phylactolaemate families, remains unclear. Our aim in the present study was to identify and document the biology of this neglected species. As the earliest branch, the reconstruction of the phylactolaemate body plan may yield essential information into the evolution of the whole phylum, as well as the evolution of coloniality among metazoans. Bryozoa is the only animal phylum in which the members are all colonial. However, since all other possible lophotrochozoan outgroups have a solitary ground pattern, it is clear that the pre-bryozoan ancestor was also a solitary, probably worm-shaped, organism. Hence the elucidation of the phylactolaemate morphological ground pattern could yield insight into the evolution of solitary into colonial forms.

Methods

Samples were collected in April 2014 and February 2015 and 2016 in Japan, either in the pond of the campus of the University of Tsukuba or in a freshwater pond in Tsuchiura-shi in close vicinity to the campus. Samples were collected from mainly wooden substrates and transferred into the laboratory for further inspection and treatment. Plants and artificial substrates (mainly garbage, shoes, etc.) were not colonized by any specimens. Living specimens were documented and filmed with a Nikon J1 camera (Nikon, Tokyo, Japan) mounted on a c-mount on a Nikon SMZ 1500.

Results & discussion

General structure of the colony and zooids

Stephanella hina colonies form an array of interconnected zooids, with each zooid generally having up to four connections to other zooids within the colony (Fig. 1a and b). These interconnections are restricted to the substrate and are, particularly in older colonies, stolon-like. Younger colonies commonly show a wider interzooidal connection. This corresponds to previous descriptions [18, 21] and is unique for a phylactolaemate bryozoan where zooids are normally interconnected by much wider spaces than in *Stephanella*. In this respect a similar tube-like connection is only found in the plumatellid genus *Stolella* [22, 23]. However, in the latter these are not as pronounced as in *S. hina*. Since there are three budding loci—one distal and two lateral—a total of four stolon-like connections are present in *S. hina* (Fig. 1a, b). Lateral buds are not present in any other phylactolaemate. This simple runner-like colony shape is also characteristic of early fossil representatives as well

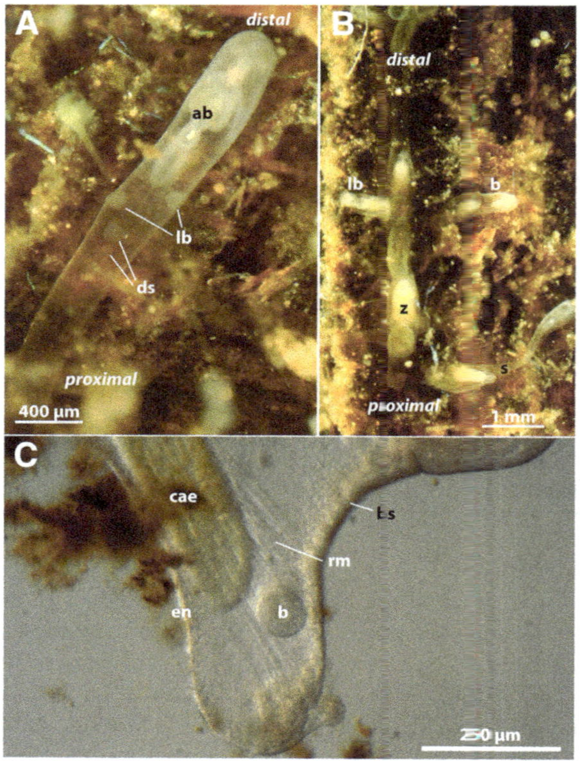

Fig. 1 Budding sites of *Stephanella hina*. **a** View of an elongated distal bud, with two lateral buds and developing statoblasts. **b** Parts of a retracted colony, showing the cruciform growth of the colonies due to lateral budding. **c** Close-up of the basal part of a detached, single zooid, showing an early two-layered bud at its base close to the retractor muscle. Abbreviations: ab – advanced bud, b – bud, bs – basal attachment side, cae – caecum, ds – developing statoblasts, en – endocyst, lb – lateral buds, rm – retractor muscle, s – stolon, z – zooid

as recent, probably early branching, taxa of the other two bryozoan taxa Stenolaemata (e.g. Corynotrypidae) [24–26] and Gymnolaemata (e.g. *Arachnidium* and *Cardoarachnidium* in Ctenostomata [27]; e.g. *Pyripora* and *Pyriporopsis* in Cheilostomata [28, 29]). Considering the primitive pattern of budding in *Stephanella* as an early branch of Phylactolaemata [14], it is possible that it has retained the early or the original budding mode, and that other Phylactolaemata have reduced it.

The main part of each zooid extends vertically from the substrate as a long slender tube that carries all the major components of the polypide, i.e., the digestive tract and the tentacle crown, the lophophore, which is used for filter-feeding (Figs. 2 and 3, see also Additional file 1: Video S1). From the lophophoral base, the intertentacular membrane extends towards the abfrontal side of the tentacles. It is rather short and surrounds the whole lophophore equally (Fig. 4b and d, see Additional file 2: Video S2). The tentacles, the epistome and the distal part of the pharynx show a reddish-brownish hue, which is more pronounced when animals are retracted and tissues are much more compact (Fig. 4a, b and d, see Additional file 2: Video S2, Additional file 3: Video S3, Additional file 4: Video S4). This coloration has been previously described by several authors [16, 18, 21]. The tentacle tinge is reportedly lost after two days of culture in pure tap water [16]. The digestive tract consists of the common parts found in other bryozoans, i.e. pharynx, oesophagus, cardia, caecum, intestine and anus. In our samples the digestive tract showed a blue-green coloration in most cases (Additional file 3: Video S3, Additional file 4: Video S4, Additional file 5: Video S5). The color of the digestive tract is commonly influenced by the food source, but the exact food contents have not been analyzed so far in *S. hina*. Previously, the coloration of the stomach was described as yellowish-green

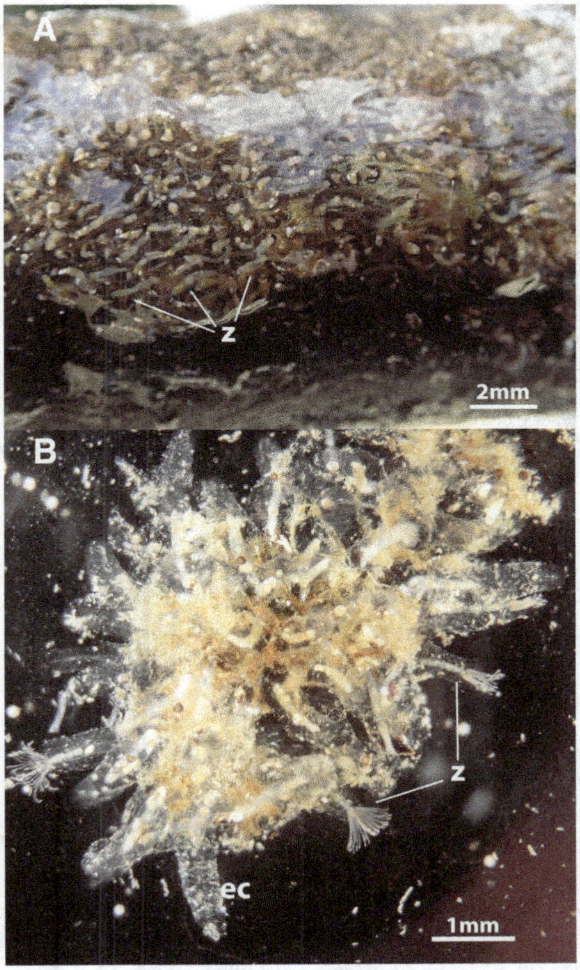

Fig. 2 Comparison of the colony of *Stephanella hina* in and out of the water, showing the non-supportive and flexible ectocyst. **a** *Stephanella hina* colony on a piece of wood taken out of the water. Note the collapsed appearance of the colony with the zooids lying flat on the substrate. **b** Piece of a colony submerged into water. Note the upright position of the zooids extending vertically from the substrate. Abbreviations: ec – ectocyst, z – zooid

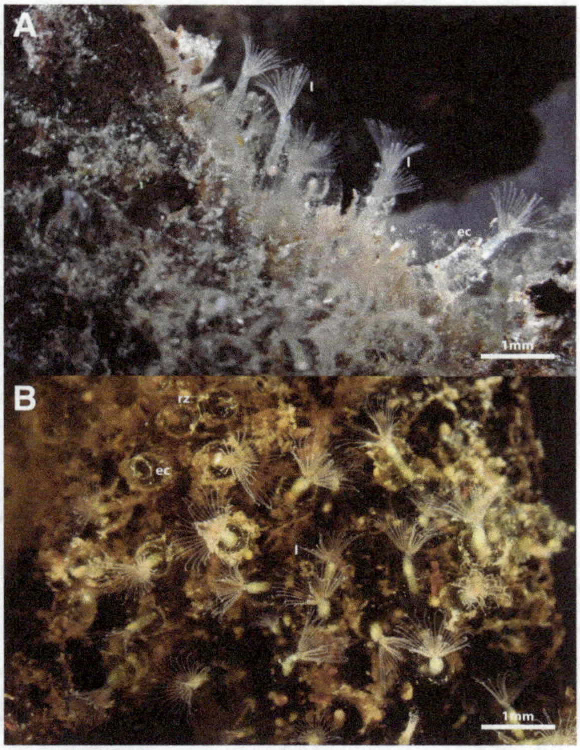

Fig. 3 General morphology of colonies of *Stephanella hina*. **a** Lateral view of a colony, showing several individual zooids extending their lophophores into the water column. Note that the polypides extend vertically from the substrate. The transparent ectocyst can only be discerned by particles adhering to it. **b** Top view of a colony showing the arrangement of the individual zooids within the colony and the typical horseshoe-shaped lophophore. Abbreviations: ec- ectocyst, l – lophophore, rz – retracted zooid

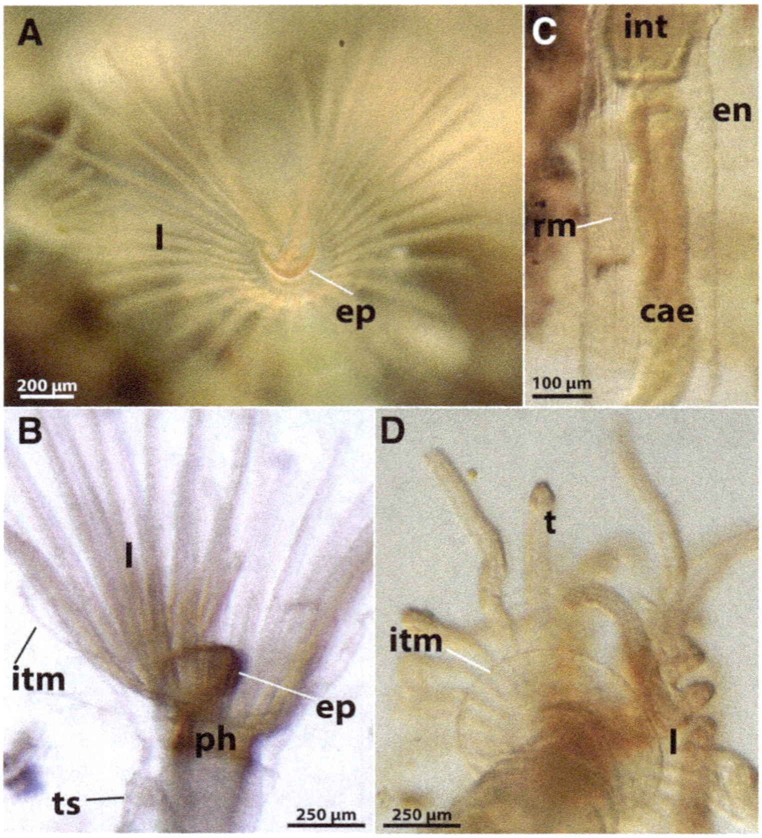

Fig. 4 Details of a single zooid of *Stephanella hina*. **a** View of the lophophore with the horseshoe-shaped tentacle arrangement and the epistome protruding over the mouth opening. Note the orange tinge of the epistome. **b** Lateral view of the lophophoral base, showing the evident coloration of the epistome as well as the pharynx. Note the intertentacular membrane on the outer side of the lophophore. **c** Lateral view through the transparent ectocyst showing parts of the digestive tract as well as retractor muscle strands traversing the visceral coelom. **d** View from the basal side of the lophophore and the intertentacular membrane extending from the lophophoral base towards the outer area of the tentacles. Abbreviations: cae – caecum, en – endocyst, ep – epistome, int – intestine, itm – intertentacular membrane, l – lophophore, ph – pharynx, rm – retractor muscle, t – tentacle, ts – tentacle sheath

[21] to pale brown [18]. Retraction of the polypide is effected by a paired retractor muscle that emanates from the basal side and inserts at the lophophoral base and parts of the digestive tract (Figs. 1c and 4c, see Additional file 3: Video S3, Additional file 4: Video S4, Additional file 5: Video S5, Additional file 6: Video S6). The general transparency of the animals allows ready observation of the contraction and relaxation of these muscles. Contraction of the regular body wall musculature increases pressure within the coelomic cavity and along with relaxation of retractor and apertural muscles pushes the polypide out of the cystid. The body wall musculature can be observed in live specimens (Fig. 5a). The funicular cord runs from the proximal pole of the caecum to the body and carries the anlagen or fully differentiated free-floating statoblasts, the floatoblasts (Fig. 6). Floatoblasts in *S. hina* are circular in shape (Fig. 6a–c). At the basal side, sessoblasts are commonly attached to the substrate (Fig. 6d and e). These often occur in certain interzooidal widenings of the stolonal network, and rarely exceed 2–3 in number. The situation thus resembles that of other phylactolaemate bryozoans [13, 30]. So far, no gonads of either sex have been encountered. A single previous description found well-developed testes and some growing ovaries in very young colonies in December at 5 °C water temperature, but could not follow its subsequent development due to frozen ponds and lakes [20].

The 'cystid' of *Stephanella hina*

The vertically extending tubes containing the polypides are embedded into a copious amount of the jelly-like, transparent ectocyst. Two different variations of this ectocyst were encountered: either it forms a connecting mass between the extended polypides, or each polypide is surrounded by its own tube, which is not connected to neighbouring ectocysts. Particularly in the latter condition, the cystid is not a firm structure that maintains the general structure of the colony, but appears to be more buoyant (Figs. 2b and 3, Additional file 1: Video S1). This

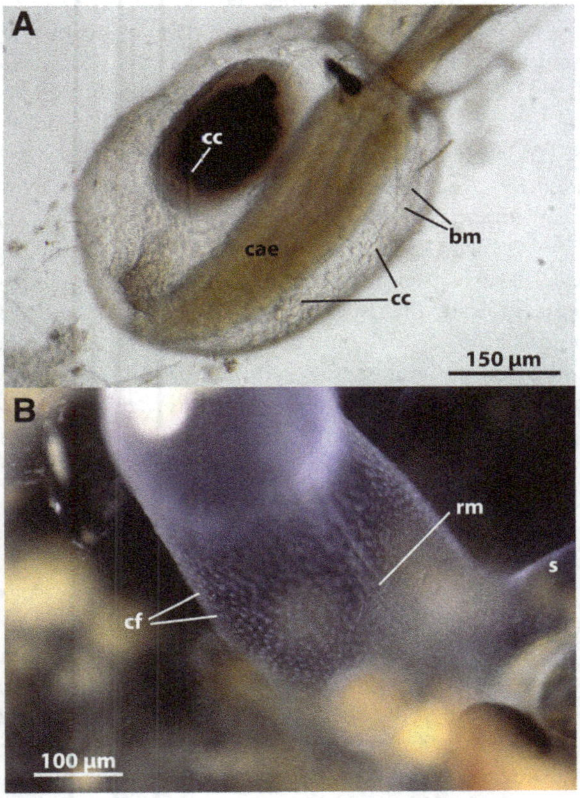

Fig. 5 Coelomic structures of *Stephanella hina*. **a** Lateral view of a solitary zooid of *Stephanella hina*, showing body-wall musculature as well as coelomocytes within the coelomic cavity. **b** Lateral view of the proximal part of a zooid, showing the regular arrangement of ciliary fields of the peritoneal lining of the coelomic cavity as well as strands of retractor muscle fibres traversing the same. Abbreviations: bm – body-wall musculature, cae – caecum, cc – coelomocyte, cf – ciliary fields of ther peritoneal epithelium, rm – retractor muscle fibres, s - stolon

Fig. 6 Statoblasts of *Stephanella hina*. **a** Statoblasts and developing statoblasts in a retracted zooid. Note the distance between the jelly-like cystid and the zooid (double arrow). **b** Single zooid with 14+ floatoblasts in the proximal cavity behind the zooid. **c** Detail of expulsed floatoblasts on the water surface. The floatoblasts are circular and show a darker annulus for floating and an inner orange fenestra, which contains the germination mass. Note a darker spot/central protuberance in the middle floatoblast indicating the deutoplasmatic side (ventral valve) of the statoblast. **d** Sessoblasts located in a distinct part of the colony not directly associated with a zooid but merely with stolons. The stolon-like processes interconnect individual zooids of the colony. **e** Detail of a sessoblast. The sessoblast is firmly attached to the substrate and lacks a floating annulus. Note on the left side sessoblast-remains. In the latter only the attaching cementus is present, indicating the probability that the frontal valve of the sessoblast was mechanically removed. Abbreviations: cp – central protuberance, ds – developing statoblast, ec – ectocyst, fb – floatoblast, l – lophophore, oc – opening of the ectocyst tube, s – stolon, sb – sessoblast, z –zooid

is most evident when colonies and their substrate are removed from the water, which results in a collapse of the colony, resulting in a flattened appearance (Fig. 2a). The soft jelly-like and barely self-supportive form of this ectocyst was previously recognized [17, 18], whereas the first more dense form has not been reported. No distinct environmental cues are known for different cystid structure among colonies, especially since colonies were collected from the same localities and sometimes also at the same time of the year.

As described above, the ectocyst is for the most part transparent – in general, however, this 'cystid' has sticky properties and particles floating in the water column easily adhere to it as previously described by Oka [21] (Figs. 2b and 3a; Additional file 3: Video S3). Cystid structures in the non-calcified Phylactolaemata are either gelatinous as in most of the larger forms (Pectinatellidae, Cristatelllidae, Lophopodidae) or sand-encrusted to chitinous in the smaller, rather branching representatives (Plumatellidae, Fredericellidae) [13, 31]. The ectocyst in *S. hina* is different from that of the above-mentioned taxa, and is probably best characterized as jelly-like. It is most comparable in appearance and mechanical properties to the protein/glycoprotein jelly-coats of frog embryos [32].

In all bryozoan zooids the cystid consists of an inner, living endocyst (the body-wall with the outer epidermis and inner peritoneum) and an extracellularly secreted cuticle, the ectocyst or sometimes called periostracum (see eg. [33] for phylactolaemates and [34] for calcareous taxa). These two components are commonly attached firmly to each other, especially in the calcified Stenolaemata and Cheilostomata. Analysis of the retraction process in *Stephanella hina* shows that the endocyst is for the most part not attached to the ectocyst and thus actually not a cuticle of the endocyst but a tube surrounding the zooid (Fig. 7). During the retraction process, the endocyst is quite distant from the ectocyst, leaving a considerable gap (Fig. 7, Additional file 7: Video S7). This kind of ectocyst is thus unique among Phylactolaemata and Bryozoa in general and superficially resembles tubes of other filter-feeding lophotrochozoans (e.g., tube-forming polychaete annelids and phoronids).

Coelomocytes

Due to its transparency, internal structures can readily be observed. One very obvious feature in *Stephanella hina* is commonly elongated to roundish coelomocytes in the body cavity of the animals (Fig. 5a, see Additional file 6: Video S6, Additional file 8: Video S8 and Additional file 9: Video S9). Their amount and density seem to be highly variable and range from very few to several dozens. Cues affecting their abundance are unknown. Ciliary tufts regularly situated at the inner peritoneal layer (cf. [30, 33]) create a current in the coelomic fluid that circulates the coelomocytes within the coelomic cavity (Fig. 5b). Previously, up to nine different coelomocytes were categorized in the Phylactolaemata [35]. None have been described as elongated/rod-like cells as found in *S. hina* (see Additional file 6: Video S6, Additional file 8: Video S8 and Additional file 9: Video S9). Coelomocytes in bryozoans appear to play different important roles that remain poorly understood. They appear to originate as detaching cells from the peritoneal layer of the body-wall. Especially vital dye experiments indicate that some of these cells have excretory functions [35–37], whereas some were commonly termed 'leucocytes' or 'lymphocytes'. An immunological function, however, could not be proven so far. Particular excretory organs/osmoregulatory organs such as proto- or metanephridial systems are absent among bryozoans (cf. [30]); their function seems to be taken over by coelomocytes, as has been described for marine bryozoans [38, 39].

In the current investigation, a few cases of the expulsion of coelomocytes were observed from specimens containing numerous coelomocytes (Additional file 10: Video S10). This phenomenon was previously documented in a few other phylactolaemates [40–42] and occurs via the vestibular pore at the anal side of the vestibular wall at the orifice. The vestibular pore is a not well-recognized structure which in addition to the expulsion of coelomocytes is also the point where statoblasts as well as sperm are commonly liberated [13, 23, 40, 42–44]. Only recently the pore was morphologically confirmed in the phylactolaemate *Lophopus crystallinus*. In the latter, muscles are present at

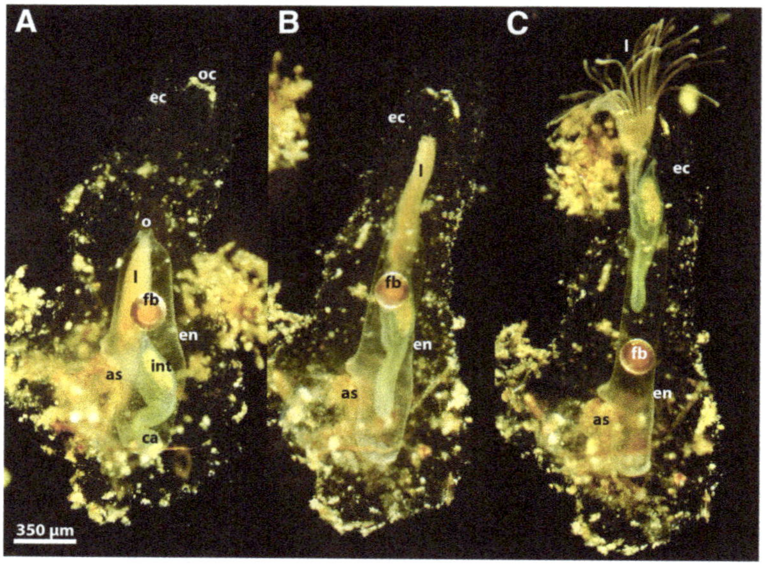

Fig. 7 The retraction process of an individual zooid of *Stephanella hina* in regard to the ectocyst. **a** Completely retracted zooid. **b** Mid-retracted/extended zooid and **c** protruded zooid. Note that in all images there is a distinct gap between the endocyst and the ectocyst, and that the endocyst slides past the ectocyst during the retraction process. Abbreviations: as – cup-shaped attachment site, ca – caecum, ec – ectocyst, en – endocyst, fb – floatoblast, int – intestine, l – lophophore, o – orifice, oc – opening of the ectocyst tube

the pore that enable active closure after the release of substances [42]. It is highly likely that *S. hina* and other phylactolaemate species possess such a pore, but this needs to be confirmed morphologically on a broader scale. Gymnolaemate bryozoans possess a comparable pore, the supraneural pore, which seems, at least partially, to fulfill similar functions. However, it is located at the lophophoral base rather than in the vestibular wall [45].

Experimental observations

De-colonisation of colonies

Some of the sample material contained single zooids not connected with others; these were thus clearly solitary. Later observations showed that this resulted mostly from the disruption of the stolon during removal of the zooids from the substrate. Consequently, individual zooids were excised from the colony to observe whether this would affect the life of the individuals. We found that de-colonisation does not seem to have any negative effect. Separated zooids that contain fully differentiated floatoblasts generally float on the top of the water column.

Considering that separation of the individuals or a colony did not seem to affect the future life of the colonies raises questions about how well zooids are integrated into colonial communication. For most bryozoans, colonial integration of the individuals is assumed (i.e. the assumption is that individuals communicate in some way with other zooids of the colony, particularly in response to threats such as predators). For this purpose a colonial nervous system is present within the body-wall (see [46]). In the more advanced and integrated taxa Stenolaemata and Gymnolaemata, the colonial plexus is localized in more direct patterns than in the Phylactolaemata, in which the nervous system in the body-wall consists merely of a diffuse plexus [30, 46–49]. Since information on the nervous system in *Stephanella hina* is lacking, the extent of neuronal integration of zooids is not known, but it is clear that zooids communicate by stolon-like connections across longer distances than in other freshwater bryozoans. Numerous observations in which individual zooids were tapped to provoke retraction found no effect on neighboring zooids (see Additional file 11: Video S11). This is not uncommon, especially for freshwater bryozoans, where stimulation of individual zooids does not necessarily lead to the retraction of more than just the affected zooid. Instead, (mechanical) stimulation of the cystid wall results in simultaneous retraction [50]. Due to the lack of a firm connection of the soft ectocyst to the endocyst, cystid vibrations often do not result in a colonial retraction process. Also, in some cases removing individual zooids from the colony did not result in any defensive response in other zooids of the colony, whereas wounding has been found to be a common stimulus for retraction in other phylactolaemates [50]. Suffice to say, this appears to be a rather ineffective protective mechanism. Moreover, *S. hina* colonies commonly do not react to regular shifts in the water column; regular movement of the dishes containing the colonies did not cause the zooids to retract, but rather caused the vertically standing tube-like extensions of the zooids, including the jelly-like ectocyst, to follow the motion of the water. This gave the colony the appearance of 'headbanging' tentacle crowns (see Additional file 12: Video S12). Currently, reduced predation during the cold months appears to be the only explanation for why zooids appear rather unresponsive to such stimuli. As aforementioned, *S. hina* is one of the few species that occurs in winter. Perhaps lower temperatures also lead to a lower metabolism in these animals, and thus slower response times. Alternatively, or additionally, predation risk may be significantly decreased during colder times in freshwater.

Ectocyst removal and cystid evolution

As described above, the ectocyst is for the most part not firmly attached to the body-wall; nonetheless, the retraction process functions properly. Investigation of decolonized, solitary forms shows that parts of the basal wall form a cup-like structure that connects with the ectocyst. This is the area at which the retractor muscle fibres attach, and thus represents the support for the retraction process. This limited attachment of the body-wall to the ectocyst gave the opportunity to release zooids from their ectocyst hull (Additional file 13: Video S13). Severing the attachment site, in most cases with a pair of forceps, gave rise to zooids devoid of an ectocyst. The animals readily survived this treatment and continued feeding. Tapping animals that lack an ectocyst does not cause retraction of the tentacle crown, only a longitudinal contraction of the trunk as, in such individuals, the retractor muscle has no support from the area normally attached to the substrate (Additional file 14: Video S14). However, such animals are capable of restoring/regenerating their ectocyst coating. At first, the new ectocyst is produced on the basal site at the attachment of the retractor fibres in order to enable the typical retraction process. The ectocyst renewal proceeds distally towards the tentacle crown that protrudes into the water column. Mechanical stimulation of such specimens leads to partial contraction of the trunk (Additional file 15: Video S15).

Experimental ectocyst removal was previously conducted only in the lophopodid *L. crystallinus* [51], in which colonies also survived the procedure without great harm. The ectocyst in *L. crystallinus* in general is just a thin gelatinous layer that is in rather loose connection with the endocyst [40]. With the current position of stephanellids and lophopodids, this may imply that the original ectocyst was not a proper cuticle attached to the body-wall, but merely a tube-like, semi-adhering protective secretion. The lophopodid genus *Lophopodella* commonly has a thin layer of ectocyst (Hirose pers. obs.), whereas the condition among the other lophopodid genus *Asajirella* hast not been analyzed recently. In the latter it appears to be more massive and gelatinous [52, 53]. It is definitely present in copious amounts on the basal side (the side facing the substrate), but it is not clear to what extent it is present on the frontal side facing the water. *Pectinatella magnifica* (sole species of Pectinatellidae) also produces copious amounts of gelatinous ectocyst, but only the basal side is covered by ectocyst, whereas the frontal one is naked [52]. Likewise, *Cristatella mucedo* (sole representative of the Cristatellidae) also produces only a thin ectocyst at its creeping sole towards the basal side [51, 52]. Pectinatellidae and Cristatellidae are generally regarded as closely related and probably constitute sister-groups [14]. *Cristatella* has motile colonies through the whole life-stages, whereas *Pectinatella* colonies are sessile gelatinous masses and only young colonies are motile [54–56]. Consequently, the three phylactolaemate families Lophopodidae, Cristatellidae and Pectinatellidae are motile, at least at one point during their ontogeny [56]. A rigid and firm ectocyst is thus obstructive for any movement. It appears that only in the most speciose family, the Plumatellidae, the separation of the ectocyst is not possible; the same applies to the Fredericellidae. However, in the latter it should be noted that zooids are capable of detaching from the cystid and leave the colony in form of swimming zooids under unfavorable conditions [57]. Given that the latter two families are currently considered as the derived form [14], it is likely that the phylactolaemate ancestor was a semi-motile, gelatinous animal with a rather loose ectocyst. *Stephanella hina* appears to be somewhat intermediate between a motile and strictly sessile form, since it is definitely sessile, but almost shows no connection to the loose ectocyst.

The *Stephanellidae*

The present study confirms that *Stephanella hina* is indeed a unique species among the Phylactolaemata. With its typical features, such as the colony structure, the jelly-like ectocyst and the mostly missing connection between endo- and ectocyst, it is clear that this species cannot be assigned to the Plumatellidae as it has been in previous studies (e.g. [17, 58]), but merits a separate family Stephanellidae as originally proposed by Lacourt [15]. Besides the clear distinguishing features presented above, Stephanellidae is, together with the Plumatellidae, the only phylactolaemate family to produce sessoblasts (sessile dormant stages/statoblasts). However, the formation of the sessoblast is essentially different in *S. hina* than in all plumatellids studied [18, 19]. In plumatellids the basal side of the sessoblast, attached to the substrate, derives from the so-called cystigenic side, whereas the opposite condition with the deutoplasmatic side forming the basal side is found in *S. hina* [18]. This substantiates separate placement into the Stephanellidae and indicates that sessoblasts probably evolved twice within the phylactolaemates. Finally, Fredericellidae and Plumatellidae show an oro-median gap within the intertentacular membrane [40], which all other families, including *Stephanella*, lack. The functional significance of this gap is not clearly understood. The lack of this particularly neglected feature again supports the placement of *Stephanella* into a separate family-level taxon. Accordingly, the features defining Stephanellidae are (i) stolon-like growth pattern with polypides mostly extending vertically from the substrate, (ii) lateral budding loci in addition to the normal buds produced on the oral side as in other Phylactolaemata, (iii) sessoblasts attaching to the substrate with the deutoplasmatic side, (iv) floatoblast circular and not oval as in Plumatellidae, (v) jelly-like ectocyst for the most part not in direct connection to the endocyst, (vi) lack of oro-median gap in the intertentacular membrane.

Conclusions and outlook

Stephanella hina is an ideal candidate for experimental approaches, as the transparency of the ectocyst as well as the easy detachment of single individuals from the colony allows insight into internal processes, such as digestion and the enigmatic coelomocytes. Other phylactolaemate families, such as the Lophopodidae, Cristatellidae or Pectinatellidae, are also transparent, but their large size and compact colony shape impede proper microscopic investigation. Moreover, they are difficult to keep in the laboratory. We kept large *Stephanella* colonies in a simple, aerated bucket at room temperature in the lab without any special care for over a week. The colonies were still in very good condition, which indicates that laboratory culture should be possible without too much effort. Some of the smaller branching forms, such as plumatellids or fredericellids, are opaque in their native habitat, but in case of *Fredericella* can be rendered transparent under proper laboratory culture [59]. The latter is particularly interesting for use in parasitological investigations, since phylactolaemate bryozoans are definitive hosts of myxozoan parasites, which can be observed within the body cavity [60]. So far no parasitic stages are known for *Stephanella*, a fact that perhaps could be correlated to their life in the cold times of the year, but it might be an interesting aspect to study in the future. Some undulating worm-like objects were encountered in some zooids. Some of these appear to be slowly circulating elongated coelomocytes, whereas others might represent ingested particles that were not properly digested and traverse the digestive tract unharmed.

What potential does this species have for phylogenetic inferences of the Bryozoa? Taking this species as one of the earliest branches, if not the earliest branch among

Phylactolaemata, the conditions found in *S. hina* are perhaps plesiomorphic. Bryozoa were often regarded being derived from worm-shaped organisms, perhaps similar to the Phoronida (e.g. [61]). Many filter-feeding annelids and also phoronids are capable of secreting protective tubes from the body-wall which the animals can retract into ([62–64] for Annelida) ([65] for Phoronida). The ectocyst of *S. hina* is more similar to these tubes rather than the firmly attached ectocyst (thus a proper cuticle) of other bryozoans. Filter-feeding worms that live in tubes commonly use the longitudinal musculature to retract into the tubes. Accordingly, where has the retractor muscle of bryozoans evolved from? The last attempt of an evolutionary interpretation regarded it as condensation and separation of longitudinal body-wall muscles (cf. [66]). This view was, however, biased by the 'lophophorate' concept, and took the phoronids as closest to the Bryozoa for granted. In this line of thought, the derivation of the retractor muscle appears likely from the prominent feather-like longitudinal muscle bundles found in phoronids. However, even without this bias, the derivation of the retractor muscle from a longitudinal body-wall musculature appears likely when a worm-shaped ancestor is assumed. The condition of zooids without an ectocyst in *Stephanella hina* also showed trunk contractions, similar to the shortening of a worm-shaped body due to contraction of a longitudinal body-wall musculature.

In the early branching lophopodids an epistome is lacking [33], which is traditionally considered to be one the phylactolaemate-specific characters (e.g. [10, 30]). Such a small flap-like protuberance over the mouth opening is also present in other 'lophophorate' phyla and was considered homologous among the three 'lophophorates'. Future studies should deal with the structure of different organ systems such as the epistome in order to assess whether this character has indeed been present at the base of all Phylactolaemata or has evolved within this group. In addition, future field work should aim at finding sexually mature colonies as well as larvae, since any kind of sexual reproduction is currently unknown in *Stephanella*.

Acknowledgements
Special thanks to all the departmental members of the AORI, Kashiwa Campus of the University of Tokyo for their kindness and hospitality.

Funding
Special thanks to the Faculty of Life Sciences and the Dean's office for financial support to TS for travelling to Japan.

Authors' contributions
TS carried out the experiments and wrote the draft. All authors contributed to the writing of the manuscript. MH coordinated necessary field work and lab work in Japan as well as contributed to data interpretation. AW supervised the study and contributed to data interpretation. All authors read and approved the final version of the manuscript.

Competing interests
The authors declare that they have no competing interests.

Author details
[1]Department of Integrative Zoology, University of Vienna, Althanstraße 14, Vienna 1090, Austria. [2]Atmosphere and Ocean Research Institute, University of Tokyo, 5-1-5 Kashiwanoha, Kashiwa-shi, Chiba 277-8564, Japan.

References

1. Halanych KM. The new view of animal phylogeny. Annu Rev Ecol Evol Syst. 2004;35:229–56.
2. Dunn CW, Hejnol A, Matus DQ, Pang K, Browne WE, Smith SA, Seaver E, Rouse GW, Obst M, Edgecombe GD, et al. Broad phylogenomic sampling improves resolution of the animal tree of life. Nature. 2008;452:745–U745.
3. Hejnol A, Obst M, Stamatakis A, Ott M, Rouse GW, Edgecombe GD, Martinez P, Baguna J, Bailly X, Jondelius U, et al. Assessing the root of bilaterian animals with scalable phylogenomic methods. Proc R Soc B Biol Sci. 2009;276:4261–70.
4. Nielsen C. Animal Evolution. Interrelationships of the Living Phyla. Oxford: University Press; 2012.
5. Hyman LH. The invertebrates. Vol. V. smaller coelomate groups. New York: McGraw-Hill; 1959.
6. Helmkampf M, Bruchhaus I, Hausdorf B. Phylogenomic analyses of lophophorates (brachiopods, phoronids and bryozoans) confirm the Lophotrochozoa concept. Proc R Soc B Biol Sci. 2008;275:1927–33.
7. Nesnidal M, Helmkampf M, Meyer A, Witek A, Bruchhaus I, Ebersberger I, Hankeln T, Lieb B, Struck T, Hausdorf B. New phylogenomic data support the monophyly of Lophophorata and an Ectoproct-Phoronid clade and indicate that Polyzoa and Kryptrochozoa are caused by systematic bias. BMC Evol Biol. 2013;13:1–13.
8. Nesnidal MP, Helmkampf M, Bruchhaus I, Ebersberger I, Hausdorf B. Lophophorata monophyletic - after all. In: Wagele JW, Bartolomaeus T, editors. Deep Metazoan Phylogeny: The Backbone of the Tree of Life: New Insights from Analyses of Molecules, Morphology, and Theory of Data Analysis. Berlin: Walter de Gruyter; 2014. 127–142.
9. Kocot K. On 20 years of Lophotrochozoa. Org Divers Evol. 2016;16:329–43.
10. Wood TS. General features of the class Phylactolaemata. In: Robinson RA, editor. Treatise on Invertebrate Palaeontology Part G: Bryozoa (Revised). Boulder and Lawrence: Geological Society of America and University of Kansas; 1983. p. 287–303.
11. Fuchs J, Obst M, Sundberg P. The first comprehensive molecular phylogeny of Bryozoa (Ectoprocta) based on combined analyses of nuclear and mitochondrial genes. Mol Phylogenet Evol. 2009;52:225–33.
12. Waeschenbach A, Taylor PD, Littlewood DTJ. A molecular phylogeny of bryozoans. Mol Phylogenet Evol. 2012;62:718–35.
13. Wood TS. Phyla Ectoprocta and Entoprocta (Bryozoans). In: Thorp JH, Rogers DC, editors. Ecology and General Biology, Vol I: Thorp and Covich's Freshwater Invertebrates, 4th Edition. London: Academic Press; 2014. 327–345.
14. Hirose M, Dick MH, Mawatari SF. Proceedings of the 14th International Bryozoology Association Conference, Boone, North Carolina, July 1–8, 2007, Virginia Museum of Natural History Special Publication No 15. In: Hageman SJ, Key MMJ, Winston JE, editors. Molecular phylogenetic analysis of phylactolaemate bryozoans based on mitochondrial gene sequences. Martinsville: Virginia Museum of Natural History; 2008. p. 65–74.
15. Lacourt AW. A monograph of freshwater bryozoa-Phylactolaemata. Zool Verh Uitgeg Rijksmus Nat Hist Leiden. 1968;93:1–159.
16. Toriumi M. Taxonomical study on fresh-water Bryozoa. XI. *Stephanella hina* Oka. Sci Rep Tokohu Imp Univ 4e serie, Biol. 1955;21:131–6.
17. Smith DG. *Stephanella hina* (Ectoprocta: Phylactolaemata) in North America, with notes on its morphology and systematics. J N Am Benthol Soc. 1988;7:253–9.
18. Mukai H. Systematic position of *Stephanella hina* (Bryozoa: Phylactolaemata), with special reference to the budding site and the attachment of sessoblasts. Zool Sci (Tokyo). 1990;7:947–54.
19. Mukai H. Comparative Morphological studies on the Statoblasts of lower phylactolaemate bryozoans, with discussion on the Systematics of the Phylactolaemata. Sci Rep Fac Educ Gunma Univ. 1999;46:51–91.
20. Mukai H. Notes on the freshwater Bryozoa in Gunma Prerecture (central Japan). Sci Rep Fac Educ Gunma Univ. 1982;31:17–32.
21. Oka A. Über eine neue Gattung von Süsswasserbryozoen. Annot Zool Jap. 1908;6:277–85.

22. Annandale N. Preliminary note on a new genus of Phylactolaematous polyzoa. Rec Ind Mus. 1909;3:279–80.
23. Marcus E. Sobre Bryozoa do Brasil. II. Boletim da Faculdade de filosofia, ciências e letras, Universidade di Sao Paolo, Zool. 1942;6:57–105.
24. Dzik J. Evolutionary relationships of the Early Paleozoic cyclostomatous Bryozoa. Palaeontology. 1981;24:827–62.
25. Taylor PD, Larwood GP. Major evolutionary radiations in the Bryozoa. In: Taylor PD, Larwood GP, editors. Major evolutionary radiations. Oxford: Clarendon Press; 1990. 209–233.
26. Taylor PD. Cyclostome systematics: phylogeny, suborders and the problem of skeletal organization. In: Herrera Cubilla A, Jackson JBC, editors. Proceedings of the 11th International Bryozoology Association Conference. Balboa: Smithsonian Tropical Research Institute; 2000. p. 87–103.
27. Taylor PD. Bioimmured ctenostomes from the Jurassic and the origins of the cheilostome Bryozoa. Palaeontology. 1990;33:19–34.
28. Taylor PD. The ancestrula and early growth pattern in two primitive cheilostome bryozoans: *Pyripora catenularia* (Fleming) and *Pyriporopsis partlandensis* Pohowsky. J Nat Hist. 1986;20:101–10.
29. Ostrovsky AN, Taylor PD, Dick MH, Mawatari SF. Pre-Cenomanian Cheilostome Bryozoa: Current State of Knowledge. In: Okada H, Mawatari SF, Suzuki N, Gautam P, editors. Proceedings of International Symposium The Origin and Evolution of Natural Diversity, 1–5 October 2007. Sapporo: Hokkaido University; 2008. p. 69–74.
30. Mukai H, Terakado K, Reed CG. Bryozoa. In: Harrison FW, Woollacott RM, editors. Microscopic anatomy of invertebrates. Chichester: Wiley-Liss; 1997. p. 45–206.
31. Kaya M, Baublys V, Satkauskiene I, Akyuz B, Bulut E, Tubelyte V. First chitin extraction from *Plumatella repens* (Bryozoa) with comparison to chitins of insect and fungal origin. Int J Biol Macromol. 2015;79:126–32.
32. Carroll EJJ, Wei SH, Nagel GM, Ruibal R. Structure and macromolecular composition of the egg and embryo jelly coats of the *Lepidobatrachus laevis*. Dev Growth Diff. 1991;33:37–43.
33. Gruhl A, Wegener I, Bartolomaeus T. Ultrastructure of the body cavities in Phylactolaemata (Bryozoa). J Morphol. 2009;270:306–18.
34. Tavener-Smith R, Williams A. The secretion and structure of the skeleton of living and fossil Bryozoa. Philos Trans R Soc Lond Ser B. 1972;264:97–159.
35. Mano R. The coelomic corpuscles and their origin in the freshwater bryozoan, *Lophopodella carteri*. Sci Rep Tokyo Kyoiku DaigakuSectB. 1964;11:211–35.
36. Harmer SF. On the nature of excretory processes in marine Polyzoa. Q J Microsc Sci. 1891;33:123–67.
37. Gordon DP. The ageing process in bryozoans. In: Woollacott RM, Zimmer RL, editors. Biology of bryozoans. New York: Academic; 1977. p. 335–76.
38. Matricon I. Étude histologique d'*Alcyonidium polyoum* (Hassall) (Bryozaire Cténostome). Cah Biol Mar. 1960;1:359–95.
39. Bobin G, Prenant M. Sur les cellules cavitaires de quelques Vésicularines (Bryozaires Cténostomes). Cah Biol Mar. 1972;12:479–510.
40. Marcus E. Über *Lophopus crystallinus* (PALL.). Zool Jb Anat. 1934;58:501–606.
41. Oda S. On the outflow of the blood in colonies of freshwater Bryozoa [japanese]. Kagaku (tokyo). 1958;28:37.
42. Gruhl A. Occurrence and Identity of "White Spots" in Phylactolaemata. In: Ernst A, Schäfer P, Scholz J, editors. Bryozoan Studies 2010. Verlin Heidelberg: Springer; 2013. p. 91–102.
43. Wiebach F. Über den Ausstoß von Flottoblasten bei *Plumatella fructicosa*. Zool Anz. 1952;149:181–5.
44. Wiebach F. Über den Ausstoss von Flottoblasten bei Plumatellen. Zool Anz. 1953;151:266–72.
45. Ostrovsky AN, Porter JS. Pattern of occurrence of supraneural coelomopores and intertentacular organs in Gymnolaemata (Bryozoa) and its evolutionary implications. Zoomorphology. 2011;130:1–15.
46. Gruhl A, Schwaha T. Bryozoa. In: Schmidt-Rhaesa A, Harzsch S, Purschke G, editors. Structure and Evolution of Invertebrate Nervous System. Oxford: Oxford University Press; 2015. p. 325–40.
47. Gerwerzhagen A. Beiträge zur Kenntnis der Bryozoen, I. Das Nervensystem von *Cristatella mucedo*. Z Wiss Zool. 1913;107:309–45.
48. Hiller S. The so-called "colonial nervous system" in Bryozoa. Nature. 1939;143:1069–70.
49. Shunkina KV, Zaytseva OV, Starunov VV, Ostrovsky AN. Comparative morphology of the nervous system in three phylactolaemate bryozoans. Front Zool. 2015;12:28.
50. Marcus E. Beobachtungen und Versuche an lebenden Süßwasserbryozoen. Zool Jb Syst. 1926;52:279–350.
51. Kraepelin K. Die deutschen Süßwasser-bryozoen. 1. Anatomisch-systematischer Teil. Abhandlungen aus dem Gebiete der Naturwissenschaften. hrsg vom Naturwissenschaftlicher Verein in Hamburg. 1887;10:168.
52. Mukai H, Oda S. Histological and histochemical studies on the epidermal system of higher phylactolaemate bryozoans. Annot Zool Jap. 1980;53:1–17.
53. Hirose M, Mawatari SF. Freshwater Bryozoa of Okinawa, Japan, with descriptions of *Rumarcanella* gen. nov (Phylactolaemata: Plumatellidae) and two new species. Zootaxa. 2011;2732:1–19.
54. Wilcox AW. Locomotion in young colonies of *Pectinatella magnifica*. Biol Bull. 1906;11:245–52.
55. Brooks CM. Notes on the statoblasts and polypides of Pectinatella magnifica. Proc Acad Nat Sci Phila. 1929;81:427–41.
56. Mukai H, Karasawa T, Matsumoto Y. Field and laboratory studies on the growth of *Pectinatella gelatinosa* Oka, a freshwater bryozoan. Sci Rep Fac Educ Gunma Univ. 1979;28:27–57.
57. Gorgoglione B, Kotob MH, El-Matbouli M. Migrating zooids allow the dispersal of Fredericella sultana (Bryozoa) to escape from unfavourable conditions and further spreading of Tetracapsuloides bryosalmonae. J Invertebr Pathol. 2016;140:97–102.
58. Koletic N, Novosel M, Rajevic N, Franjevic D. Bryozoans are returning home: recolonization of freshwater ecosystems inferred from phylogenetic relationships. Ecol Evol. 2015;5(2):255–64.
59. Kumar G, Abd-Elfattah A, Soliman H, El-Matbouli M. Establishment of medium for laboratory cultivation and maintenance of *Fredericella sultana* for in vivo experiments with *Tetracapsuloides bryosalmonae* (Myxozoa). J Fish Dis. 2013;36:81–8.
60. Tops S, Hartikainen H, Okamura B. The effects of infection by *Tetracapsuloides bryosalmonae* (Myxozoa) and temperature on *Fredericella sultana* (Bryozoa). Int J Parasitol. 2009;39:1003–10.
61. Farmer JD. An adaptive model for the evolution of the ectoproct life cycle. In: Woollacott RM, Zimmer RL, editors. Biology of Bryozoans. New York: Academic; 1977. p. 487–518.
62. Vinn O, Ten Hove HA, Mutvei H. On the tube ultrastructure and origin of calcification in sabellids (Annelida, Polychaeta). Paleontology. 2008;51:295–301.
63. Vinn O, Ten Hove HA, Mutvei H, Kirsimae K. Ultrastructure and mineral composition of serpulid tubes (Polychaeta, Annelida). Zool J Linnean Soc. 2008;154:633–50.
64. Shah DU, Vollrath F, Stires J, Deheyn DD. The biocomposite tube of a chaetopterid marine worm constructed with highly-controlled orientation of nanofilaments. Mater Sci EngC Mater Biol Appl. 2015;48:408–15.
65. Herrmann K. Phoronida. In: Harrison FW, Woollacott RM, editors. Microscopic anatomy of invertebrates. Chichester: Wiley-Liss; 1997. p. 207–35.
66. Jebram D. Arguments concering the basal evolution of the Bryozoa. Z zool Syst Evoltut-forsch. 1986;24:266–90.

3

An exceptionally preserved Eocene shark and the rise of modern predator–prey interactions in the coral reef food web

Federico Fanti[1,2†], Daniela Minelli[1], Gabriele Larocca Conte[1] and Tetsuto Miyashita[3*†]

Abstract

Background: Following extreme climatic warming events, Eocene Lagerstätten document aquatic and terrestrial vertebrate faunas surprisingly similar to modern counterparts. This transition in marine systems is best documented in the earliest teleost-dominated coral reef assemblage of Pesciara di Bolca, northern Italy, from near the end of the Eocene Climatic Optimum. Its rich fauna shows similarities with that of the modern Great Barrier Reef in niche exploitation by and morphological disparity among teleost primary consumers. However, such paleoecological understanding has not transcended trophic levels above primary consumers, particularly in carcharhiniform sharks.

Results: We report an exceptionally preserved fossil school shark (*Galeorhinus cuvieri*) from Pesciara di Bolca. In addition to the spectacular preservation of soft tissues, including brain, muscles, and claspers, this male juvenile shark has stomach contents clearly identifiable as a sphyraenid acanthomorph (barracuda). This association provides evidence that a predator–prey relationship between *Galeorhinus* and *Sphyraena* in the modern coral reefs has roots in the Eocene. A growth curve of the living species of *Galeorhinus* fitted to *G. cuvieri* suggests that all specimens of *G. cuvieri* from the lagoonal deposits of Bolca represent sexually and somatically immature juveniles.

Conclusion: The modern trophic association between higher-degree consumers (*Galeorhinus* and *Sphyraena*) has a counterpart in the Eocene Bolca, just as Bolca and the Great Barrier Reef show parallels among teleost primary consumers. Given the age of Bolca, trophic networks among consumers observed in modern coral reefs arose by the exit from the Climatic Optimum. The biased representation of juveniles suggests that the Bolca Lagerstätte served as a nursery habitat for *G. cuvieri*. Ultraviolet photography may be useful in probing for exceptional soft tissue preservation before common acid preparation methods.

Keywords: Early eocene climatic optimum, Carcharhinuformes, Triakidae, *Galeorhinus*, *Sphyraena*, Bolca, Soft tissue, Stomach content, Nursery habitat, von Bertalanffy

Introduction

Fossil record documents profound changes in both marine and terrestrial ecosystems across the Paleocene-Eocene Thermal Maxima (55.8 and 53.7 million years ago) [1–4]. These changes are preludes to the modern vertebrate faunas, with the increasing dominance of acanthomorph teleosts in marine systems and the emergence of artiodactyls, perissodactyls, and primates in terrestrial systems, among other radiation and extinction events [1, 3]. It still remains an open question as to whether, how, and exactly what modern ecological traits and feeding interactions arose in such biodiversity hotspots as coral reefs during this time. These fundamental questions demand case studies of ecological traits in the fossil counterparts against those in components of living coral reef vertebrate fauna, but direct evidence that allows such one-to-one comparison is rare.

Pesciara di Bolca Konservat–Lagerstätte from northern Italy occupies a special place in understanding the shift toward modern marine ecosystems. This late Ypresian Lagerstätte with more than 250 vertebrate species (approximately 90 families) [5–7] coincides chronologically with the latest phase of the Early Eocene Climatic

* Correspondence: tetsuto@ualberta.ca
†Equal contributors
[3]Department of Biological Sciences, University of Alberta, Edmonton, Alberta T6G 2E9, Canada
Full list of author information is available at the end of the article

Optimum [2, 8, 9] and that of the 'reef gap' in which foraminiferal/algal banks and shoals replaced coral-dominated reefs globally across low latitudes (57–42 mya) [2, 10–12]. The locality also sat in the Tethys Sea connecting the Atlantic and Indo–Pacific oceans [13, 14]. Documenting the earliest occurrences of many acanthomorph lineages, the Bolca fishes represent the earliest of clearly defined coral reef fish assemblages [6, 15–17]. This assemblage resembles modern coral reef faunas in its remarkably high functional diversity of primary consumers, which in large part was facilitated by significant increase in morphological disparities among acanthomorph teleosts during the Paleocene-Eocene interval [3, 18–21].

Despite the wealth of ecomorphological data on primary consumers, chondrichthyans from this latest phase of the Climatic Optimum have received little attention. Most chondrichthyans are secondary or tertiary consumers or apex predators, and those from the Bolca Lagerstätte should be no exception [5, 22]. However, the overall morphology of chondrichthyans cannot be readily compared with that of teleosts using same anatomical landmarks in a single ecomorphological analysis, and direct evidence of trophic interactions has been lacking. Consequently, the teleost-centered paleoecological understanding of the Early Eocene vertebrate fauna has not readily transcended trophic levels above primary consumers. Several factors present a challenge to incorporating sharks into the paleoecological scheme, including documentation of head and body morphology, test of sexual dimorphism, and inference of trophic relationships. We solve these problems with the finely preserved specimens of the extinct carcharhiniform shark *Galeorhinus cuvieri* from Pesciara di Bolca Konservat-Lagerstätte. One of the specimens (MGGC 1976) receives a special focus for its exquisite preservation of soft tissues and for its stomach content.

List of institutional abbreviations

BM, Museo dei Fossili di Bolca, Bolca (Verona), Italy; **MCSNV**, Museo Civico di Storia Naturale, Verona, Italy; **MGGC**, Museo Geologico Giovanni Capellini, Bologna, Italy; **MGP-PD**, Museo di Geologia e Paleontologia, Padova, Italy.

Methods

The restoration project for damage caused by a 2012 earthquake allowed close examination of MGGC 1976 (slab and counterslab) using natural color and ultraviolet photography, stereomicroscopy, X-ray computed tomography, and SEM of extracted samples. The systematic use of UV light distinguished preserved tissues from cryptic reconstruction with fine grout and organic pigments in the 18th century. Furthermore, black light exposure revealed tissue–specific fluorescence, which delineated individual cartilages, muscles, brain, and visceral structures. We tested these interpretations under natural light, using differences in texture, topographical features, and colors.

We set three criteria for identification of the preserved elements: (1) clearly delineated edges under both visible and ultraviolet light, which are not obscured by more than 50 % of the outline; (2) area distinguished by texture, topography, and/or color enclosed within the outline; and (3) topographical and morphological consistency with corresponding elements in the living sharks (Fig. 1). As *G. galeus* is a FAO-listed vulnerable species, we used males of the living carcharhiniforms *Squalus acanthias* and *Mustelus mustelus* for dissection.

Ultraviolet photography could improve screening for soft tissue preservation in specimens from Lagerstätten like Pesciara di Bolca. Although it is a common practice to prepare such specimens using acid solution [23], dissolution of matrix can lead to loss of preserved soft tissues or impressions.

Results

Systematic paleontology

Carcharhiniformes Compagno, 1973

Triakidae Gray, 1851

Galeorhinus Blainville, 1816

Emended diagnosis

A carcharhiniform selachian with a unique combination of characters assembled from descriptions of living and fossil species [24, 25]: mouth arcuate, with sides of lower jaw convex; pectoral fin greater in area than first dorsal fin; distance from snout tip to first dorsal origin much greater than interdorsal space; supraorbital crest moderately concave; teeth up to 5 mm tall; cusp bent toward the rear (from anterior files distally); upper anterolateral teeth low–crowned; mesial cutting edge much greater than distal cutting edge; distinct cusplets of decreasing size on distal heel, well–separated from cusp; thin root with a distinct furrow; basal edge weakly concave.

Galeorhinus cuvieri (Agassiz, 1835)

Figs. 2, 3, 4, 5, 6, 7 and 8

Referred specimens

MNHN F.Bol.516 (holotype) (Fig. 8a), MCSNV T 1124 (Fig. 8d–f), MCSNV VII B 96-97 (Fig. 8c), MGGC 1976 (Figs. 2, 3, 4, 5, 6 and 7), MGP-PD 8871C/8872C (Fig. 8b), BM B70.

Locality and horizon

Pesciara di Bolca, Veneto, Italy (upper Ypresian; SBZ 11; NP 14).

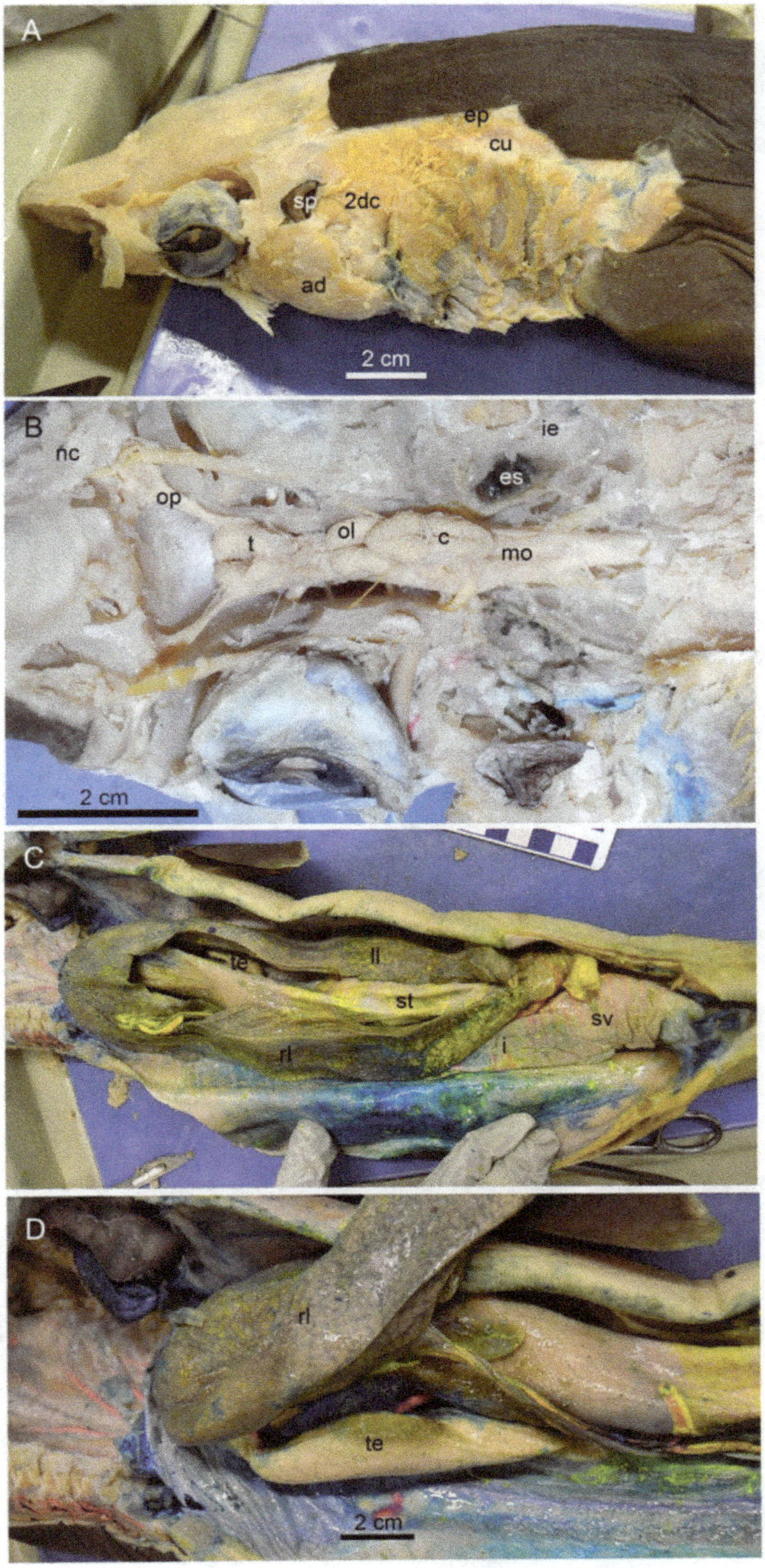

Fig. 1 (See legend on next page.)

(See figure on previous page.)
Fig. 1 A dissected male spiny dogfish (*Squalus acanthias*), showing the soft tissues identified in MGGC 1976 (*G. cuvieri*). **a** The head in left dorsolateral view with the skin removed. **b** The endocranial cavity in dorsal view with the chondrocranial roof removed. **c** The abdominal region in ventral view. **d** The abdominal region in right ventrolateral view. Abbreviations: *2 dc*, second dorsal constrictors; *ad*, adductor mandibulae; *cu*, cucullaris; **d**, diencephalon; *ep*, epaxial muscles; *es*, endolymphatic sac; *i*, intestine; *ie*, inner ear; *ll*, left liver; *mo*, medulla oblongata; *nc*, nasal cavity; *ol*, optic lobe; *op*, olfactory peduncle; *rl*, right liver; *sp*, spiracle; *st*, stomach; *sv*, spiral valve; *t*, telencephalon; *te*, testis

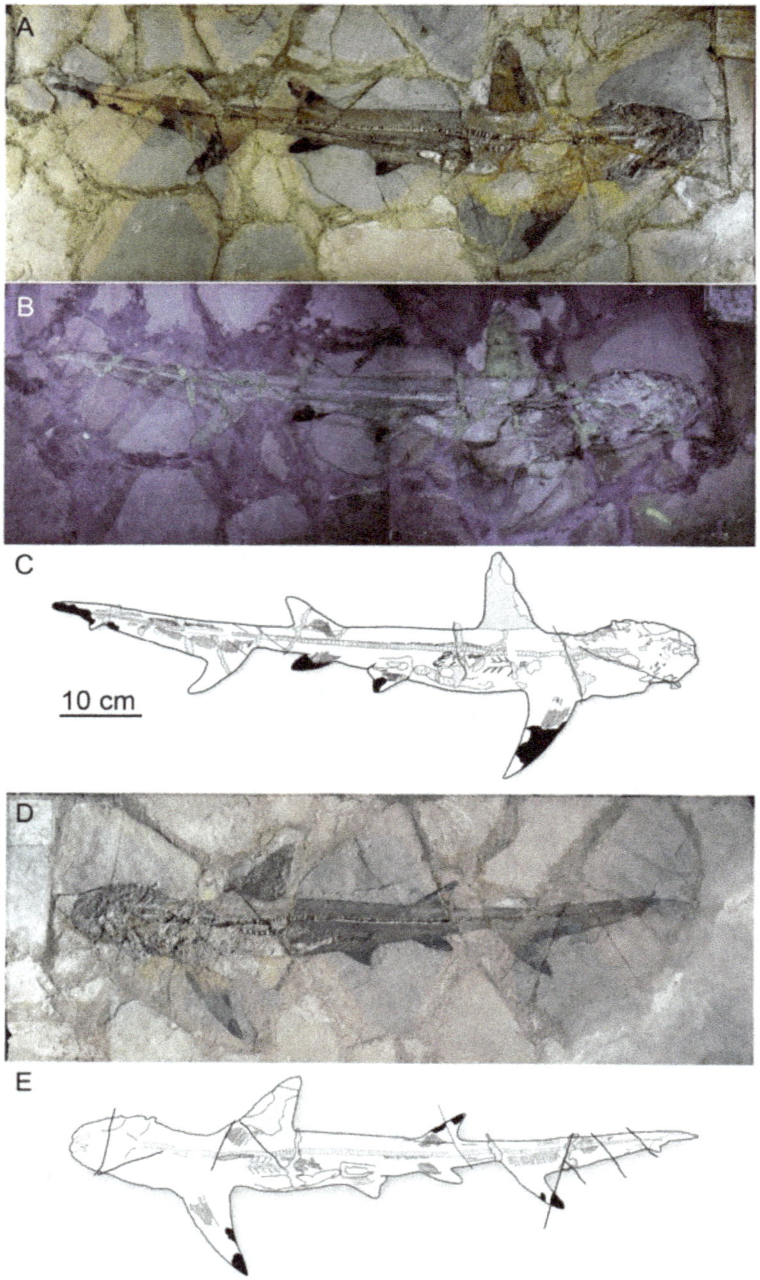

Fig. 2 MGGC 1976, a juvenile male of *Galeorhinus cuvieri* in dorsal to right lateral view. Photographs of the main slab under natural (**a**) and UV light (**b**) and an interpretive drawing of anatomical structures (**c**) with artifacts in light grey and areas preserved with dermal denticles in black. Photographs of the counterslab under natural light (**d**) and an interpretive drawing of anatomical structures (**e**) with artifacts in light grey and areas preserved with dermal denticles in black

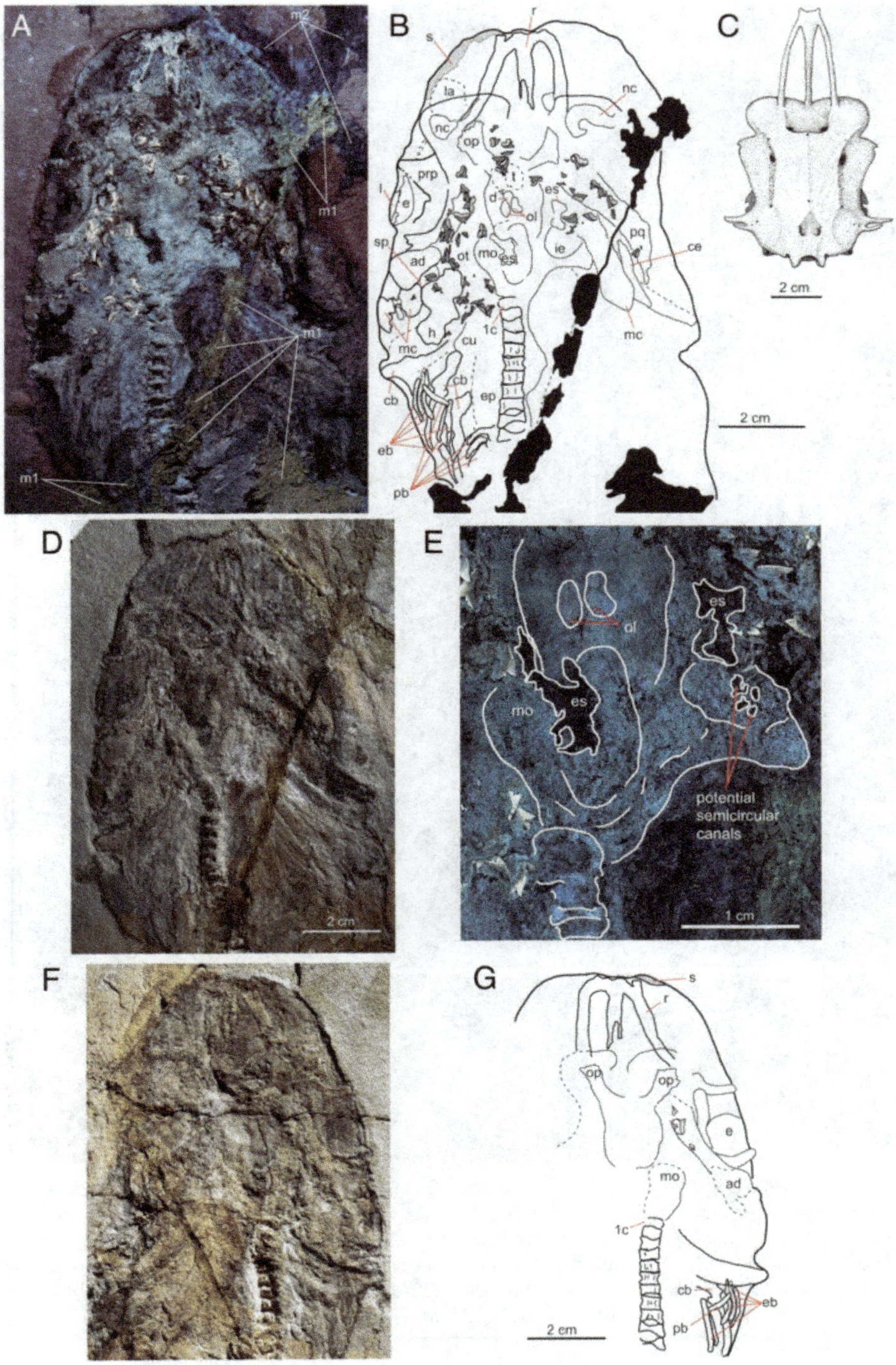

Fig. 3 Head anatomy of MGGC 1976 (*G. cuvieri*) with exceptional preservation of soft tissues. Photograph of the head region of the main slab under UV light (**a**) and interpretive drawing of anatomical structures (**b**) in dorsal view, compared to the chondrocranium of the living *G. galeus* (**c**) in dorsal view. The chondrocranium (**c**) is modified from [25] and scaled to the same anteroposterior length. The slab was assembled from multiple parts using mortar glue (*m1*) and reconstructed using pigmented mortar (*m2*). **d** Photograph of the head region of the main slab under natural light. **e** Details of the endocranial cavity and the braincase under UV light. **f, g** The head region of the counter slab under natural light (**f**) and outline (**g**). Abbreviations: *1c*, first cervical vertebra; *ad*, adductor mandibulae; **c**, cerebellum; *cb*, ceratobranchial; *ce*, ceratohyal; *cu*, cucullaris; **e**, eye; *eb*, epibranchials; *ep*, epaxial muscle; *es*, endolymphatic sac; **h**, hyomandibular overlapped by second dorsal constrictors; *ie*, inner ear; **l**, lens; *la*, sensory area with the ampullae of Lorenzini; *m1*, mortar used to glue parts of the slab; *m2*, pigmented mortar used to reconstruct anatomical parts; *mc*, Meckel's cartilage; *mo*, medulla oblongata; *nc*, nasal cavity; *ol*, optic lobes; *op*, olfactory peduncle; *ot*, otic capsule; *pb*, pharingobranchial; *pq*, palatoquadrate; *prp*, preorbital process; *r*, rostrum; *s*, patches of dermal denticles; *sp*, spiracle

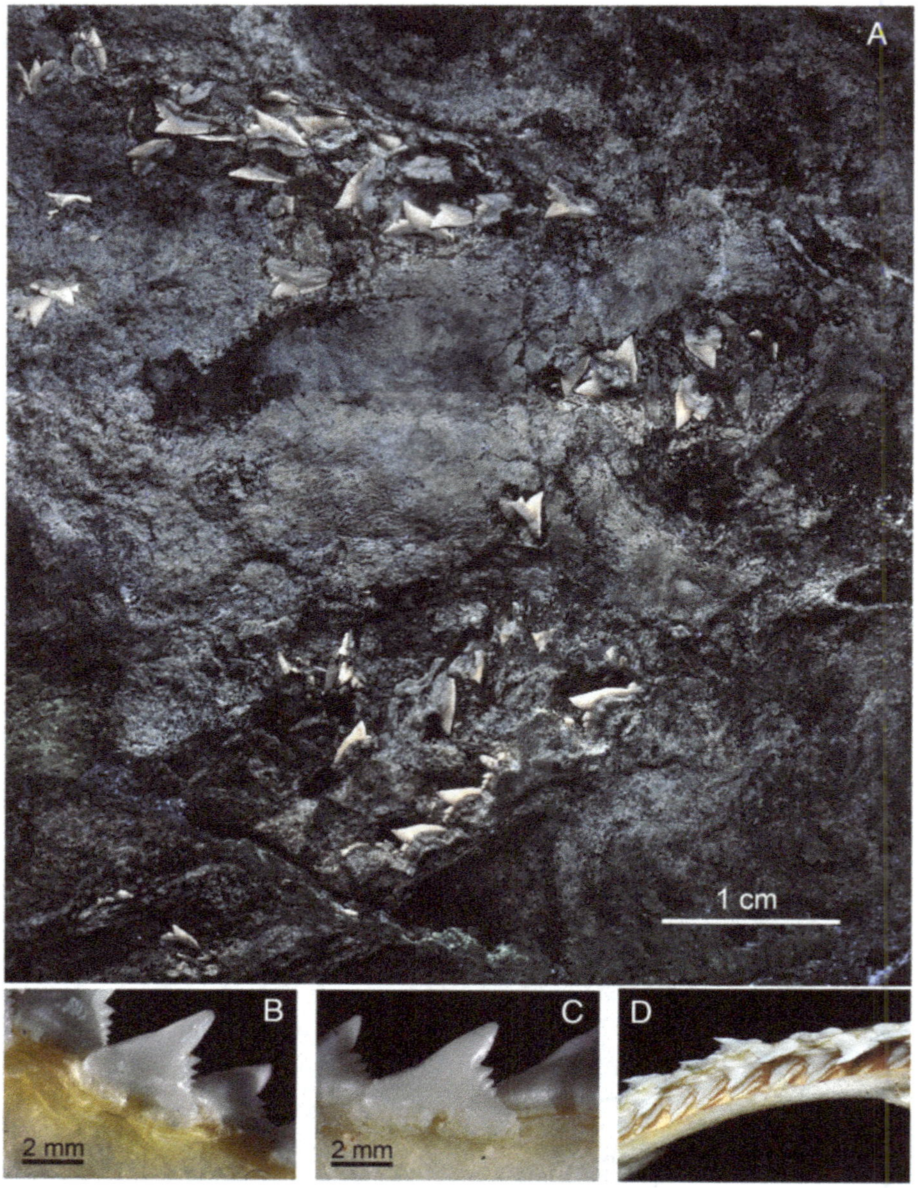

Fig. 4 Tooth morphology of the extinct and living species of *Galeorhinus*, *G. cuvieri* MGGC 1976 (**a**) and *G. galeus* (**b-d**). **a** A composite image of UV photographs showing the upper and lower tooth arcades of MGGC 1976. **b-d** Detailed photographs of the dentition of *G. galeus*, showing **b** a right upper anterior tooth (tooth position: #3), **c** right upper anterolateral tooth (#5), and left upper anterolateral to lateral tooth files (**d**). Arrow indicates some of the characters differentially diagnostic to *G. cuvieri* (see Diagnosis). Ch. 1: tooth cusp upturned near tip (from anterior files distally); Ch. 2: distal and mesial shoulders of tooth at similar horizontal levels; Ch. 3: distal heel separated from cusp by a notch deeper than cusplets measured along distal cutting edge; Ch. 4, relatively wide basal furrow

Emended diagnosis

Galeorhinus with the following unique combination of features: relatively short rostrum (not significantly longer than orbital length) (Figs. 2 and 9); tooth cusp upturned near tip (from anterior files distally) (Fig. 5; Character [Ch.] 1); distal and mesial shoulders of tooth at similar horizontal levels (Fig. 5; Ch. 2); distal heel separated from cusp by a notch deeper than cusplets measured along distal cutting edge (Fig. 5; Ch. 3); relatively wide basal furrow (Fig. 5; Ch. 4); vertebral count greater than 200 (Fig. 2); second dorsal fin taller than half the height of first dorsal (Figs. 2 and 9); dorsal edge of caudal fin as long as or longer than interdorsal distance (Figs. 2 and 9); accessory lobe of caudal fin less than a quarter of area of distal rest of main lobe (Figs. 2 and 9).

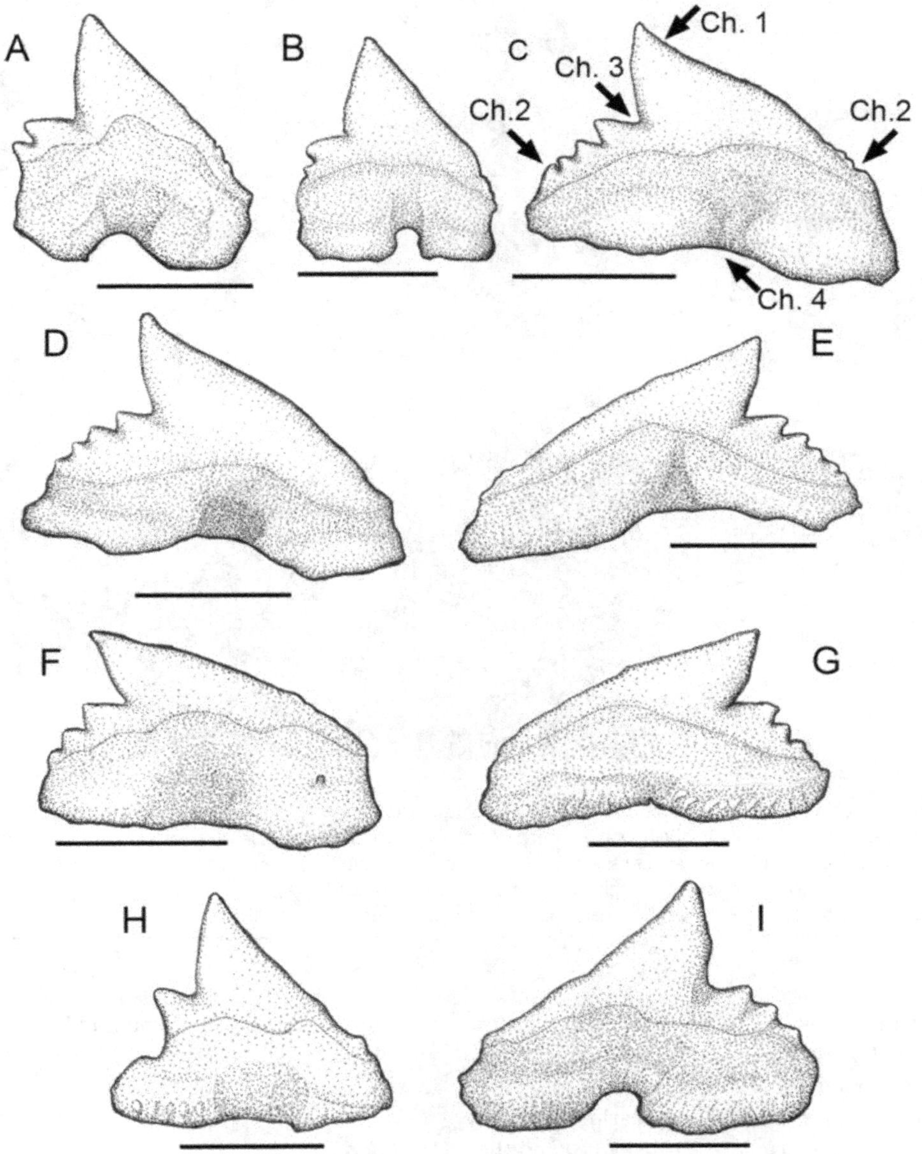

Fig. 5 Representative teeth of MGGC 1976 (*G. cuvieri*). Interpretive drawings of two left lower anterior teeth (**a**, **b**), two left lower anterolateral teeth (**c**, **d**), three left lower lateral teeth (**e-g**), a right upper anterior tooth (**h**) and a left upper anterior tooth (**i**). MGGC 1976 incompletely preserves both right and left pairs of upper and lower tooth series. Scale bar = 2 cm

Description

Overall morphology

MGGC 1976 (the best-preserved specimen of *G. cuvieri*) lies on the right side of the body (main slab: upper; counterslab: lower), with the head twisted clockwise (Fig. 2). The pectoral fin has approximately twice the area of the first dorsal fin, and this ratio is greater than in the living species of *Galeorhinus* [25]. The caudal fin occupies a third of the entire body length, and its dorsal lobe has an accessory lobe. In comparison to the living species of *Galeorhinus*, the caudal fin is slender and the accessory lobe is small. The pelvic girdle extends beyond the pelvic fin posteriorly into a clasper, which indicates that MGGC 1976 is a male (Fig. 6b).

MGGC 1976 (92 cm in total length) has a vertebral count of between 208 and 213 (114 to 118 in the trunk, 94–95 in the tail region). In comparison, MCSNV T 1124 (Fig. 8d–f) is a female of the same species (pelvic girdle clearly lacking claspers) and has a vertebral count between 200 and 204 (104 to 106 in the trunk, 96 to 98 in the tail). The vertebral count exceeding 200 in *G. cuvieri* is significantly higher than that in the living galeorhines (*G. galeus*:

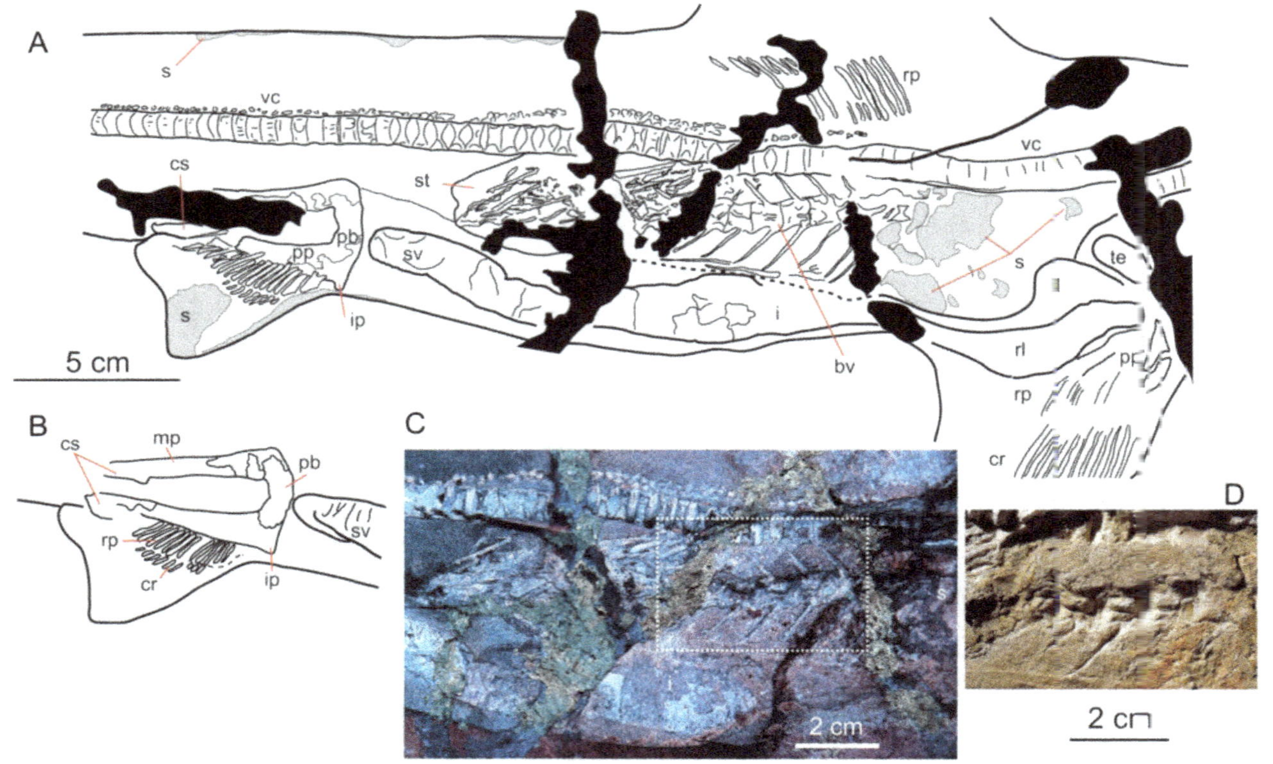

Fig. 6 Visceral anatomy and stomach content of MGGC 1976 (*G. cuvieri*) with exceptional preservation of soft tissues. **a** An interpretive drawing of the abdominal region of MGGC 1976 on the main slab, showing the visceral tissues, pelvic girdle, and stomach content. **b** An interpretive drawing of the pelvic girdle of MGGC 1976 on the counterslab. **c** Photograph under UV light of the stomach and intestine, revealing an undigested partial skeleton of the barracuda *Sphyraena bolcensis*. **d** Photograph under natural light to show the detailed morphology of the caudal vertebrae of *S. bolcensis*. Abbreviations: *bv*, caudal vertebrae of *S. bolcensis*; *cr*, ceratotrichia; *cs*, claspers; *i*, intestine; *ip*, iliac process; *ll*, left liver; *mp*, metapterygium; *pb*, puboischiatic bar; *pp*, propterygium; *rl*, right liver; *rp*, radial pterigyophores; *s*, dermal denticle patches; *st*, stomach; *sv*, spiral valve; *te*, testis; *vc*, vertebral column

136 centra, 83 in the trunk, 53 in the tail; *G. japonicus*: 160 centra, 106 in the trunk, 54 in the tail). Across the living selachians, vertebral counts of greater than 200 are rare. The exceptions are several carcharinid genera (*Carcharhinus*, *Galeocerdo*, *Prionace*, *Triaenodon*) and the leptochariid *Leptocharias* [25]. Dermal denticles are preserved in patches across the body (Fig. 7).

Chondrocranium and branchial skeleton

The tripartite rostrum of MGGC 1976 is relatively short among triakid sharks, not significantly longer than the orbital length (Fig. 3). The neurocranium is preserved with a clear outline on the left side of the head. The sigmoidal nasal cavity is entirely exposed on the right side and the sensory area for the ampullae of Lorenzini on the left side. Both right and left Meckel's cartilages can be delineated along the tooth series, meeting in roughly 70° (Fig. 3). The ceratohyals sit close to the proximal ends of the lower jaws. The palatoquadrates are preserved nearly in parallel to the Meckel's cartilages. On the left side of the head, a spiracle sits between the palatoquadrate and the hyomandibula. A full set of pharyngobranchials and epibranchials overlaps with at least three exposed ceratobranchials.

Dentition

The teeth of MGGC 1976 are small (3–4 mm crown height), cuspidate, and not bladelike (Figs. 4 and 5). Unlike the coeval *Eogaleus* [26], the crown height is never greater than the fore-aft basal length across the tooth series, even near the symphysis. From anterior files distally, the teeth have a cusp bent more posteriorly than dorsally so that the mesial cutting edge is greater than the distal edge. The cusp is upturned near the tip, which differs from other extinct species of *Galeorhinus* such as *G. gomphorhiza*, *G. minutissimus*, and *G. ypresiensis* and from other triakid genera such as *Khouribgaleus*, *Palaeogaleus*, and *Triakis* [24, 25]. A similarly upturned outline is also present in *G. duchaussoisi*, *G. goncalvesi*, *G. louisi*, and *Iago angustidens* [24, 27, 28]. However, the teeth of *G. cuvieri* have dorsoventrally taller roots than the teeth of *G. goncalvesi*; have smaller cusplets relative to the

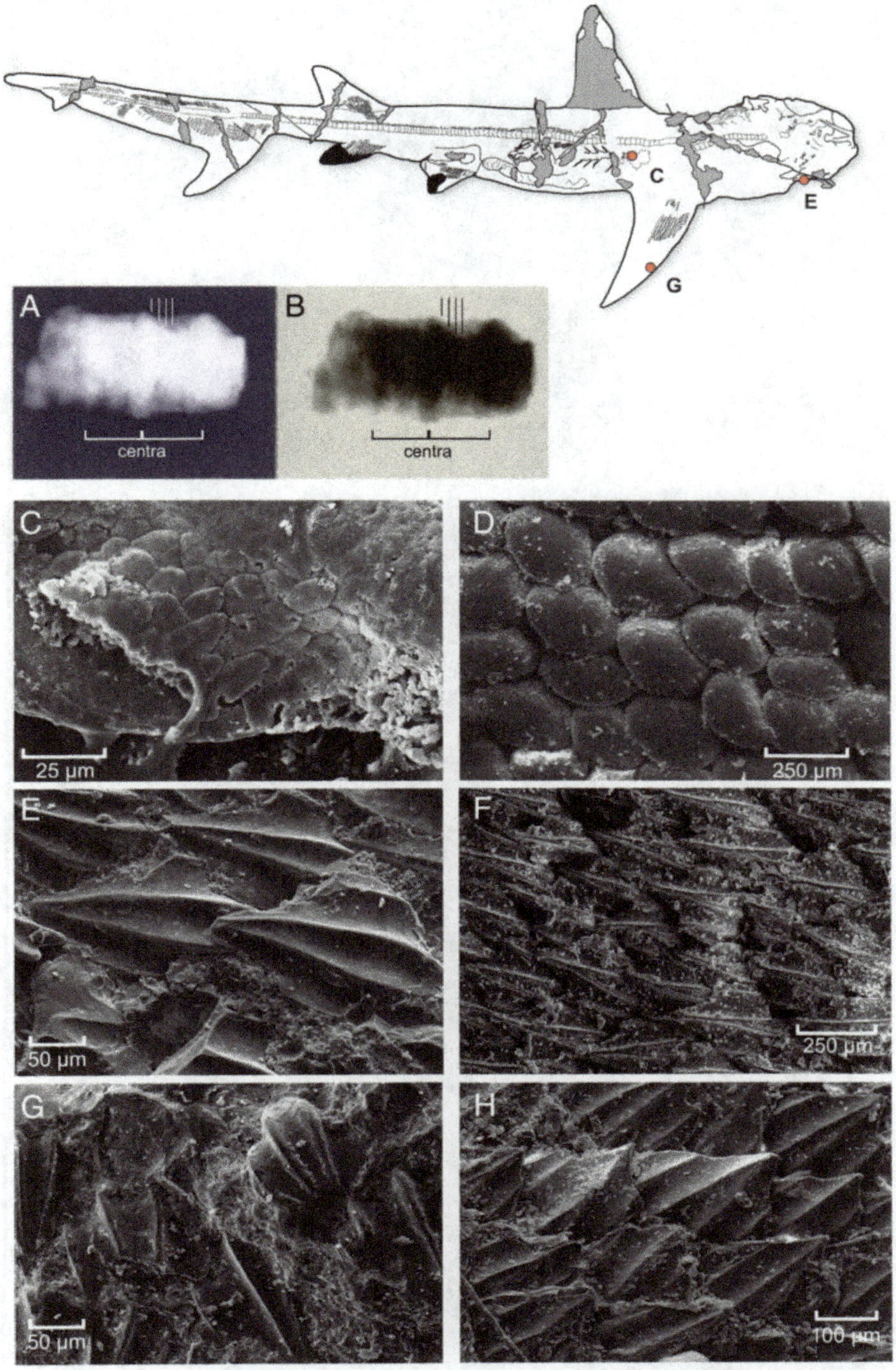

Fig. 7 Additional information on the morphology of *G. cuvieri*, with emphasis on dermal denticles in the extinct and living species of *Galeorhinus*. **a, b**, X-ray radiograph of two vertebral centra sampled from a young female of *G. cuvieri* (MCSNV T 1124), showing four vertebral bands. *G. cuvieri* MGGC 1976 (**c**, **e**, **g**) and *G. galeus* (**d**, **f**, **h**). An interpretive drawing of MGGC 1976 shows areas of preserved dermal denticles sampled for scanning electron microscopy (SEM). **c, d** SEM micrographs of dermal denticles from the abdominal region in *G. cuvieri* (**c**) and *G. galeus* (**d**). **e, f** SEM micrographs of dermal denticles from the head region in *G. cuvieri* (**e**) and *G. galeus* (**f**). **g, h** SEM micrographs of dermal denticles from the pectoral fin in *G. cuvieri* (**g**) and *G. galeus* (**h**)

Fig. 8 Specimens of *G. cuvieri* with whole body preservation from Pesciara di Bolca. **a**, MNHN F Bol516 (holotype, sex undetermined; photograph provided by G. Clément). **b**, MGP-PD 8871C (sex undetermined). This specimen is the smallest among those we examined and preserved with digested and unidentifiable stomach contents. **c**, MCSNV VII B96 (sex undetermined). **d**-**f**, MCSNV T.1124 (juvenile female) in left lateral view (**d**) and detailed view of the pelvic girdle under natural (**e**) and UV (**f**) light. [planned for single column width: 3.54 inches]

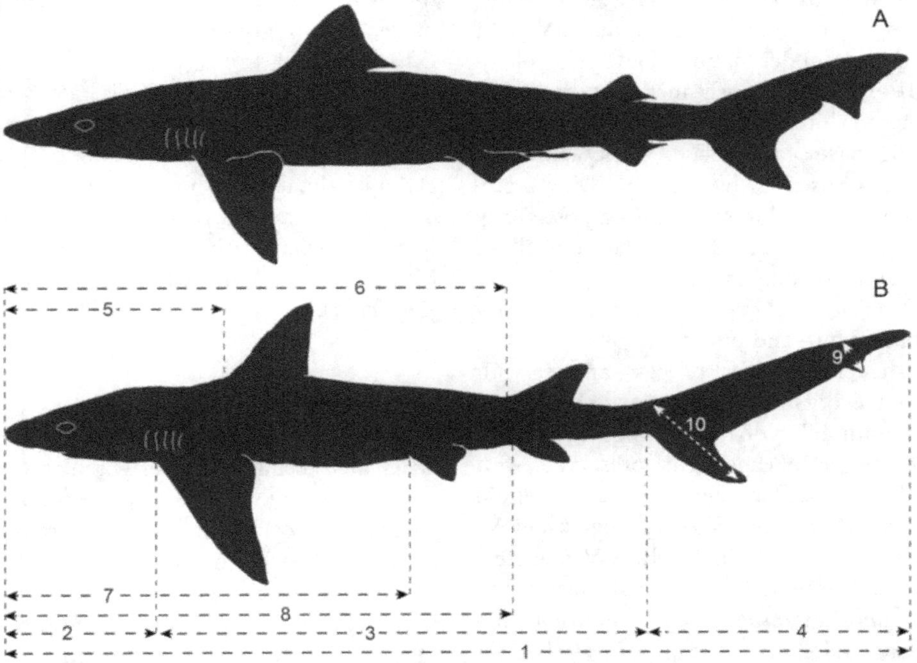

Fig. 9 Comparison of body proportions and outlines between the living and extinct species of *Galeorhinus* in left lateral view. **a** Mature male of the living *G. galeus* (modified from [25, 41]). **b** Juvenile male of the extinct *G. cuvieri* reconstructed on the basis of MGGC 1976, showing standard metric traits for Additional file 1: Table S1. Relative to *G. galeus*, *G. cuvieri* has a shorter rostrum, larger pectoral fins, taller second dorsal fin, shorter interdorsal distance, longer caudal fin, smaller accessory lobe, and smaller ventral lobe of caudal fin. Not to scale

cusp than in *G. duchaussoisi* and *G. louisi*; and are overall lower apicobasally than those of *I. angustidens*. Unlike *G. duchaussoisi* and *Hypogaleus* [25, 28], the mesial and distal shoulders are at similar horizontal levels across the tooth series. From the anterior files distally, the distal heel develops distinct cusplets of decreasing sizes. The number and size of the cusplets vary from tooth to tooth (one in symphyseal to five in lateral; Figs. 4 and 5), but they are invariably oriented posterodorsally, irrespective of the orientation of the cusp. There are no more than five distal cusplets. The count of distal cusplets in *G cuvieri* is greater than in *G. mesetaensis* and *G. minutissimus* (less than three), but smaller than in *G. duchaussoisi* (up to six). The distal series of cusplets is separated from the cusp by a notch deeper than the distal cutting edge of the first cusplet — a unique feature among triakids with the exceptions of *Hypogaleus* and *Khouribgaleus* [25, 29]. *G. duchaussoisi* has a similarly deep notch in some of the anterior teeth, but these cusplets are larger relative to the cusp in this taxon than in *G. cuvieri* [28]. The teeth of *G. duchaussoisi* are also more asymmetric mesiodistally than those of *G. cuvieri*. In the anterior to anterolateral teeth of *G. cuvieri*, the mesial cutting edge has weakly developed denticles, but none of them is a distinct cusplet. This feature distinguishes *G. cuvieri* from other species of *Galeorhinus* such as *G. ypressiensis* [28]. The basal edge is weakly concave, and the basal furrow is as wide as the first distal cusplet.

These features distinguish the dentition of *G. cuvieri* from that of other triakids [24–31]. *G. cuvieri* differs from all other species of *Galeorhinus* in having the deep notch separating the cusp and the first distal cusplet, among other features that differentially sets *G. cuvieri* apart from each of the species of *Galeorhinus* in various combinations. The cusp is taller than at least three times the height of the first cusplet in *G. cuvieri* unlike *Hemitriakis*, *Hypogaleus*, *Mustelus*, *Pachygaleus*, and *Triakis*. The presence of multiple distinct distal cusplets in all but symphyseal teeth rules out *Furgaleus*, *Iago*, *Paratriakis*, *Scylliogaleus*, and *Triakis*, whereas the absence of parallel apicobasal ridges precludes *Archaeotriakis*, *Mustelus*, *Palaeogaleus*, *Squatigaleus*, *Triakis*, and *Xystrogaleus*.

Nervous system

Centered in the head of MGGC 1976, an area of intense UV reflection in contrast to the less reflective surrounding matrix indicates that the preserved structure is not simply an infilling of the endocranial cavity. Instead, it likely represents permineralization in the cerebrospinal space between the meninx and brain (M. I. Coates, pers. commn. 2015). Remarkable preservation allows identification of major external features in the central and peripheral nervous systems (Fig. 3), which can be delineated with both highly reflective areas under UV light and texture/topography under natural light. The olfactory tracts (CN I) extend anteriorly in the approximate angle of 90° from each other. The right cerebral hemisphere sits above the base of the olfactory tract, although the left counterpart is obscured by the underlying teeth. These parts of the telencephalon are followed posteriorly by a pair of the optic lobes, and finally the cerebellum that sits on the medulla oblongata.

Within the left orbit, the eye and lens are clearly demarcated both in texture and outline.

Just anterior to the left preorbital process, an area with strongly UV-reflective dots may represent the chemosensory field of the ampullae of Lorenzini. On the right side of the head, the otic capsule is preserved with possible traces of semicircular canals. These potential semicircular canals are distinguished as narrow UV-nonreflective loops within an intensely UV-reflective area. They appear to represent the proximal third of the anterior, posterior, and horizontal loops. Anterior to the otic capsule, an area filled with the non-UV reflective matrix represents the endolymphatic sac. Its potential counterpart on the left side sits on top of the medulla oblongata and cerebellum, presumably shifted as the chondrocranium was crushed.

Musculature

In MGGC 1976, several major muscles are visible both under natural and UV light on the left side of the head and in the branchial region (Fig. 3). The UV-reflective structure overlapping the left palatoquadrate and Meckel's cartilage represents the adductor mandibulae, a major jaw adductor that attaches to these two cartilages and occupies a large area of the head behind the postorbital process in lateral view. A similarly UV-reflective structure overlapping the hyomandibula posterior to the spiracle represents a second dorsal constrictor (levators hyomandibulae and hyoideus). Near the left series of the gill arches, two unclearly defined structures extend in parallel with the vertebral column. If these are muscles, the lateral element would represent the cucullaris, whereas the medial one would be the epaxial musculature.

Visceral tissues

The abdomen of MGGC 1976 is preserved in right lateral view on the main slab (Fig. 6). The stomach overlaps the large left liver, whereas the smaller right liver partly parallels the ventral margin of the stomach. The testis sits near the proximal end of the livers. The intestine follows the stomach in the sigmoidal twist on the left side of the body. Once at the midline, it parallels the ventral margin of the stomach posteriorly. The helical contents reveal that it was a spiral valve, as in the living selachians. Count of the spiral valves is six or more, slightly greater than that in the living species of *Galeorhinus* at 4 or 5 [25].

The pelvic girdle of MGGC 1976 has claspers. The stout, slightly anteriorly convex puboischiatic bar bridges the iliac process. The propterygium extends at approximately 80 degrees with respect to the puboischiatic bar. The metapterygium posteriorly extends into a clasper on both sides. This is not the case in MCSNV T 1124, which, therefore, is a female (Fig. 8d-f).

Stomach content

The stomach content of MGGC 1976 consists of six articulated distal caudal vertebrae and dismembered caudal fin rays (Fig. 6). The amphicoelous centra are as long anteroposteriorly as tall dorsoventrally (approximately 7 and 8 mm each). In absolute size, the consumed prey had centra as large as those of the consumer (dorsoventral height of 8.2 mm for the centra below the first dorsal fin in MGGC 1976). The neural and haemal spines are semi–equal in length and oriented posterodorsally at approximately 40°. The spine length and the associated body outline show no evidence of tapering toward the caudal fin.

At this body size, candidates for the consumed fish are limited in the Bolca fauna. We considered scombrids and sphyraenids, both well represented in the locality [5]. The stomach content is not a scombrid, because neural and haemal spines do not decrease significantly in length from anterior to posterior positions along the vertebral column, because the spine orientation is at substantially lower angle around 40°, and because dorsal finlets are absent [32]. The vertebral centra are longer anteroposteriorly than tall dorsoventrally. These morphological features are consistent with the sphyraenid *Sphyraena bolcensis* [33, 34] (Fig. 10). This association does not entirely rule out scavenging, but the intact preservation of an articulated skeleton in the stomach indicates that the preserved portion of *S. bolcensis* was ingested at or near death in a single large bite. Consistent with this, there is no evidence of erosion or weathering on the vertebrae of the prey.

Age and size estimates

Curiously, all six specimens of *G. cuvieri* fall into a range of similar body sizes from the total length of 54 to 92 cm (Figs. 2 and 8). Perhaps this small variation reflects similar ontogenetic ages among the individuals. We used the von Bertalanffy growth equation [35] to calculate age and mass estimates as routinely applied to modern chondrichthyans [36, 37]:

Fig. 10 Two specimens of *Sphyraena bolcensis* (Sphyraenidae) from Pesciara di Bolca. This taxon formed part of the diet for *G. cuvieri.* **a** MCSNV VII B18 in right lateral view. **b**, **c** MCSNV VII B12 in left lateral view, with a close up of the caudal region for comparison with the stomach content of MGGC 1976

$$t_1 = \left(\frac{1}{k}\right) \ln \frac{L_\infty}{L_\infty - L_t} + t_0 \qquad (1)$$

where t_1 is age of the individual, k is a growth coefficient (rate of change of length increment), L_∞ is the mean maximum body length for the population, L_t is body length of the individual, and t_0 is hypothetical postnatal age extrapolated from a growth curve when length equals zero. We substituted for the parameters k, L_∞, and t_0 with values taken from growth curves of multiple living carcharhiniform populations (Table 1). These calculations provide a range of numerical estimates. For example, a dataset for males of *G. galeus* from Australia [38] has the following life history parameters:

$$k = 0.1675; L_\infty = 158.33\ (\text{cm}); t_0 = -1.2545\ (\text{years})$$

As $L_t = 92$ (cm) for MGGC 1976, the estimated age of the individual is:

$$t_1 = \left[\frac{1}{0.1675} \times \ln \frac{158.33}{158.33\text{-}92}\right] \text{-}1.2545 \cong 3.9\ (\text{years})$$

With the estimated age, another von Bertalanffy growth equation provides a mass estimate of MGGC 1976:

$$w = P_\infty \left[1\text{-}e^{-k(t_1 - t_0)}\right]^3 \qquad (2)$$

where P_∞ is mean maximum weight for the population. Again, we use known values taken from a growth curve based on a dataset for males of *G. galeus* in Australia [38] in this example:

$$P_\infty = 44.7\ (\text{kg}); k = 0.1675$$

At $t_1 = 3.940$ for MGGC 1976:

$$w = 44.7 \times \left[1\text{-}e^{-0.1675(3.940+1.2545)}\right]^3 \cong 8.77\ (kg)$$

We fitted multiple von Bertalanffy growth curves derived from various living carcharhiniforms to six

Table 1 Age estimates for specimens of *Galeorhinus cuvieri* using the von Bertalanffy functions for living relatives. All of the specimens are predicted to represent juveniles well before — or in some cases near — sexual and somatic maturity

Taxon	S.	k	L_∞	t_0	Age at maturity	MGGC 1976[M]	MCSNV T.1124[F]	MGP-PD 8871C	MCSNV VII B96	MNHN F Bol516	BM 370	Source
Galeorhinus galeus												
*Galeorhinus galeus**	M	0.1675	158.33	-1.2545	8+	3.94	3.94	2.16	2.87	2.23	1.24	[38]
*Galeorhinus galeus**	F	0.1600	161.83	-1.2818	10+	3.97	3.97	2.19	2.90	2.26	1.26	[38]
*Galeorhinus galeus**	C	0.1693	160.04	-1.2699	8-10+	3.95	3.95	2.17	2.88	2.24	1.24	[38]
Galeorhinus galeus§ R1	M	0.154	142.9	-1.64	12-17	5.06	5.06	2.64	3.59	2.73	1.44	[39]
Galeorhinus galeus§ R1	F	0.086	179.2	-2.68	13-15	5.70	5.70	2.97	4.08	3.08	1.49	[39]
Galeorhinus galeus§ R1	C	0.104	165.8	-2.37	12-17	5.41	5.41	2.80	3.85	2.90	1.42	[39]
Galeorhinus galeus§ R2	C	0.086	180.4	-2.48	12-17	5.81	5.81	3.13	4.22	3.23	1.65	[39]
Galeorhinus galeus§ LF	C	0.131	154.9	-1.91	12-17	4.97	4.97	2.59	3.54	2.68	1.35	[39]
Triakidae												
Mustelus mustelus	M	0.12	145.1	-2.14	6-9	6.24	6.24	3.24	4.41	3.35	1.74	[63]
Mustelus mustelus	F	0.06	204.96	-3.55	12-15	6.38	6.38	3.29	4.56	3.41	1.55	[63]
Mustelus mustelus	C	0.06	198.94	-3.82	6-15	6.53	6.53	3.28	4.61	3.41	1.46	[63]
Triakis semifasciata	M	0.089	149.9	-2.31	7+	8.38	8.38	4.62	6.10	4.76	2.71	[64]
Triakis semifasciata	F	0.073	160.2	-2.31	10+	9.39	9.39	5.41	7.00	5.56	3.32	[64]
Triakis semifasciata	C	0.082	153.6	-2.31	7-10+	8.83	8.83	4.96	6.50	5.11	2.97	[64]
Non-triakid carcharhiniforms												
Carcharrhinus leucas	C	0.076	285	-3	M: 14-15 F: 18+	2.13	2.13	0.65	1.27	0.71	-0.24	[65]
Carcharhinus plumbeus	C	0.046	186	-6.45	30	8.39	8.39	3.63	5.57	3.81	1.0[illegible]	[66]
Negaprion brevirostris	C	0.057	317.65	-2.302	M: 11.6+ F: 12.7+	3.70	3.70	1.99	2.71	2.07	0.97	[67]
Sphyrna lewini	M	0.131	336.4	-1.09	4.3+	1.35	1.35	0.66	0.95	0.69	0.25	[68]
Sphyrna lewini	F	0.156	353.3	-0.633	5.8+	1.30	1.30	0.76	0.99	0.78	0.43	[68]
Non-triakid carcharhiniforms with unrealistic age estimates												
Galeus sauteri	M	0.081	70.1	-0.527	9+	—	—	50.76 x	—	80.37 x	17.63	[69]
Galeus sauteri	F	0.089	69.8	-0.307	7+	—	—	49.90 x	—	—	16.39	[69]
Rhizoprionodon taylori	M	1.337	65.2	-0.41	1	—	—	—	—	—	0.91	[70]
Rhizoprionodon taylori	F	1.013	73.2	-0.455	1	—	—	2.37	—	2.63	0.87	[70]
Scyliorhinus canicula	M	0.118	87.42	-1.09	6.6+	—	—	12.11	18.74 x	12.58	7.06	[71]
Scyliorhinus canicula	F	0.15	75.14	-0.96	7.9+	—	—	15.74	—	16.92	7.49	[71]

The von Bertalanffy growth curves for small non-triakid carcharhiniforms do not fit readily to the specimens of *G. cuvieri*

They either return unrealistic age estimates that are significantly older than the longevity of the living counterpart (x) or are unable to calculate the estimates because the specimens of *G. cuvieri* are larger than the mean maximum length (L_∞) of the original population (—). k = growth coefficient (rate of change in length increment); L_∞ = mean maximum length for the population; t_0 = hypothetical postnatal length (y intercept of the growth curve). S. = sex: C = male and female combined; F = female; M = male. Age at maturity is in years

*Australian population; §New Zealand population; LF Length-frequency data; R1 Reader 1 (for vertebral bands); R2 Reader 2 (for vertebral bands)

specimens of *G. cuvieri* (Table 1). The data sets for different populations of *G. galeus* suggest the age of 3.9 to 5.8 years for the largest of the specimens (MGGC 1976 and MCSNV T 1124) and that of 2.2 to 3.1 years for the smallest (MGP-PD 8871C). This age estimate roughly agrees with the count of hypermineralized major vertebral bands of MCSNV T 1124 in X-ray radiograph ($n = 4$; Fig. 7a, b).

The biological implication of these estimates lies not in the absolute numerical values, but in the predicted relative age within each of the surrogate populations. When fitted against specimens of *G. cuvieri*, the von Bertalanffy growth curve for the Australian population of *G. galeus* suggests that all specimens of *G. cuvieri* with whole body preservation from Bolca, including MGGC 1976, represent sexually and somatically immature juveniles (Table 1;

total body length at sexual maturity in the Australian population of *G. galeus*: males > 120 cm; females > 130 cm [38]). Although *G. galeus* has regional variations in growth rates, the body size at sexual maturity is largely consistent across the species. With the data set for the slow-growing New Zealand population of *G. galeus* [39], the age estimates are numerically different (MGGC 1976: approximately 5.8 years old). However, the specimens of *G. cuvieri* still fell into the range of juveniles (sexual maturity reached after 12 years of age in this population; Table 1). The total body length of 92 cm in MGGC 1976 would indicate age of 7-8 years in the Brazilian population of *G. galeus*, again well before sexual maturity (reached after 12 years of age at body length > 113 cm in males) [40]. Although numerical estimates for *G. cuvieri* vary geographically and from warm to cool waters, the inference of immaturity for all specimens of *G. cuvieri* remains consistent across different data sets for the living *Galeorhinus*.

To our knowledge, this study represents the first application of von Bertalanffy equation to fossil selachians. An important inference borne out in this exercise is probable sexual and somatic immaturity, not numerical values of the estimates. This is because of the assumptions underlying our fitting of the growth curves. These estimates assume that *G. cuvieri* had similar growth parameters to those of *G. galeus*. There is no direct evidence to support this assumption other than the congeneric status. However, the implication for sexually immature status would still be robust to violation of this assumption. When fitting the von Bertalanffy growth curves of other triakids and non-triakid carcharhiniforms, the specimens of *G. cuvieri* represent either sexually immature individuals or individuals near the sexual maturity in those respective taxa (Table 1). The only exceptions are those in which the mean maximum size is substantially smaller than the specimens of *G. cuvieri*. In these small carcharhiniforms, the plateau of the growth curve is well below the sizes of the specimens of *G. cuvieri* so the age estimates are unrealistically high (denoted with 'x' or '—' in Table 1).

Hemitriakis, *Mustelus*, and *Triakis* may reach sexual maturity in the range of body size exhibited by the specimens of *G. cuvieri* (body length from 54 to 92 cm) among the living triakids (Additional file 1: Table S2). This is intriguing because triakids like *Hemitriakis* and *Hypogaleus* inhabit warmer waters than many of the sampled triakid populations (Table 1) [25]. Unfortunately, there is no robust data set for *Hemitriakis* and *Hypogaleus* to estimate von Bertalanffy growth curve. For this group of relatively small triakids, only the two largest specimens of *G. cuvieri* (MGGC 1976 and MCSNV T.1124) barely overlap the onset of sexual maturity in males of the documented populations of *Mustelus* and *Triakis* (Table 1). MCSNV T.1124 is a female (Fig. 8d-f) so this inference of sexual maturity does not apply to this specimen at the very least. In any case, none of these genera shares morphological characters with *G. cuvieri* that other species of *Galeorhinus* do not, both in the teeth and body proportions [24–26]. It is therefore difficult to consider them as a better analogue to estimate the relative age of *G. cuvieri* than the living *Galeorhinus*.

Discussion

Taxonomy and sexual dimorphism

The alpha taxonomy of chondrichthyans from Pesciara di Bolca has a convoluted history. In brief, the six specimens referred to *G. cuvieri* (and to its junior synonyms) were distinguished from the coeval carcharhinid *Eogaleus bolcensis* on the basis of tooth and body morphology [24, 26, 41] despite some attempts for synonymization [42]. Subsequently, many tooth-based extinct species of *Galeorhinus* from other early Cenozoic localities were attributed to the tooth-based extinct carcharhinid *Physogaleus*, and *G. cuvieri* was also swept into that latter genus [22, 28, 43]. Finally, the fossil teeth classically assigned to *Galeorhinus* and *Physodon* were considered as female and male morphs of *Physogaleus* spp. [24, 26]. This suspicion was raised because the type specimen of *Physodon muelleri* was re-identified as a male of *Scoliodon laticaudus* [44].

Our observations contradict the referral of *G. cuvieri* to *Physogaleus* and the implied sexual dimorphism in dentition in this species. *G. cuvieri* is best referred to the living genus *Galeorhinus* on the basis of a diagnostic combination of tooth and body characters. The living *Galeorhinus* does not exhibit distinguishable sexual dimorphism in tooth forms [30, 38, 45], and neither does *G. cuvieri*. This is borne out by the fact that a male (MGGC 1976) and a female (MCSNV T 1124) of *G. cuvieri* could be identified using the same diagnostic combination of tooth and body characters. However, the status of *G. cuvieri* does not determine whether or not other tooth-based species of *Galeorhinus* represent the carcharhinid *Physogaleus* [28] or whether or not the teeth of *Physogaleus* represent discrete sexual morphotypes [24, 26].

The assignment of *G. cuvieri* to the living genus *Galeorhinus* reflects the observation that this taxon is morphologically closer to *G. galeus* than to any other living triakids. A suite of characters in *G. cuvieri* — both in the tooth morphology and body proportions — follows that of the genus *Galeorhinus* (listed in **Diagnosis**) [25] and differs from the type *G. galeus* only in those diagnostic to the species (listed in **Diagnosis**; Figs. 4, 5 and 9). These characters diagnostic to *G. cuvieri* also differentially distinguish the taxon against other species of *Galeorhinus* (outlined in **Dentition**). The congeneric status with *G.*

galeus implies the existence of the genus for at least 50 million years. If *G. glickmani* from the Cenomanian of Russia also represents a congeneric species [28, 31], the chronological range may even extend well into the Late Cretaceous times. With mosaic distribution of tooth characters among the living and fossil species of *Galeorhinus*, it would be difficult to distinguish *G. cuvieri* at the generic level to the exclusion of all other species of *Galeorhinus*. Optimally sorting the observed morphological variations at the generic level would require a comprehensive systematic reassessment of the fossil *Galeorhinus* spp. At this point, the conservative approach is to reflect apparent morphological proximity with *G. galeus* in the generic assignment of *G. cuvieri*.

The bolca as a possible nursery habitat for *G. cuvieri*

The extant *G. galeus* inhabits warm temperate and tropical waters on continental shelves [25]. Juveniles tend to occupy inshore 'nursery' habitats such as protected bays or estuaries, whereas adults occur across the continental shelf. [30, 38, 45–50] *G. cuvieri* may have had similar ecology to *G. galeus*. All six specimens of *G. cuvieri* with whole body preservation from Bolca likely represent sexually and somatically immature juveniles (discussed in **Age and Size Estimates**; Table 1). MGGC 1976 (length: 92 cm) is the largest, and five other coeval specimens range from 54 cm (BM B70) to 92 cm (MCSNV T 1124) in total length (Figs. 2 and 8; Additional file 1: Table S1). If all of the specimens of *G. cuvieri* are indeed sexually immature, the Pesciara di Bolca locality may have been the nursery habitat.

Consistent with the nursery interpretation, the Pesciara di Bolca assemblage deposited in a shallow lagoonal setting bordered by active fluvial and coastal systems and by coral reefs [8, 9, 51]. These environmental features agree well with the modern nursery areas (generally protected, nearshore/inshore habitats) for various coastal selachians, including triakids [52, 53]. In such nearshore/inshore habitats, juvenile sharks tend to form a diffuse community of mesopredators feeding on similarly sized preys [54]. A nursery area may be identified on the basis of three criteria: (1) relative occurrence of the species in the habitat/area versus outside; (2) duration of occupying the habitat/area; and (3) repeated use of the habitat/area across years [52]. None of these criteria can be evaluated quantitatively for *G. cuvieri* with the current sample. Although the occurrence of likely juvenile individuals alone does not constitute direct support for the Bolca locality as a nursery for *G. cuvieri*, circumstantial evidence is consistent with that interpretation.

Trophic interactions in the Eocene coral reefs

The stomach content of MGGC 1976 indicates that *G. cuvieri* fed on *S. bolcensis*. The extant *Galeorhinus* frequently preys upon *Sphyraena* as a significant component of its diet that heavily relies on reef fish communities [25, 30, 38, 45, 46]. *Sphyraena* itself is a secondary to tertiary consumer in low-latitude coral reefs [47, 55]. Therefore, the stomach content provides a rare glimpse of feeding relationships near top of the trophic network in this coral reef fish community near the exit of the Early Eocene Climatic Optimum [2, 15].

The *Galeorhinus-Sphyraena* feeding interaction at higher trophic levels underscores modern features of the fish community of Bolca. The high functional diversity of herbivorous Bolca fishes is linked to post-Cretaceous increase in morphological disparity of acanthomorphs, and these are considered as important factors in shaping the Cenozoic coral reef communities [3, 20]. Disparity analyses suggest that the colonization followed a classic niche–filling scenario [3, 18–21, 56]. The invasion of herbivorous acanthomorphs into the coral reef habitats must have been accompanied by correlated shifts at higher trophic levels, and the *Galeorhinus-Sphyraena* association supports that prediction.

Coupled with the ecomorphological comparison of primary consumers [18–20, 56], the *Galeorhinus-Sphyraena* association provides evidence for deep conservation of the food web structure in modern coral reef communities. The late Ypresian age for the Bolca locality suggests that this particular feeding interaction occurred under the shift of marine vertebrate fauna across the Paleocene–Eocene thermal maxima. The challenges to testing this interpretation are: (a) poor age constraints for both the predator and prey lineages; and (b) the paucity of pre-Ypresian marine Lagerstätten to link the modern features of the Bolca vertebrate fauna unambiguously to the climatic pattern. As for the *Galeorhinus-Sphyraena* association, the trophic relationship cannot be presumed earlier than the earliest occurrence of either predator or prey lineage. The uncertain identification notwithstanding, teeth referable to the Triakidae occur in the Cenomanian of Ukraine and, possibly, the Hauterivian of England [31, 57]. It is difficult to constrain the origin of *Galeorhinus* or that of the Triakidae further. This is partly because the tooth morphology is highly heterogeneous among triakids [24, 28] and partly because molecular evidence is equivocal about whether or not *Galeorhinus* is nested within the Triakidae [58–60]. As for *Sphyraena*, *S. bolcensis* provides the hard minimum calibration point for the split between sphyraenids and their sister groups [61]. So the trophic association is as deep into fossil record as the Sphyraenidae. This is consistent with the hypothetical link to the Climatic Optimum, but the causal relationship is non-falsifiable with the data available at hand.

At the faunal level, the paucity of pre-Ypresian Lagerstätten poses a challenge to associate modern features of the Bolca fauna with the exit of the Climatic

Optimum [3, 20]. Elsewhere in the world, the stomach content of the scombrid teleost *Auxides huberti* from the contemporaneous late Ypresian locality of Senegal indicates that this tuna–like extinct scombrid fed on another extinct scombrid, the mackerel-like *Eoscomber senegalicus* [62]. Although this association has no congeneric parallel in modern marine ecosystem as in between *Galeorhinus* and *Sphyraena*, it reinforces the occurrence of modern feeding relationships at higher trophic levels near the exit of the Climatic Optimum. However, this interpretation has its basis in the marine Lagerstätten from the post–Climatic Optimum; little evidence exists in the immediately preceding stages to support or reject it. As such, the link to global climatic patterns awaits a test through increased sampling in the Paleocene-Eocene assemblages.

Additional file

Additional file 1: Table S1. A table of measurements for selected specimens of *Galeorhinus cuvieri*. Numbers for metric traits refer to those in Fig. 8. Measurements were made in cm. Specimen housed in the Museo Geologico Giovanni Capellini, Bologna, Italy: MGGC 1976a, b (slab and counter slab): a juvenile male (Figs. 2, 3, 4, 5, 6, and 7). Specimen housed in the Muséum national d'Histoire naturelle, Paris, France: MNHN F Bol516 (holotype specimen of *G. cuvieri*) (Fig. 8A). Specimen housed in the Museo di Geologia e Paleontologia, Padova, Italy: MGP-PD 8871C, 8872C (slab and counter slab): a juvenile (Fig. 8B). Specimens housed in the Museo Civico di Storia Naturale in Verona, Italy: MCSNV VII B96, B97 (slab and counter slab): a juvenile (Fig. 8C); MCSNV T.1124: a juvenile female (Fig. 8D-F). **Table S2.** Body lengths (in cm) of triakid taxa that are known to reach sexual maturity in the size range represented by specimens of *Galeorhinus cuvieri*. The data are based on ref. [1].

Competing interests
The authors declare that they have no competing interests.

Authors' contributions
FF and DM conceived the project; FF, DM, GLC, and TM contributed to data collection; TM and FF interpreted the data, prepared figures, and wrote the manuscript. All authors read and approved the final manuscript.

Acknowledgments
We thank C. Sarti and G.B. Vai (MGGC), R. Zorzin (MCSNV), M. Fornasiero (MGP), G. Clément and P. Janvier (MNHN) for access to specimens in their care. G. Clément provided a photograph of MNHN F.Bol.516. We thank B. Barr (University of Alberta) for a supply of *Squalus acanthias* and use of lab space, M. Coates (University of Chicago) for discussion on soft tissue preservation, and K. Miyashita for logistical support. T.M. was supported by Vanier Canada Graduate Scholarship.

Author details
[1]Dipartimento di Scienze Biologiche, Geologiche e Ambientali, Alma Mater Studiorum, Università di Bologna, Via Zamboni 67, Bologna 40126, Italy. [2]Museo Geologico Giovanni Capellini, Alma Mater Studiorum, Università di Bologna, Via Zamboni 63, Bologna 40126, Italy. [3]Department of Biological Sciences, University of Alberta, Edmonton, Alberta T6G 2E9, Canada.

References
1. Compagno PD. Environment and evolution through the Paleocene–Eocene thermal maximum. Trends Ecol Evol. 2006;21:246–53.
2. Norris RD, Turner SK, Hull PM, Ridgwell A. Marine ecosystem responses to Cenozoic global change. Science. 2013;341:492–8.
3. Friedman M. Explosive morphological diversification of spiny-finned teleost fishes in the aftermath of the end-Cretaceous extinction. Proc R Soc Lond B Biol Sci 2010:rspb20092177.
4. Near TJ, Eytan RI, Dornburg A, Kuhn KL, Moore JA, Davis MP, Wainwright PC, Friedman M, Smith WL. Resolution of ray-finned fish phylogeny and timing of diversification. Proc Natl Acad Sci U S A. 2012;109:13698–703.
5. Blot J. La faune ichthyologique des gisements du Monte Bolca (Province de Vérone, Italie). Bull Muséum Natl D'Histoire Nat 4e Sér Sect C. 1980;2:339–96.
6. Patterson C. An overview of the early fossil record of acanthomorphs. Bull Mar Sci. 1993;52:29–59.
7. Patterson C. Osteichthyes: Teleostei. In: Benton MJ, editor. The Fossil Record, vol. 2. Oxford: Blackwell; 1993. p. 621–65.
8. Papazzoni CA, Trevisani E. Facies analysis, palaeoenvironmental reconstruction, and biostratigraphy of the "Pesciara di Bolca" (Verona, northern Italy): An early Eocene fossil-Lagerstätte. Palaeogeogr Palaeoclimatol Palaeoecol. 2006;242:21–35.
9. Trevisani E, Papazzoni CA, Ragazzi E, Roghi G. Early Eocene amber from the "Pesciara di Bolca" (Lessini Mountains, Northern Italy). Palaeogeogr Palaeoclimatol Palaeoecol. 2005;223:260–74.
10. Scheibner C, Speijer RP. Late Paleocene–early Eocene Tethyan carbonate platform evolution — A response to long- and short-term paleoclimatic change. Earth-Sci Rev. 2008;90:71–102.
11. Payros A, Pujalte V, Tosquella J, Orue-Etxebarria X. The Eocene storm-dominated foralgal ramp of the western Pyrenees (Urbasa–Andia Formation): An analogue of future shallow-marine carbonate systems? Sediment Geol. 2010;228:184–204.
12. Perrin C, Kiessling W. Latitudinal trends in Cenozoic reef patterns and their relationship to climate. In: Mutti M, Piller W, Betzler C, editors. Carbonate Systems during the Oligocene–Miocene Climatic Transition. Oxford, UK: Wiley-Blackwell; 2012. 17–33.
13. Cowman PF, Bellwood DR. The historical biogeography of coral reef fishes: global patterns of origination and dispersal. J Biogeogr. 2013;40:209–24.
14. Cowman PF, Bellwood DR. Coral reefs as drivers of cladogenesis: expanding coral reefs, cryptic extinction events, and the development of biodiversity hotspots. J Evol Biol. 2011;24:2543–62.
15. Bellwood DR. The Eocene fishes of Monte Bolca: the earliest coral reef fish assemblage. Coral Reefs. 1996;15:11–9.
16. Carnevale G, Pietsch TW. An Eocene frogfish from Monte Bolca, Italy: The earliest known skeletal record for the family. Palaeontology. 2009;52:745–52.
17. Friedman M. The evolutionary origin of flatfish asymmetry. Nature. 2008; 454:209–12.
18. Bellwood DR, Goatley CHR, Brandl SJ, Bellwood O. Fifty million years of herbivory on coral reefs: fossils, fish and functional innovations. Proc R Soc Lond B Biol Sci. 2014;281:20133046.
19. Bellwood DR. Origins and escalation of herbivory in fishes: a functional perspective. Paleobiology. 2003;29:71–83.
20. Price SA, Schmitz L, Oufiero CE, Eytan RI, Dornburg A, Smith WL, Friedman M, Near TJ, Wainwright PC. Two waves of colonization straddling the K–Pg boundary formed the modern reef fish fauna. Proc R Soc Lond B Biol Sci. 2014;281:20140321.
21. Lobato FL, Barneche DR, Siqueira AC, Liedke AMR, Lindner A, Pie MR, Bellwood DR, Floeter SR. Diet and diversification in the evolution of coral reef fishes. PLoS One. 2014;9:e102094.
22. Carnevale G, Bannikov AF, Marramà G, Tyler JC, Zorzin R. The Pesciara-Monte Postale Fossil-Lagerstätte: 2. Fishes and other vertebrates. Rendiconti Della Soc Paleontol Ital. 2014;4:37–63.
23. Friedman M. Osteology of †*Heteronectes chaneti* (Acanthomorpha, Pleuronectiformes), an Eocene stem flatfish, with a discussion of flatfish sister-group relationships. J Vertebr Paleontol. 2012;32:735–56.
24. Cappetta H. Handbook of Paleoichthyology/Chondrichthyes · Mesozoic and Cenozoic Elasmobranchii: Teeth. München: Verlag Dr. Friedrich Pfeil; 2012.
25. Compagno LJV. Sharks of the Order Carcharhiniformes. Caldwell: The Blackburn Press; 2003.
26. Cappetta H. Handbook of Paleoichthyology Vol 3B: Chondrichthyes II Mesozoic and Cenozoic Elasmobranchii. Stuttgart, New York: Gustav Fischer; 1987.
27. Antunes MT, Balbino AC, Cappetta H. A new shark, *Galeorhinus goncalvesi* nov. sp. (Triakidae, Carcharhiniformes) from the latest Miocene of Portugal. Tert Res. 1999;19:101–6.
28. Adnet S, Cappetta H. New fossil triakid sharks from the Early Eocene of Prémontré, France, and comments on fossil record of the family. Acta Palaeontol Pol. 2008;53:433–48.

29. Noubhani A, Cappetta H. Les Orectolobiformes, Carcharhiniformes et Myliobatiformes (Elasmobranchii, Neoselachii) des bassins a phosphate du Maroc (Maastrichtien-Lutetien basal). Palaeo Ichthyol. 1997;8:1–327.
30. Compagno LJV. FAO Species Checklist Volume 4. Sharks of the world. An annotated and illustrated catalogue of shark species known to date. Part 2. Carcharhiniformes. 1984.
31. Popov EV, Lapkin AV. A new shark species of the genus *Galeorhinus* (Chondrichthyes, Triakidae) from the Cenomanian of the lower Volga River Basin. Paleontol J. 2000;34:435–8.
32. Monsch KA. A Revision of scombrid fishes (Scombroidei, Perciformes) from the Middle Eocene of Monte Bolca, Italy. Palaeontology. 2006;49:873–88.
33. Agassiz L. Recherches Sur Les Poissons Fossiles. Tome V, Part I. Neuchâtel: Petitpierre; 1844.
34. De Sylva DP. Systematics and Life History of the Great Barracuda, *Sphyraena barracuda* (Walbaum), vol. 1. Coral Gables: University of Miami Press; 1963 [*Studies of Tropical Oceanography*].
35. von Bertalanffy L. A quantitative theory of organic growth (inquiries on growth laws. II.). Hum Biol. 1938;10:181–213.
36. Allen KR. A method of fitting growth curves of the von Bertalanffy type to observed data. J Fish Res Board Can. 1966;23:163–79.
37. Chen Y, Jackson DA, Harvey HH. A comparison of von Bertalanffy and polynomial functions in modelling fish growth data. Can J Fish Aquat Sci. 1992;49:1228–35.
38. Olsen AM. Synopsis of biological data on the school shark *Galeorhinus australis* (Macleay 1881). FAO Fish Synop. 1984;139:1–49.
39. Francis MP, Mulligan KP. Age and growth of New Zealand school shark, *Galeorhinus galeus*. N Z J Mar Freshw Res. 1998;32:427–40.
40. Ferreira BP, Vooren CM. Age, growth, and structure of vertebra in the school shark *Galeorhinus galeus* (Linnaeus, 1758) from southern Brazil. Fish Bull. 1991;89:19–31.
41. Cappetta H. Les Sélaciens éocènes du Monte-Bolca. I - Les Carcharhinidae. Misc Paleontol Studi E Ric Sui Giacimenti Terziari Bolca. 1975;2:279–305.
42. Applegate SP. Phyletic studies; part I; tiger sharks. Univ Nac Autón México Inst Geol. 1978;2:55–64.
43. Cappetta H. Modification du statut générique de quelques espèces de sélaciens crétacés et tertiaires. Palaeovertebrata. 1980;10:29–42.
44. Springer VG. A revision of the carcharhinid shark genera *Scoliodon*, *Loxodon*, and *Rhizoprionodon*. Proc U S Natl Mus. 1964;115:559–632.
45. Walker TI, Cavanagh RD, Stevens JD, Carlisle AB, Chiaramonte GE, Domingo A, Ebert DA, Mancusi CM, Massa A, McCord M, Morey G, Paul LJ, Serena F, Vooren CM. IUCN Redlist. *Galeorhinus galeus*. 2006.
46. Ebert DA, Stehmann MFW. FAO Species Catalogue for Fishery Purposes No. 7. Sharks, Batoids and Chimaeras of the North Atlantic. Rome: Food and Agriculture Organization of the United Nations; 2013.
47. Fischer W, Bianchi G. FAO Species Identification Sheets for Fishery Purposes. Western Indian Ocean. Fishing Area 51, vol. 4. Rome: Food and Agriculture Organization of the United Nations; 1984.
48. Peres MV, Vooren CM. Sexual development, reproductive cycle, and fecundity of the school shark *Galeorhinus galeus* off southern Brazil. Fish Bull. 1991;89:655–67.
49. Hurst RJ, Baglet NW, McGregor GA, Francis MP. Movements of the New Zealand school shark, *Galeorhinus galeus*, from tag returns. N Z J Mar Freshw Res. 1999;33:29–48.
50. Holden MJ, Horrod RG. The migrations of tope, *Galeorhinus galeus* (L), in the eastern North Atlantic as determined by tagging. J Cons. 1979;38:314–7.
51. Giusberti L, Bannikov A, Boscolo Galazzo F, Fornaciari E, Frieling J, Luciani V, Papazzoni CA, Roghi G, Schouten S, Sluijs A, Bosellini FR, Zorzin R. A new Fossil-Lagerstätte from the Lower Eocene of Lessini Mountains (northern Italy): A multidisciplinary approach. Palaeogeogr Palaeoclimatol Palaeoecol. 2014;403:1–15.
52. Heupel MR, Carlson JK, Simpfendorfer CA. Shark nursery areas: concepts, definition, characterization and assumptions. Mar Ecol: Prog Ser. 2007; 337:287–97.
53. Knip DM, Heupel MR, Simpfendorfer CA. Sharks in nearshore environments: models, importance, and consequences. Mar Ecol: Prog Ser. 2010;402:1–11.
54. Heupel MR, Knip DM, Simpfendorfer CA, Dulvy NK. Sizing up the ecological role of sharks as predators. Mar Ecol: Prog Ser. 2014;495:291–8.
55. Senou H. *Sphyraenidae*. Rome: Food and Agriculture Organization of the United Nations; 2001. p. 3685–97 [*FAO WCP*].
56. Bellwood DR, Hoey AS, Bellwood O, Goatley CHR. Evolution of long-toothed fishes and the changing nature of fish–benthos interactions on coral reefs. Nat Commun. 2014;5:3144.
57. Underwood CJ, Mitchell SF, Veltcamp KJ. Shark and ray teeth from the Hauterivian (Lower Cretaceous) of north-east England. Palaeontology. 1999; 42:287–302.
58. Naylor GJP, Caira JN, Jensen K, Rosana KAM, Straube N, Lakner C. Elasmobranch Phylogeny. In Biology of Sharks and Their Relatives, Second Edition. Oxford, UK: CRC Press; 2012. p. 31–56. [Marine Biology].
59. López JA, Ryburn JA, Fedrigo O, Naylor GJP. Phylogeny of sharks of the family Triakidae (Carcharhiniformes) and its implications for the evolution of carcharhiniform placental viviparity. Mol Phylogenet Evol. 2006;40:50–60.
60. Vélez-Zuazo X, Agnarsson I. Shark tales: A molecular species-level phylogeny of sharks (Selachimorpha, Chondrichthyes). Mol Phylogenet Evol. 2011;58:207–17.
61. Santini F, Carnevale G, Sorenson L. First timetree of Sphyraenidae (Percomorpha) reveals a Middle Eocene crown age and an Oligo–Miocene radiation of barracudas. Ital J Zool. 2014;0:1–10.
62. Claeson KM, Sarr R, Hill RV, Sow EH, Malou R, O'Leary MA. New fossil scombrid (Pelagia: Scombridae) fishes preserved as predator and prey from the Eocene of Senegal. Proc Acad Nat Sci Phila. 2015;164:133–47.
63. Goosen AJJ, Smale MJ. A preliminary study of age and growth of the smoothhound shark *Mustelus mustelus* (Triakidae). South Afr J Mar Sci. 1997;18:85–91.
64. Kusher DI, Smith SE, Cailliet GM. Validated age and growth of the leopard shark, *Triakis semifasciata*, with comments on reproduction. Environ Biol Fishes. 1992;35:187–203.
65. Branstetter S, Stiles R. Age and growth estimates of the bull shark, *Carcharhinus leucas*, from the northern Gulf of Mexico. Environ Biol Fishes. 1987;20:169–81.
66. Casey JG, Natanson LJ. Revised estimates of age and growth of the sandbar shark (*Carcharhinus plumbeus*) from the western North Atlantic. Can J Fish Aquat Sci. 1992;49:1474–7.
67. Brown CA, Gruber SH. Age Assessment of the Lemon Shark, *Negaprion brevirostris*, Using Tetracycline Validated Vertebral Centra. Copeia. 1988; 1988:747–53.
68. Anislado-Tolentino V, Robinson-Mendoza C. Age and growth for the scalloped hammerhead shark, *Sphyrna lewini* (Griffith and Smith, 1834) along the central Pacific coast of Mexico. Cienc Mar. 2001;27:501–20.
69. Liu K-M, Lin C-P, Joung S-J, Wang S-B. Age and growth estimates of the blacktip sawtail catshark *Galeus sauteri* in northeastern waters of Taiwan. Zool Stud. 2011;50:284–95.
70. Simpfendorfer CA. Age and growth of the Australian sharpnose shark, *Rhizoprionodon taylori*, from north Queensland, Australia. Environ Biol Fishes. 1993;36:233–41.
71. Ivory P, Jeal F, Nolan CP. Age determination & growth and reproduction in the lesser-spotted dogfish *Scyliorhinus canicula* (L.). J Northwest Atl Fish Sci. 2005;35:89–106.

4

Functional morphology of giant mole crab larvae: a possible case of defensive enrollment

Nicole R. Rudolf*, Carolin Haug and Joachim T. Haug

Abstract

Background: Mole crabs (Hippidae) are morphologically distinct animals within Meiura, the "short-tailed" crustaceans. More precisely, Hippidae is an ingroup of Anomala, the group which includes squat lobsters, hermit crabs, and numerous "false" crabs. Within Meiura, Anomala is the sister group to Brachyura, which includes all true crabs. Most meiuran crustaceans develop through two specific larval phases. The first, pelagic one is the zoea phase, which is followed by the transitory megalopa phase (only one stage). Zoea larvae are rather small, usually having a total size of only a few millimeters. Zoea larvae of some hippidan species grow significantly larger, up to 15 mm in size, making them the largest known zoea larvae of all anomalan, and probably all meiuran, crustaceans. It has been suggested that such giant larvae may be adapted to a specific defensive strategy; i.e., enrollment. However, to date such giant larvae represent a rarity.

Methods: Eight specimens of large-sized hippidan larvae from museum collections were photographed with a Canon Rebel T3i digital camera under cross-polarized light. Additionally, one of the specimens was documented with a Keyence BZ-9000 fluorescence microscope. The specimen was subsequently dissected to document all appendages in detail. UV light (377 nm) was used for illumination, consistent with the specimen's autofluorescence capacities. For high-resolution images, composite imaging was applied.

Results: All specimens differ in important aspects from all other known hippidan zoea larvae, and thus probably represent either previously unreported larvae or stages of known species, or larvae of unknown species. The sixth pleon segment articulates off the telson, a condition not previously reported in hippidan zoea larvae, but only for the next larva phase (megalopa). The larvae described here thus most likely represent the ultimate pelagic larval stages, or rare cases of 'early megalopae'. The morphological features indicate that giant hippidan larvae perform defensive enrollment.

Conclusions: Our investigation indicates a larger morphological diversity of hippidan larvae than was known previously. Moreover, their assumed functional morphology, similar to the condition in certain stomatopod larvae, indicates a not yet directly observable behavior by these larvae, namely defensive enrollment. In a wider context, we are only just beginning to understand the ecological roles of many crustacean larvae.

Keywords: Giant larva, Zoea, Hippidae, Defensive behavior, Museum material

Abbreviations: ANT, Antenna; ATL, Antennula; BA, Basipod; CE, Compound eye; CX, Coxa; ED, Endit; EN, Endopod; EX, Exopod; FL, Flagellum; GE, Gnathic edge; GI, Gills; LB, Labrum; MXP, Maxilliped; PD, Peduncle; PL, Pleon; PLS, Postero-lateral spine; RST, Rostral spine; TE, Telson; TP, Thoracopod; VG, Ventral gape

* Correspondence: nicole.rudolf.nr@googlemail.com
Ludwig-Maximilians-Universität München, Fakultät für Biologie, Biozentrum, Großhaderner Str. 2, 82152 Planegg-Martinsried, Germany

Background

Within the diverse group of Eucrustacea, Hippidae is a rather small ingroup with a distinct adult morphology; its representatives are known as sand or mole crabs [1]. Hippidae is an ingroup of Anomala (often also termed Anomura), the group uniting hermit crabs, false crabs, and squat lobsters. Anomala and Brachyura (true crabs) together form Meiura.

Within Hippidae, three species groups are generally differentiated: *Emerita* Scopoli, 1777, *Hippa* Fabricius, 1787 and *Mastigochirus* Miers, 1878 [2]. As with other meiurans, representatives of Hippidae develop through two distinct larval phases: a zoea phase with 3–6 pelagic zoea stages, followed by a critical metamorphic molt into a single megalopa stage representing a still-swimming transitory form [3]. The juvenile and adult stages have a benthic mode of life in intertidal and upper subtidal sandy marine environments [4].

Representatives of Hippidae are special among Meiura in that some of their zoea larvae may achieve impressive sizes. These can reach shield lengths of over 6 mm and, together with the long and slender pleon, may be more than 15 mm long when outstretched [4], whereas most meiuran megalopae are significantly smaller. In fact, these probably represent the largest zoea larvae of all anomalan crustaceans, and possibly all meiurans.

Martin and Ormsby ([4], their Fig. 1b) depicted one such super-sized specimen positioned with a strongly anteriorly flexed pleon. They furthermore pointed out how well the "opercular-like" telson (term from [4]) fits the ventral shape of the shield. While not further discussed in this original work, the function of this tight fit seems most likely to be a specific defensive strategy; more precisely, these larvae appear able to perform defensive enrollment.

Enrollment is a defensive mechanism which apparently evolved several times independently within Metazoa, often combined with morphological specializations, such as hard plates or large sclerotized spines (e.g., [5–7]). In enrollment, the body is strongly curved ventrally, forming a nearly perfect ball, and the anterior and posterior end lie adjacent to each other. As a result, sclerotized or hardened dorsal structures protect the softer, ventral side of the body and all appendages.

Within vertebrates, armadillos (Xenarthra, Mammalia) are able to bend their body to such an extent that they form a ball (e.g. [8]); their "armor" of dorsal overlapping plates composed of bone with a covering of keratin [9], provides protection in this position. Among mollusks, polyplacophorans (chitons) roll up their bodies to the ventral side when detached from the substrate. In this posture, their dorsal shell plates protect the broad and fleshy foot ([10, 11]; see also Fig. 1a).

In arthropods, enrollment of the body as a protective mechanism against predators and other threats is widespread and primarily known from terrestrial arthropods, e.g., pill bugs and pill millipedes. However, some extinct marine species, e.g., trilobites, also performed enrollment [12–15]. Here again, as described for the other groups, the body is strongly curved ventrally and the tergites (dorsal sclerotisations of the segments) protect the softer ventral side of the body and all the appendages.

More recently it has been reported that certain larval representatives of mantis shrimps (Stomatopoda, Eucrustacea) are also able to tightly enroll their bodies. The pleon is bent forward, constituting a sclerotized protection for the entire body with no major gaps ([7]; see also Fig. 1b).

While the description in Martin and Ormsby [4] indicated the possibility of morphological adaptations for enrollment in hippidan larvae, this appears not to have been investigated further. In the present report, we present new specimens of giant hippidan larvae and provide a detailed description of the general morphology using modern imaging techniques. We discuss morphological details which support the interpretation that these larvae can indeed perform defensive enrollment. We furthermore document an unexpected morphological diversity among giant hippidan larvae.

Methods

Materials

Eight hippidan larval specimens were the basis for the present investigation. Six of the specimens came from the zoological collections of the Natural History Museum of Denmark, Copenhagen (ZMUC), registered under the numbers ZMUC-CRU-8679 to 8684. These specimens were collected during the Dana expeditions (1921–22 and 1928–30; Schmidt 1926, 1931; Broch 1936). One specimen came from the crustacean collections of the Senckenberg Naturmuseum Frankfurt (Mu_267), and one from the Muséum national d'Histoire naturelle Paris (MNHN-IU-2014-5468). All are currently stored in 70 % ethanol, probably after previous fixation in formalin. For ventral and dorsal documentation (always within the storage liquid), some specimens were carefully outstretched and fixed with a cover slip. For large specimens, posterior and anterior ends were each fixed with separate cover slips. In other orientations, specimens were either propped against glass or metal objects, or placed into depressions. Specimens in unusual positions were not altered, but kept in this specific position. A single specimen (ZMUC-CRU-8679) was dissected directly in 70 % ethanol using a dissection microscope.

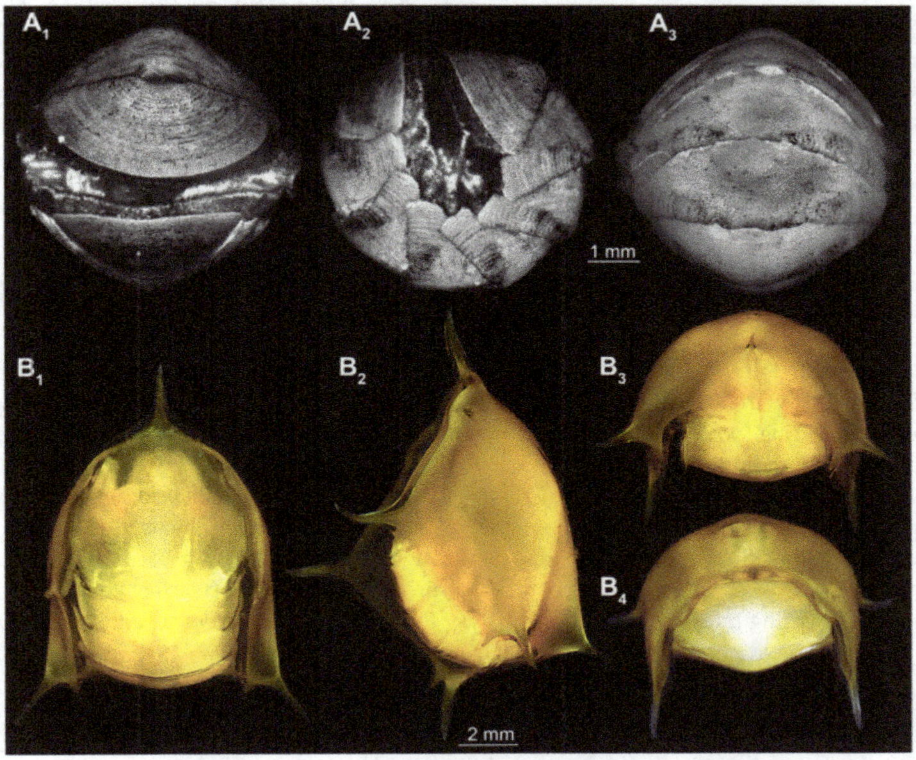

Fig. 1 a–b Commonly known species that exhibit defensive enrollment. A_1–A_3 Autofluorescence images of *Chiton* spec. (Polyplacophora). A_1 Ventral view. A_2 Lateral view. A_3 Dorsal view. B_1–B_4 Composite images under cross-polarized light of a mantis shrimp larva (Stomatopoda, Erichthus-type, see [7]). B_1 Ventral view. B_2 Lateral view. B_3 Posterior view. B_4 Frontal view

Documentation

All eight specimens were photographed using a Canon Rebel T3i camera with a MP-E 65 mm macro lens. Light was provided by a Canon Macro Twin Flash MT 24EX or a MeiKe FC 100 LED ring light. Light sources were equipped with polarization filters. A cross-polarized filter was placed in front of the lens. Cross-polarized light reduces reflections and enhances colour contrast (e.g., [16] and references therein). Additionally, one of the eight specimens (ZMUC-CRU-8679) was documented in 70 % ethanol using a Keyence BZ-9000 fluorescence microscope with either a 2×, 4× or 10× objective (resulting in approximately 20×, 40×, and 100× magnification, respectively; in a few cases the zoom function of the camera was also employed) depending on the different sizes of the body parts. UV light (377 nm) was used for illumination, using the autofluorescence capacities of the specimens (e.g. [17]). For high-resolution images, composite imaging was applied [18, 19].

Image processing

Image stacks were fused with the computer software CombineZP into sharp images. Adobe Photoshop CS3 was used to merge different sharp image details resulting in large panorama images. Finally, images were edited in Adobe Photoshop CS6 (optimization of the histogram and sharpness, manual removing of dirt particles etc. e.g. [7]).

Drawings

For better comparison, the different telson shapes of the specimens ZMUC-CRU-8679, 8682, and 8683 were drawn in Adobe Illustrator CS 3.

Presentation

The description is provided as a descriptive matrix (Additional file 1) [20]. This allows a more direct comparison of corresponding structures, which may facilitate future detailed descriptions of other larvae.

Terminology

Most terms applied are standard crustacean terms (e.g. [21, 22]). However, we have sought to keep terminology neutral to the extent possible, in the interests of allowing comparisons across a wider (arthropod) range. Special terminology of malacostracan or decapod-type is provided in brackets.

Results

In the following, we describe one of largest specimens (specimen A) in detail. Furthermore, we provide a morphological description of comparable features of the additional seven specimens (specimen B–H). As the latter ones were not dissected, only features that were available in the intact specimens are described.

Specimen A (ZMUC-CRU-8679):

Habitus (Fig. 2). Small arthropod larva with a globose shield, bearing a long, anteriorly directed, rostral spine (slightly shorter than shield length) and lateral spines (similar length as rostral spine).

Body (Fig. 2) differentiated into cephalothorax, pleon and non-somitic telson. Body with 20 segments, ocular segment plus 19 appendage-bearing (post-ocular) segments.

Ocular segment incorporated into the cephalothorax, dorsal area contributes to the shield.

Post-ocular segment 1-13 (Fig. 2) incorporated into the cephalothorax, dorsal area contributes to the shield.

Post-ocular segment 14-19 (Fig. 2) are separate pleon segments, each dorsally forming a tergite.

Cephalothorax (Fig. 2) shield more or less spherical, without setae; large, shield-like, cuticular structure formed by dorsal region of cephalothoracic segments. Anterior rim of the shield drawn out into prominent rostral spine. Posterior rim of the shield slightly convex, with a confined gape, as wide as the posterior gape of the shield. Rostrum unpaired anterior extension of shield, elongated, without spines; anterior region slightly bent upwards. Shield length about 8.5 mm (measured with rostral spine) and 5.1 mm without rostral spine, maximum shield width (measured without spines), about 5.2 mm (about 60 % of shield length with rostral spine). Rostral spine about 40 % of the shield length with rostral spine.

Post-ocular segment 14 (Fig. 2) anterior-posterior dimension about 25 % of the shield length (without rostral spine); total width of the segment 25 % of the maximum shield width, as wide as the posterior gape of the shield; tergo-pleura not developed; anterior region of post ocular segment 15 slightly convex.

Post-ocular segment 15 (Fig. 2) anterior-posterior dimension about 5 % of the shield length (without rostral spine). Total width of the segment 25 % of the maximum shield width, as wide as the posterior gape of the shield. Tergo-pleura not developed. Post-ocular segment 15 armed with one cone-shaped spine in the middle of anterior rim of the segment.

Post-ocular segment 16 (Fig. 2) anterior-posterior dimension about 20 % of the shield length (without rostral spine). Total width of the segment about 25 % of the maximum shield width. Tergo-pleura not developed.

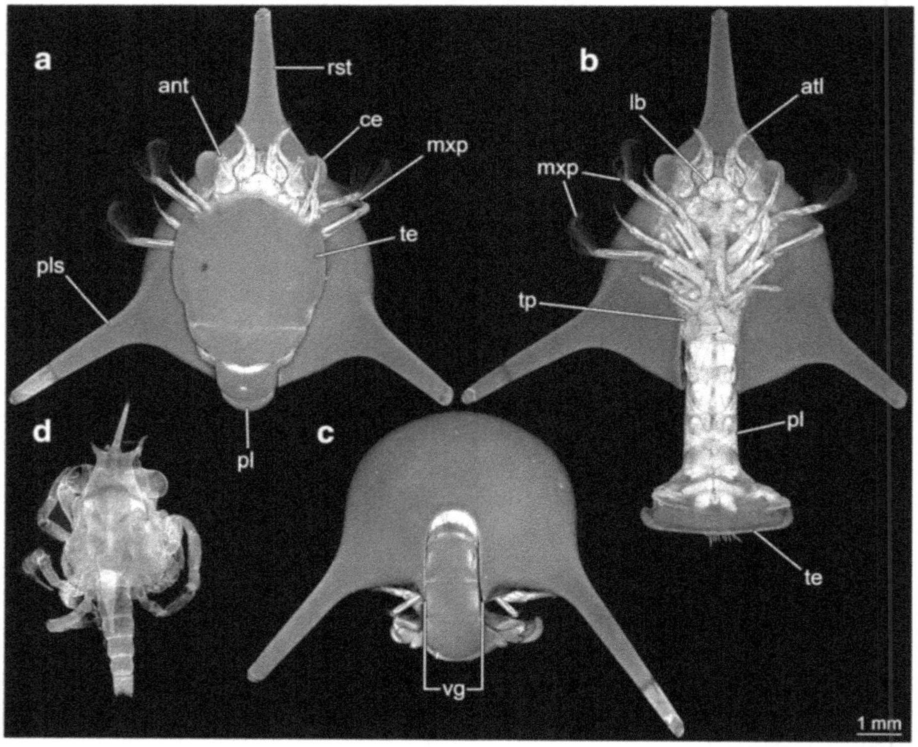

Fig. 2 a–d Autofluorescence images of a hippidan larva (ZMUC-CRU-8679) and a spider crab larva (*Maja* sp.). **a** Ventral view, fully enrolled. **b** Ventral view, fully outstretched. **c** Posterior view. **d** Dorsal view. Abbreviations: ant = antenna; atl = antennula; ce = compound eye; lb = labrum; mxp = maxilliped; pl = pleon; pls = postero-lateral spine; rst = rostral spine; te = telson; tp = thoracopod; vg = ventral gape

Post-ocular segment 17 (Fig. 2) anterior-posterior dimension about 20 % of the shield length (without rostral spine). Total width of the segment about 25 % of the maximum shield width. Tergo-pleura not developed.

Post-ocular segment 18 (Fig. 2) anterior-posterior dimension about 15 % of the shield length. Total width of the segment 50 % of the maximum shield width, measured on posterior rim of the segment. Axial region 25 % of the maximum shield width (without rostral spine). Tergo-pleura about 40 % of the axial region, on each side.

Post-ocular segment 19 (Fig. 2) anterior-posterior dimension about 10 % of the shield length (without rostral spine). Total width of the segment about 50 % of the maximum shield width. No clear differentiation between axial region and tergo-pleura.

Telson (Fig. 2) in dorsal view more or less rectangular. About 45 % of shield length (without rostral spine) and about 30 % wider than long. Anterior rim slightly concave, posterior rim convex. The lateral rim on each side slightly convex, telson width suddenly increased after about 20 % from anterior to posterior rim. Telson shape in lateral view distally tapering. Tip of telson more or less triangular-shaped from dorsal view, with a flattened tip. Forty-seven simple setae on tip of telson. Further lateral setae shorter than distal ones. The 20th setae counted from each terminal rim are the longest ones, the most central one is about 50 % shorter than the longest. Telson armed with two spines on distal rim as protrusion of lateral rim on each side.

Lateral eyes (Fig. 2) compound eyes, with numerous ommatidia covered by cornea; stalked.

Hypostome-labrum complex (Fig. 2) with more or less triangular-shaped labrum in ventral view, anteriorly surrounded by hypostome.

Appendage 1 (Antennula) (Fig. 3) differentiated into peduncle and one flagellum. Antennula with aesthetascs. Peduncle more or less tube-shaped and curved to outer lateral rim of the shield. With spine-like protrusion on distal part of the inner lateral rim. Not yet divided into

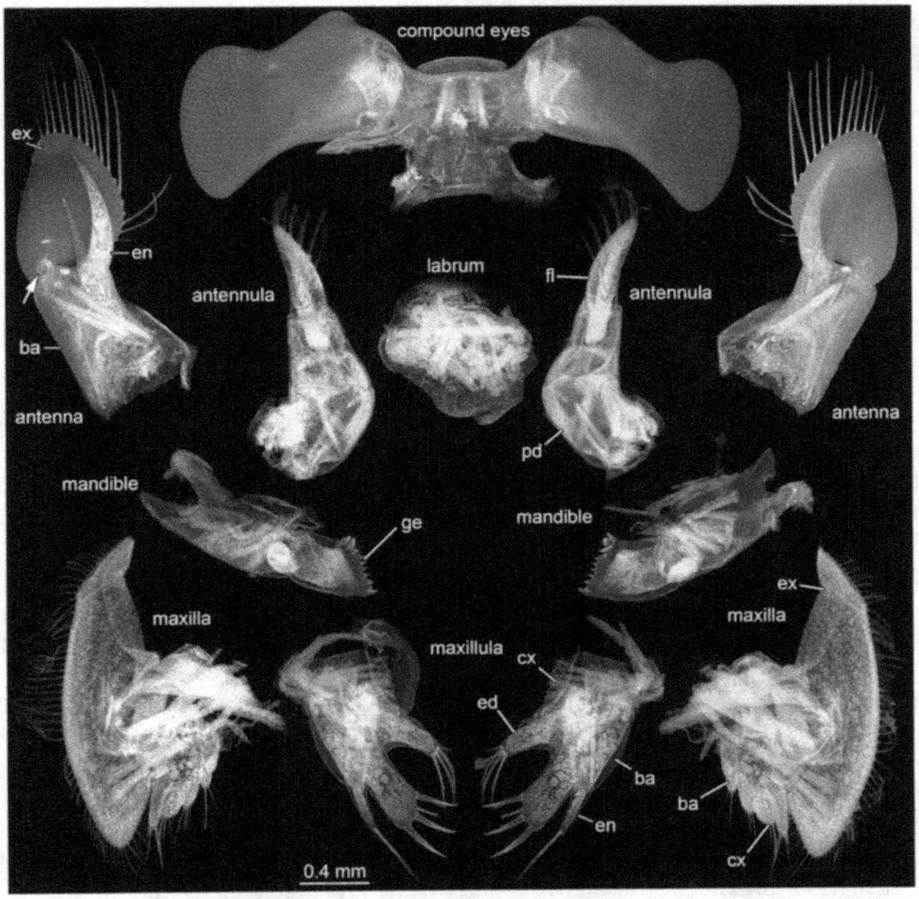

Fig. 3 Autofluorescence images of compound eyes, labrum, antennula, antenna, mandible, maxillula, and maxilla of the hippidan specimen (ZMUC-CRU-8679). Abbreviations: ba = basipod; cx = coxa; ed = endit; en = endopod; ex = exopod; fl = flagellum; ge = gnathic edge; pd = pedunde. Arrow: excretory opening

elements, future subdivision into three elements visible. Width of broadest part about 50 % of maximum length. Flagellum 1 not yet developed. Flagellum 2 tube-shaped with a rounded tip. About 35 % shorter than peduncle, with a slightly curved inner lateral rim with numerous setae (aesthetascs) arranged in six tiers.

Appendage of post-ocular segment 2 (Antenna) (Fig. 3) differentiated into coxa, basipod (peduncle), endopod and a paddle-shaped exopod; bears opening of antennal gland on basipod. Peduncle not yet divided into elements, with one spine on distal rim of basipod, where endopod arises from it. Endopod pointed and curved, not yet divided into elements, without setae. Exopod paddle-shaped, with 17 plumose setae on the rounded tip and the outer lateral rim.

Appendage of post-ocular segment 3 (Mandible) (Fig. 3) differentiated into coxa with endite and mandibular palp. Coxa elongate in medio-lateral axis, medially ending in a row of about 12 teeth. Row consisting of more or less lobate teeth, short triangular and longer elongate teeth with a pointed tip. Mandibular palp not yet developed, but future palp visible. Sternal protrusion of mandibular segment (paragnaths) u-shaped with two lateral elongate paddle-shaped setae bearing protrusions on distal rim. About 35 % wider than maximum length and about as large as hypostome-labrum complex.

Appendage of post-ocular segment 4 (Maxillula) (Fig. 3) differentiated into coxa with coxal endite and basipod with basipodal endite and endopod. Coxal endite more or less triangular-shaped from proximal to distal, with a rounded tip, with four elongated plumose setae at the tip. Basipodal endite paddle-shaped, elongate, with four spines at the tip, armed with tiny spines; about 30 % longer than coxal endite. Endopod pointed extension on basipod, not subdivided; one elongate, plumose seta, and one smaller seta on the tip.

Appendage of post-ocular segment 5 (Maxilla) (Fig. 3) differentiated into coxa and basipod, both drawn out into two pronounced lobate endites each and exopod. Coxa with two lobate endites with four setae on each lobe. Distal lobe smaller, than proximal one. Basipod with two lobate endites with four setae on each lobe. Distal lobe larger, than proximal one. Endopod not yet developed. Exopod of appendage 5 largest element, bilobed with a distal and proximal lobe; with numerous plumose setae around the rim.

Appendage of post-ocular segment 6 (Maxilliped 1) (Fig. 4) with coxa and basipod, from which endopod and exopod arise. Coxa and basipod with endites. Coxa more or less tube-shaped; with endite. Coxal endite small, more or less triangular from proximal to distal, without setae. Basipod more or less rectangular, with heart-shaped protrusion of anterior part; bears endopod and exopod; about 73 % longer than coxa, about twice as long as wide. Basipodal endite very prominent, slightly curved; with nine plumose setae, and four spines (armed with tiny spines) on inner lateral rim. Endopod with five elements (ischium, merus, carpus, propodus, dactylus); about as long as maximum length of basipod. Endopod element 1 more or less tube-shaped; about 45 % longer than wide, with four plumose setae on distal rim. Endopod element 2 more or less tube-shaped; about as long as preceding element; about 45 % longer than wide, with two plumose setae on distal rim. Endopod element 3 more or less tube-shaped, about 30 % shorter than preceding element; about 40 % longer than wide, with two plumose setae on distal rim. Endopod element 4 more or less tube-shaped; about 20 % shorter than preceding element, about 30 % longer than wide, with three plumose setae on distal rim. Endopod element 5 pointed, about 16 % shorter than preceding element; about 60 % longer than wide, with five plumose setae on distal rim. Exopod tube-shaped, tapering; not yet subdivided into elements; bent backwards. 85 % longer than wide, with about 12 plumose setae on the tip; laterally with one lobate protrusion slightly beyond the inner proximal rim and one lobate protrusion slightly below the tip.

Appendage of post-ocular segment 7 (Maxilliped 2) (Fig. 4) with coxa and basipod, from which endopod and exopod arise. Coxa and basipod with endites. Coxa more or less tube-shaped, without setae; coxal endite small, more or less triangular from proximal to distal, without setae. Basipod more or less rectangular-shaped, with heart-shaped protrusion of anterior part; bears endopod and exopod, about 70 % longer than coxa, about twice as long as wide. Basipodal endite very prominent, slightly curved, with 3 plumose setae, on inner lateral rim. Endopod with four elements, about as long as maximum length of basipod. Endopod element 1 more or less tube-shaped, about 40 % longer than wide, with three plumose setae on inner distal rim. Endopod element 2 more or less tube-shaped; about 45 % longer than wide; about 10 % longer than preceding element; with two plumose setae on inner distal rim. Endopod element 3 more or less tube-shaped; about 50 % longer than maximum width, and about the same length than preceding element; with two plumose setae on inner distal rim. Endopod element 4 tapering with a rounded tip; about 60 % longer than wide; about 25 % shorter than preceding element; with three plumose setae on tip. Exopod of appendage 7 tube-shaped, tapering; not yet subdivided into elements; bent backwards, about 10 % shorter than endopod and about 85 % longer than wide; with about 12 plumose setae on the tip; setae bearing tip bent to the inner lateral side; laterally with one lobate protrusion slightly beyond the inner proximal rim and one lobate protrusion slightly below the tip.

Appendage of post-ocular segment 8 (Maxilliped 3) (Fig. 4) with coxa and basipod, from which endopod

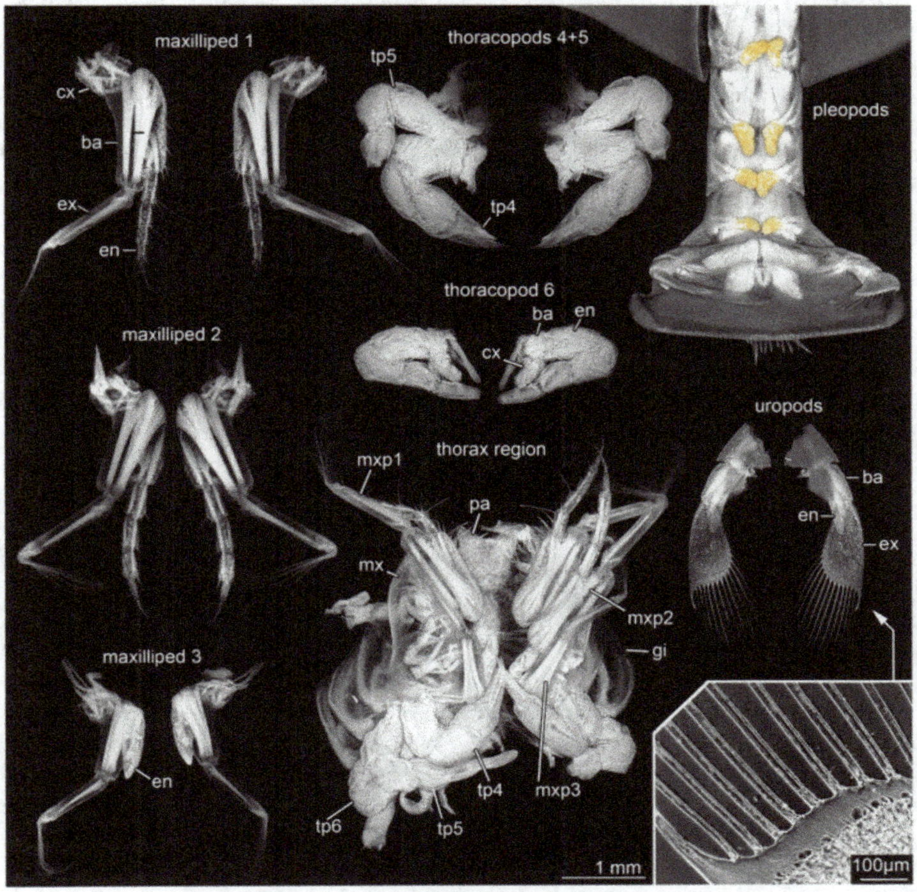

Fig. 4 Autofluorescence images of maxillipeds 1–3; thoracopods 4, 5, 6; thorax region; pleopods (orange); uropods (arrow = details of the exopod with setulae-bearing setae). Abbreviations: ba = basipod; cx = coxa; en = endopod; gi = gills; mxp = maxilliped; tp = thoracopod

and exopod arise; without endites. Coxa more or less rectangular-shaped, without setae; coxal endite not developed. Basipod more or less tube-shaped; bears endopod and exopod. Endopod about 55 % longer than coxa; about twice as long as wide. Endopod with four elements, separation indicated by faint lines, about 15 % shorter than maximum length of basipod and about 70 % longer than wide. Endopod element 1 more or less tube-shaped, about 20 % longer than maximum width; without setae. Endopod element 2 more or less rectangular shaped; about 30 % wider than maximum length; about 40 % shorter than preceding element; without setae. Endopod element 3 more or less tube-shaped; about 50 % longer than maximum width, and about the same length than preceding element; with two plumose setae on inner distal rim. Endopod element 4 tapering, about as long as maximum width; about 15 % longer than preceding element, with two simple setae on tip. Exopod tube-shaped, tapering. Not yet subdivided into elements; slightly bent backwards; about 15 % longer than endopod and about 80 % longer than wide; with about 18 plumose setae on the tip. Setae bearing tip bent to the inner lateral side. Laterally with one lobate protrusion slightly below the tip.

Appendage of post-ocular segment 9 (Thoracopod 4) (Fig. 4) with coxa and basipod (difficult to differentiate in this developmental stage) and endopod; without setae. Endopod of appendage 9 consists of five visible elements (difficult to identify at this early developmental stage) separated by faint lines; distal part of this appendage is modified to a prominent chela. Endopod elements 1–2 probably corresponding to ischium and merus, not yet separated; more or less tube-shaped; about 25 % longer than maximum width. Endopod element 3 (carpus) more or less tube-shaped, curved to the inner side; about 15 % longer than maximum width; about 12 % shorter than preceding element.

Endopod element 4 (propodus) about 45 % longer than maximum width, with outgrowth, which represents the complement of the following element; outgrowth about 35 % shorter than mainpart of the element. Endopod element 5 (dactylus) movable against outgrowth of propodus; tapering, slightly curved; about 50 % of maximum length of preceding element (propodus). Chela is formed by the articulation of element 5 (dactylus) against an outgrowth of element 4 (propodus).

Appendage of post-ocular segment 10 (Thoracopod 5) (Fig. 4) with coxa and basipod (difficult to differentiate in this early developmental stage) and endopod. Without setae. Coxa difficult to differentiate in this early developmental stage; without setae. Basipod difficult to differentiate in this early developmental stage; without setae. Endopod consists of five elements (difficult to identify in this early developmental stage); distal part without chela. Endopod element 1 (ischium) more or less triangular-shaped from proximal to distal; about 25 % longer than maximum width; without setae. Endopod element 2 (merus) more or less tube-shaped, slightly curved to inner side; about 40 % longer than maximum length and 40 % longer than preceding element; without setae. Endopod element 3 (carpus) tube-shaped; about 30 % longer than maximum width and about 8 % shorter than preceding element; without setae. Endopod element 4 (propodus) tube-shaped, about as long as maximum width and about 30 % shorter than preceding element; without setae. Endopod element 5 (dactylus) tapering, with a slightly rounded tip; about 80 % longer than maximum width, and about 60 % longer than maximum width; without setae.

Appendage of post-ocular segment 11 (Thoracopod 6) (Fig. 4) with coxa and basipod (difficult to differentiate in this early developmental stage) and endopod; without setae.

Appendage of post-ocular segment 12 (Thoracopod 7) (Fig. 4) with coxa and basipod and endopod (difficult to differentiate in this early developmental stage, separation indicated by faint lines); without setae.

Appendage of post-ocular segment 13 (Thoracopod 8) (Fig. 4) with coxa and basipod and endopod (difficult to differentiate in this early developmental stage, separation indicated by faint lines); without setae.

Appendage of post-ocular segment 14 (Pleopod 1) (Fig. 4) not found, not documented.

Appendage of post-ocular segment 15 (Pleopod 2) (Fig. 4) differentiated into basipod and endopod. Separation indicated by a faint line. Basipod elongate tube-shaped, about 50 % longer than maximum width. Endopod of appendage 15 tube-shaped, with a rounded tip; about 40 % longer than maximum width, and about 20 % shorter than basipod.

Appendage of post-ocular segment 16 (Pleopod 3) (Fig. 4) differentiated into basipod and endopod. Separation indicated by a faint line. Basipod elongate tube-shaped, about 35 % longer than maximum width. Endopod of appendage 16 tube-shaped, with a rounded tip; about 25 % longer than maximum width, and about 35 % shorter than basipod.

Appendage of post-ocular segment 17 (Pleopod 4) (Fig. 4) differentiated into basipod and endopod. Separation indicated by a faint line. Basipod elongate tube-shaped, about 35 % longer than maximum width. Endopod of appendage 17 tube-shaped, with a rounded tip; about 40 % longer than maximum width, and about 10 % shorter than basipod.

Appendage of post-ocular segment 18 (Pleopod 5) (Fig. 4) differentiated into basipod and endopod. Separation indicated by a faint line. Basipod elongate tube-shaped, about 47 % longer than maximum width. Endopod of appendage 18 tube-shaped, with a rounded tip; about 40 % longer than maximum width, and about 35 % shorter than basipod.

Appendage of post-ocular segment 19 (Uropod) (Fig. 4) differentiated into basipod and endopod and exopod. Basipod tube-shaped, 55 % longer than maximum width. Endopod tube-shaped with a rounded tip, about 40 % longer than maximum width. Without setae. Exopod paddle-shaped, with about 16 plumose setae on the distal rim, and the inner lateral rim of appendage. With one spine representing the delineation of setae and extention of the outer rim of exopod

Morphological description of comparable features of specimens B–H

Specimen B (ZMUC-CRU-8681) (Fig. 5):

Shield. Shield length about 8.5 mm (measured with rostral spine). Maximum width (measured without spines) about 3.5 mm (about 40 % of shield length).

Rostral spine. Anterior part slightly bent upwards. About 45 % of the shield length.

Telson. Anterior and posterior rim slightly convex. The lateral rim on each side slightly convex, width slowly increasing from anterior to posterior. Rim of telson more or less curly brace-shaped, with a rounded tip. Telson armed with two spines on distal rim as protrusion of lateral rim on each side.

Specimen C (ZMUC-CRU-8680) (Fig. 5):

Shield. Shield length about 7.5 mm (measured with rostral spine). Maximum width (measured without spines, about 3.4 mm (45 % of shield length).

Rostral spine. Anterior part slightly bent upwards. About 35 % of the shield length.

Telson. Anterior rim slightly convex, posterior rim slightly concave. The lateral rim on each side slightly convex, width slowly increasing from anterior to posterior. Rim of telson more or less curly brace-shaped, with a rounded tip. Telson armed with two spines on distal rim as protrusion of lateral rim on each side.

Specimen D (ZMUC-CRU-8683) (Fig. 5):

Shield. Shield length about 5 mm (measured with rostral spine). Maximum width (measured without spines, about 2.3 mm (45 % of shield length).

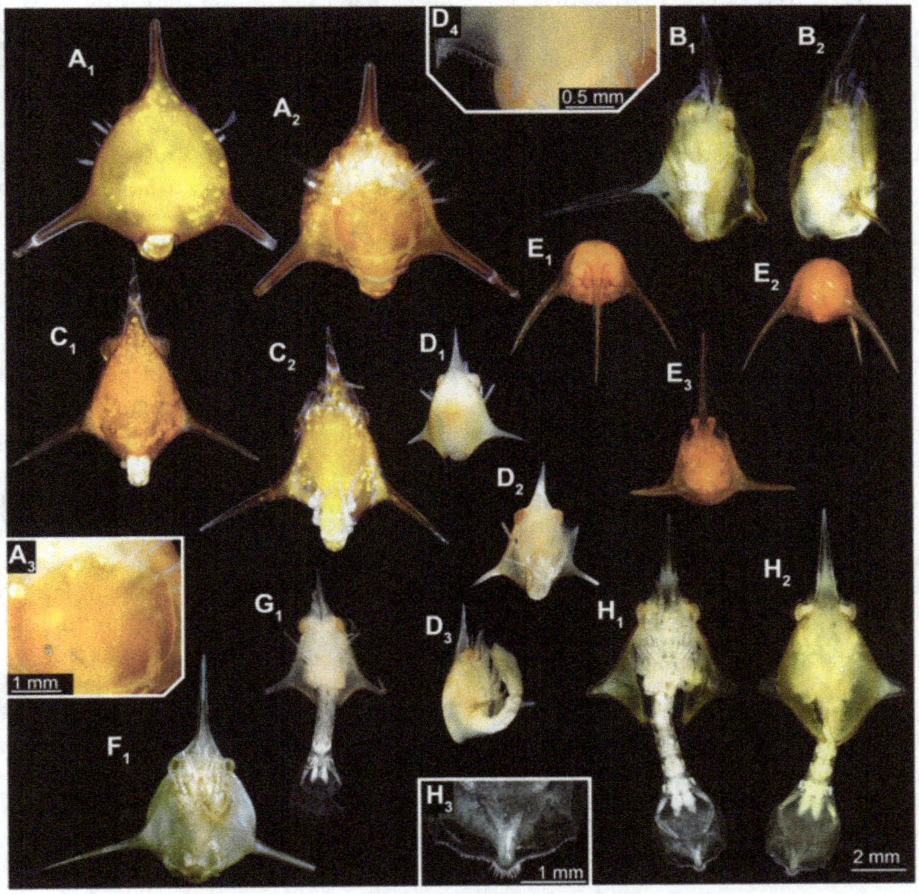

Fig. 5 a–h Composite images under cross-polarized light of hippidan specimens. **A_1–A_2** Dorsal and ventral view (entirely enrolled) of specimen ZMUC-CRU-8679; morphotype 1. **A_3** Detail of telson of morphotype 1. **B_1–B_2** latero-ventral and lateral view of specimen ZMUC-CRU-8681 (entirely enrolled); morphotype 3. **C_1–C_2** Dorsal and ventral view of specimen ZMUC-CRU-8680 (entirely enrolled); morphotype 3. **D_1–D_3** Dorsal, ventral and lateral view of specimen ZMUC-CRU-8683 (enrolled); morphotype 2. **D_4** Detail of telson of morphotype 2. **E_1–E_3** Frontal, postero-lateral, and ventral view of specimen MNHN-IU-2014-5468 (entirely enrolled); morphotype 4. **F_1** Ventral view of specimen Mu_267 (entirely enrolled); morphotype 1. **G_1** Ventral view of specimen ZMUC-CRU-8684 (entirely outstretched); morphotype 2. **H_1–H_2** Ventral and dorsal view of specimen ZMUC-CRU-8682 (entirely outstretched); morphotype 3. **H_3** Detail of telson of morphotype 3

Rostral spine. Anterior part slightly bent upwards. About 35 % of the shield length.

Telson. Anterior rim slightly convex, posterior rim slightly concave. The lateral rim on each side slightly convex, width slowly increasing from anterior to posterior. Rim of telson more or less triangular-shaped from dorsal view, with a flattened tip. Telson armed with two spines on distal rim as protrusion of lateral rim on each side.

Specimen E (MNHN-IU-2014-5468) (Fig. 5):

Shield. Shield length about 7.4 mm (measured with rostral spine) and about 3 mm (measured without rostral spine). Maximum width (measured without spines, about 2.5 mm (30 % of shield length).

Rostral spine. Anterior part strongly bent downwards. About 120 % of the shield length.

Telson. Anterior rim slightly convex, posterior rim slightly concave. Lateral rim difficult to recognize; apparently slightly convex, probably with lobate structure. Telson width suddenly increasing after about 45 % from anterior to posterior rim. Rim of telson more or less triangular-shaped from dorsal view, with a slightly flattened tip. Telson probably armed with two spines on distal rim as protrusion of lateral rim on each side.

Specimen F (Mu_267) (Fig. 5):

Shield. Shield length about 8.3 mm (measured with rostral spine). Maximum width (measured without spines), about 4 mm (50 % of shield length).

Rostral spine. No bending visible. Not documented from lateral, or ventro-lateral view. About 50 % of the shield length

Telson. Anterior rim slightly convex, posterior rim slightly concave. Lateral rim slightly convex, with lobate structure. Telson width suddenly increased after about 40 % from anterior to posterior rim. Rim of telson more or less triangular-shaped from dorsal view, with a flattened tip. Telson armed with two spines on distal rim as protrusion of lateral rim on each side.

Specimen G (ZMUC-CRU-8684) (Fig. 5):

Shield. Shield length about 5.2 mm (measured with rostral spine). Maximum width (measured without spines), about 2.5 mm (50 % of shield length).

Rostral spine. No bending visible. Not documented from lateral, or ventro-lateral view. About 40 % of the shield length.

Telson. Anterior rim slightly convex, posterior rim slightly concave. The lateral rim on each side slightly convex, width slowly increasing from anterior to posterior. Rim of telson more or less triangular-shaped in dorsal view, with a flattened tip. Telson armed with two spines on distal rim as protrusion of lateral rim on each side.

Specimen H (ZMUC-CRU-8682) (Fig. 5):

Shield. Shield length about 9 mm (measured with rostral spine). Maximum width (measured without spines, about 3.6 mm (40 % of shield length).

Rostral spine. Anterior part slightly bent upwards. About 40 % of the shield length.

Telson. Anterior rim slightly convex, posterior rim slightly concave. The lateral rim on each side slightly convex, width slowly increasing from anterior to posterior. Rim of telson more or less curly brace-shaped, with a rounded tip. Telson armed with two spines on distal rim as protrusion of lateral rim on each side.

Discussion

Identification of the specimens

The specimens described here show some inter-individual differences, but are sufficiently similar to be discussed together. The overall morphology immediately identifies them as reptantian zoea larvae; the embryonic-like posterior thoracopods identify them as an ingroup of Meiura (cf. e.g. [3, 23]). Most meiuran zoeae possess a forked telson (e.g. [3, 23–25]), while the posterior rim of the telson of the here described specimens has a roughly convex shape in dorsal view (with additional lobate protrusions in some specimens). This shape is known from zoea larvae of hippidan species (Figs. 5, 6). Hippidan larvae, similar to the present specimens, have been described with a more or less spherical shield with a rostral spine and two large spines on each postero-lateral margin, but no postero-dorsal spine. Finally, representatives of Hippidae achieve significantly larger sizes as zoea larvae [3, 4] compared to other meiurans. This is also true for the specimens described here (Figs. 2, 5).

After metamorphosis to the megalopa stage in hippidans, antennula and antenna include a long, setose flagellum, the mandible is divided into two parts, thoracopods are divided into elements and largely resemble the setae-bearing juvenile and adult ones; the pleopods as well as the exopods of the uropods bear setae [4]. The new specimens are therefore interpreted as zoea larvae of species of Hippidae. A late zoea stage is indicated by a differentiation of the non-setae-bearing uropods into an endopod and exopod, by more than three tiers of aesthetascs on the antennula, comparably large maxillipeds, and the presence of primordial thoracopods and pleopods. The details of the appendages of specimen A also strongly resemble known late zoea features (besides their size) [26]. Although the other specimens have not been documented in detail, all are interpreted as representing rather late or final zoea stages.

This is supported in particular by a notable difference to already known zoea larvae of Hippidae: in all specimens the sixth pleomere is already set off from the telson (Figs. 2, 4, 5, 6). Hippidan species for which the larval sequence is known are considered to have the sixth pleomere conjoined to the telson in all zoea stages (forming a pleotelson), becoming finally articulated in the megalopa stage [27].

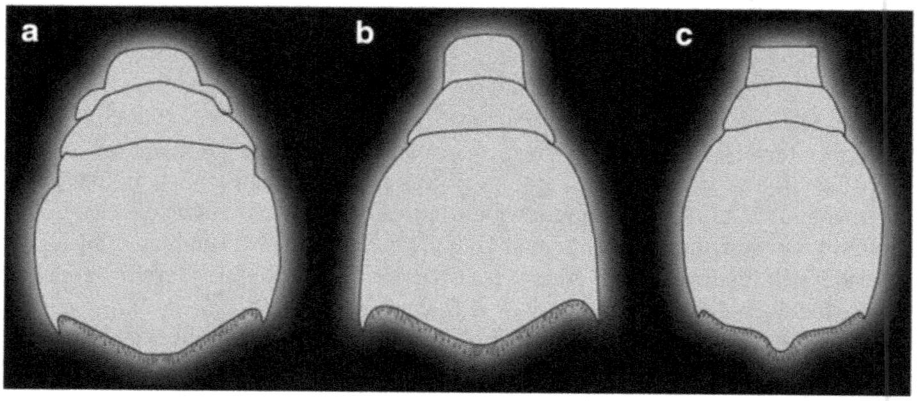

Fig. 6 Drawings of the different telson-shapes in dorsal view of the investigated species. **a** Morphotype 1. **b** Morphotype 2. **c** Morphotype 3

Hence, in the specimens described here, we probably have ultimate zoea larvae, which could be also interpreted as early megalopae [28], retaining some (in fact most) zoea characters, but already having some megalopa characters. Hippidans are able to vary the number of larval stages (e.g. [29]); these specimens could represent such a case of a prolonged pelagic phase, most likely due to the lack of a settling trigger, which would induce transformation to the megalopa.

Larvae of mole crabs – what is known so far

Hippoidea includes the groups Blepharipodidae, Albuneidae and Hippidae, the latter two representing sister groups (e.g. [30]). Larval representatives of Albuneidae and Hippidae feature a more or less spherical shield equipped with one prominent lateral spine on each side (absent in larvae of blepharipodids) and one elongate rostral spine [25, 31, 32]. Despite obvious general similarities due to their close relationships, larvae of Albuneidae and Hippidae differ in many aspects [4, 26, 31, 33, 34], assuming the here described specimens are hippidans. A specimen observed by Gurney [25] was referred to as *Albunea* sp. It strongly resembles one of the specimens described herein (Fig. 5h), and is probably also a hippidan larva.

Within Hippidae, there are currently only three species groups recognised: *Emerita*, *Hippa* and *Mastigochirus* (e.g. [4, 26, 33]). Species of *Hippa* and *Emerita* have very similar zoea larvae, with only a few differences. For comparison we refer to the last zoea stage before metamorphosis as the megalopa, since the detailed described larva (specimen A, ZMUC-CRU-8679) is most likely a very late stage zoea.

In larvae of *Hippa* species the rostral spine is slightly curved downwards, but upwards in larval representatives of *Emerita*. Also, the former bear fewer aesthetascs on the antennula, a shorter flagellum on the antenna, fewer setae on the exopod of the maxilla, and fewer setae on the telson. Additionally, larval forms of *Emerita* (as far as known) do not achieve the impressive size of larval forms of *Hippa*. The specimens in the 6th zoeal stage of *Hippa* can reach a shield length of up to 6 mm, whereas zoea larvae in the same stage of *Emerita* only reach a shield length of about 2 mm (see [4, 26, 33]).

Grouping of material described here

Based on gross morphological aspects, the specimens described here can be sorted into four more or less distinct groups or morphotypes. This will allow an easier comparison with existing descriptions instead of treating each specimen separately.

Morphotype 1 includes specimens A and F. Both share a more spherical shield, a straight rostral spine, a telson with lobate extensions laterally and a flattened tip.

Morphotype 2 includes specimen D and probably G. Both possess a more elongate shield, a straight rostral spine and a telson lacking a lateral protrusion.

Morphotype 3 includes specimens B, C and H. They all possess a shield comparable to morphotype 2, yet the shape of the posterior margin of the telson is not simple and convex (as in morphotype 2), but with a curled, brace-shaped distal margin with a small and rounded lobate tip.

Morphotype 4 is represented by specimen E, which features a more or less spherical shield with a curved rostral spine. The telson of the specimen features lateral protrusions and a flattened tip as in morphotype 1.

Comparison to the new material

Specimen A (morphotype 1) is quite large, almost reaching the size of a specimen described by Martin and Ormsby [4], which was interpreted as representing the larva of a species of *Hippa*. The rostral spine of the latter specimen is curved downwards, whereas in specimen A the rostral spine is curved slightly upwards.

The antennula of specimen A bears six, rather than the typical five, tiers of setae. It is divided into strongly curved peduncle and flagellum, instead of being not subdivided into a peduncle and flagellum (Fig. 3). The antenna also differs in many aspects. The basipod bears one spine on its distal rim, and there is no flagellum developed as described by Martin and Ormsby [4]. A marked difference constitutes the pointed and strongly curved endopod without setae and the presence of a well-developed paddle-shaped exopod with numerous plumose setae, instead of an endopod with spines and no exopod on the described species of Martin and Ormsby [4] (Fig. 3).

The maxilla has two coxal and two basipodal endites and no endopod, instead of no endites and a setae-bearing endopod. The bilobed exopod and the number of setae of the specimen described here largely resemble the description of Martin and Ormsby [4] (Fig. 3).

The number of telson setae is about twice the number described by Martin and Ormsby [4]. Additionally, the telson is equipped with two lobate structures on each lateral rim (Fig. 4).

Concerning the large size, one might assume that the observed larva is a representative of *Hippa*. However, the considerable morphological differences between the described structures of the specimen of Martin and Ormsby [4] and specimen A (and morphotype 1 in general) does not support its interpretation as a representative of *Hippa*.

Specimen A also differs in many aspects from larvae of species of *Emerita* described by Knight [26]. Concerning the larger number of aesthetascs on the antennula, and the higher number of setae on the exopod of the maxilla and the telson, specimen A resembles larvae of an *Emerita* species. Additionally, the rostral spine of specimen A is bent upwards, as presented in the drawings of Knight [26].

In specimen A, there is no flagellum on the antennula, the antenna bears a paddle-shaped exopod, the telson differs morphologically concerning the lobate structure on the lateral rim, and most strikingly, larvae of *Emerita* are significantly smaller; they achieve a mean size of only about 2 mm shield length [26].

Specimen E (morphotype 4), due to the downwards curvature of the rostral spine and the size of about 3 mm (Fig. 5), matches earlier descriptions of larvae of *Hippa* (e.g. [4]). Hence, specimen E is likely a larval representative of *Hippa*.

The additionally documented specimens, although not investigated in detail, do not show many similarities with the known larvae of *Hippa* or *Emerita* either. They also differ among each other (see also below). While based on the size we can hypothesize that the specimens represent late, but different larval stages, not all specimens can be attributed to a single developmental sequence.

Since all specimens described here, except specimen E, differ greatly from earlier described species (Fig. 5), we are unable to determine whether they are larvae of a species of *Emerita*, *Hippa* or *Mastigochirus*. These few specimens indicate that there is still an unknown morphological diversity within larval hippidans. The morphological differences are probably not caused by ontogenetic factors. As discussed above, the larvae most likely either represent ultimate zoea stages or even early megalopa stages. Therefore, they probably do not represent subsequent stages of a single species. The unexpected diversity may indicate that these larvae are individuals of species for which larvae are wholly unknown, differing more significantly than expected from other larval sequences. Yet, it is also possible that they represent special cases of developmental plasticity, which means that they are morphological variations of already known larvae caused by specific environmental conditions. In any case, the morphological diversity of hippidan larvae appears to be higher than that of hippidan adults.

Enrollment

The specimens described here have not been observed when alive. Still, we can make inferences about their original behavior based on their functional morphology, as recently suggested by Haug and Haug [7]. The basic idea is to employ approaches from paleontology; e.g., recording the preserved position of specimens, identifying specialized morphological features, and comparing these to those of animals with similar features in which behavior can be directly observed. Using these approaches, while any conclusions remain a matter of conjecture, they nonetheless represent an important tool for understanding. So far it has been impossible to breed giant hippidan larvae or to observe them directly in the field. Hence, the approach discussed here is currently the only possibility. Ideally the prediction made here can be tested in the field (see [7] for more details and a comparable example, also corroborated by field observation).

All specimens were originally preserved in an enrolled position, indicating the possibility of the animals to achieve this position. Further morphological adaptations are given for each morphotype separately.

Morphotype 1 (Figs. 2, 5, 6): Specimen A (ZMUC-CRU-8679), and specimen F (Mu_267). In these specimens the shield appears more or less spherical and the posterior gape of the shield has the same width as the pleonal tergites. This combines a spherical shield with maximum mobility of the pleon. The pleon can be stretched out or flexed forward, without any limitations. The large and lid-like telson features a lobate structure on each lateral margin, which ends up in a spine on the distal rim, and the rim of the telson is more or less triangular-shaped in dorsal view and has a slightly flattened tip. The width of the telson and the length of the pleon appear not to be entirely adapted to the ventral gape of the shield. The telson is broader, and due to a comparatively short pleon, the telson does not reach the anterior rim of the shield, even if the pleon is fully flipped forward. Therefore, antennula and antenna as well as the compound eyes and the distal parts of the maxillipeds are not fully concealed.

Morphotype 2 and morphotype 3 (Fig. 5, 6): Specimen D (ZMUC-CRU-8683) and probably specimen G (ZMUC-CRU-8684); respectively specimen B (ZMUC-CRU-8681), specimen C (ZMUC-CRU-8680) and specimen H (ZMUC-CRU-8682). The two morphotypes differ from each other only in the shape of the posterior rim of the telson, but otherwise are quite similar and therefore treated together.

Similar to morphotype 1, both morphotypes feature a posterior gape which has the same width as the pleon. In both morphotypes, the width of the telson is not adapted to the ventral gape of the shield. The telson is wider than the ventral gape and has a different shape. Yet, when flipped forward, the ventral gape is entirely covered across its width.

The large telson has slightly convex lateral margins, which end up in a spine on each side, and the telson width increases slightly from anterior to posterior. Additionally, due to a longer pleon the larvae are able to flex the telson far anteriorly so that the ventral gape of the shield is entirely closed and almost all parts of the appendages are covered by the telson, which perfectly protects the entire body. Only the compound eyes and the distal parts of antennula and antenna remain exposed.

Morphotype 4 (Fig. 5, 6): Specimen E (MNHN-IU-2014-5468) mainly resembles morphotype 1 in relation to the more or less spherical shield shape, the ventral gape and the shape of the telson. As in morphotypes 2 and 3, the pleon is flexed far anteriorly. Additionally, the telson is positioned inside the shield and reaches the anterior rim (Fig. 5). All appendages, except the eyes and the distal tip of the antennulae are protected by the shield, pleon and telson. The fully enrolled specimen has the appearance of a compact ball, armed with spines.

Here the position of the telson is most important. The telson appears to perfectly fit into the shield, with "rail-like" protrusions of the shield keeping it in place. In this position, the telson is tightly locked in place, and the dorsal area of the pleon perfectly closes the posterior gape. With this, the lateral rim of the telson and the ventral rim of the shield apparently form coaptative structures that tightly enclose the enrolled body. The animal probably achieves this position by 1) flexing its pleon forward, 2) pressing it towards its ventral side right in front of the mouthparts and maxillipeds, and 3) sliding it back. During the last step, the lid-like telson is pulled inside the ventrally curved rims of the shield. As a result, the animal secures the enrolled position and achieves full protection of the ventral appendages. Interestingly, this mechanism appears to be arranged "the-other-way-round" as compared to the coaptative structures in stomatopod larvae where the telson is pushed forward in order to lock it [7].

So do giant hippidan larvae perform defensive enrollment?
Based on our observations we can state that:

1) all specimens are preserved in an enrolled position, indicating that the animals can achieve this position;
2) the shield is large and drawn out ventrally for some distance, unlike in many other decapod zoea larvae, and it is able to house most of the appendages;
3) the width of the shield and the width of the pleon are perfectly adapted to one another; this is not a widespread phenomenon (see Discussion in [7]) and can hence be interpreted as an adaptation for performing enrollment;
4) a large lid-like telson covers most of the ventral gape if flipped forward; also this morphology is rather unusual for meiurans (where the pleon is usually forked); thus it is also possible that this morphology represents a further adaptation for enrollment;
5) at least for morphotype 4 (specimen E) we have indications that there are coaptative structures; such structures are strong indicators of defensive enrollment, as these are developed in other groups for which enrollment seems now corroborated, such as trilobites ([15], their figs. 1E–F; [35, 36]) or stomatopod larvae ([7]).

Hence, for morphotype 4 there should be little doubt about whether it was able to perform enrollment. For the other three, the stronger argument of the coaptative structures cannot be used. Yet, as in the discussion about different morphotypes of stomatopod larvae [7], the presence of only some of the adaptations cannot be used as an argument to exclude this behavior. Also among other animals which are known to perform defensive enrollment coaptative structures appear to be absent (e.g. polyplacophorans; stonefly larvae of *Pteronarcys dorsata* [37]).

With this, we consider it likely that all specimens were able to perform defensive enrollment, but to differing degrees of specialization. The ventral "softer" part of the body is in all cases concealed by the spine-bearing shield and the sclerotized pleonal tergites. Yet, in morphotype 1 (specimens A, F) there is a larger unconcealed region remaining up to the anterior margin of the ventral gape of the shield (Fig. 5; similar to a stomatopod larva morphotype, [7]). Morphotypes 2 (specimens D, G) and 3 (specimens B, C, H) achieve the same defensive effectiveness since they flex the telson further anteriorly, and the ventral gape of the shield is almost entirely closed (Fig. 5). Morphotype 4 (specimen E) achieves the most effective degree of defense since there is no unconcealed region left up to the anterior margin of the ventral gape, and due to the position of the telson inside the shield (Fig. 5).

Comparison to Brachyura and Stomatopoda
Martin and Ormsby [4] have stated that hippidan larvae appear very similar to brachyuran larvae. This seems to be largely attributable to the shield morphology. Brachyuran zoeae also feature a spherical shield; yet here the rostral spine is directed more ventrally (e.g. [38]) instead of being anteriorly directed as in most hippidan larvae (Fig. 5). Additionally, brachyuran zoea larvae have no uropods [39, 40], whereas hippidan larvae feature uropods from their early zoeal stages onwards [26] (Figs. 2, 4, 5). Brachyuran zoea larvae have two lateral spines and the rostral spine (as hippidan zoeae), but additionally a long postero-dorsal spine, which is absent in hippidan larvae (Figs. 2, 5). Also the telson differs morphologically. There is a pronounced furca with a medial cleft in brachyurans (at least in early zoea stages; [23, 34]. Hippidan larvae never have a forked telson (Figs. 2, 3, 4 and 5), but the posterior rim is convex.

Most strikingly, brachyuran larvae do not achieve the giant size of hippidan larvae; brachyuran larvae with small body lengths seem to be very common (e. g. [41, 42]). Discrete lengths are rarely stated for brachyuran larvae. A shield length (without the rostral spine) of 0.88 mm in an advanced zoeal stage has been reported [34]. The largest late stage zoea in our investigation (specimen F, Mu_267) reaches a shield length of about 5.5 mm (Fig. 5); 6 mm has been reported [4].

There have been no reports to date of defensive enrollment in brachyuran zoea larvae, nor do we see morphological adaptations for it, but that may change if such features are searched for directly. Currently, only a roughly spherical shield with three spines seems to be similar between brachyuran and hippidan zoeae (Fig. 2).

Stomatopods (mantis shrimps) are also discussed here in light of their similarities to hippidan zoea larvae, since some mantis shrimp larvae can also tightly enroll their bodies [7] (Fig. 1). Interestingly, the specimens described here were found between mantis shrimp larvae in two museum collections (see Methods for details; also in other collections, pers. obs.) in roughly pre-sorted samples. This shows nicely how similar mantis shrimp and hippidan larvae appear at first sight.

Also here especially the shield appears similar, even more similar than to brachyuran larvae as stomatopod larvae, like hippidan zoea larvae, lack the pronounced postero-dorsal spine of brachyuran zoeae (Figs. 1, 5). A future, more intensive functional comparison of hippidan and stomatopod larvae could reveal the evolutionary mechanisms leading to the morphological adaptation coupled to defensive enrollment.

Conclusions and Prospects

Our investigation indicates a broader morphological diversity of hippidan larvae than has been described previously. The functional morphological aspects of these larvae suggest a behavior by these larvae that has not been directly observable to date. It thus appears that we are just starting to understand the ecological roles played by many crustacean larvae. Hence, we expect to continue to uncover hidden morphological diversity among these larvae, and will seek to reconstruct their functional morphology and evolutionary history.

Acknowledgements

We are grateful to numerous persons and institutions that made this investigation possible. Jason Dunlop, Berlin, kindly made linguistic improvements. Enrico Schwabe, ZSM Munich, Jørgen Olesen, Tom Schiøtte, Danny Eibye-Jacobsen, and Jens T. Høeg, ZMUC Copenhagen, the curating people from the crustacean collection at the Senckenberg museum Frankfurt, and Laure Corbari, MNHN Paris, provided access to the collections and assisted with the technical equipment. All authors would like to thank J. Matthias Starck, LMU Munich, for his support. This study was also supported by people providing free or extremely low-cost software, such as CombineZM/ZP, Microsoft Image Composite Editor, and OpenOffice.

Funding

This contribution is part of JTH's research project kindly funded by the German Research Foundation (DFG) under Ha 6300/3-1. CH is grateful for support via the Bavarian Equal Opportunities Sponsorship of the LMU. The research visits at ZMUC Copenhagen and MNHN Paris of CH and JTH have been made possible by grants from the European Commission's (FP 6) Integrated Infrastructure Initiative programme SYNTHESYS (DK-TAF-2591, FR-TAF-5175, FR-TAF-5181).

Authors' contributions

JTH and CH designed the study. JTH and CH documented specimens ZMUC-CRU-8680 to 8684, MNHN-IU-2014-5468, and Mu_267. NRR performed the preparation and documentation of specimen ZMUC-CRU-8679 and drafted the manuscript. All authors read and approved the final manuscript.

Competing interests

The authors declare that they have no competing interests.

References

1. Haye PA, Tam YK, Kornfield I. Molecular phylogenetics of mole crabs (Hippidae: *Emerita*). J Crustac Biol. 2002;22:903–15.
2. Haig J. The status of *Remipes testudinarius* Latreille, and designation of a neotype for *Hippa adactyla* J.C. Fabricius (Decapoda, Hippidae). Crustaceana. 1970;19:288–96.
3. Harvey A, Boyko CB, McLaughlin P, Martins JW. Anomura. In: Martin JW, Olesen J, Høeg JT, editors. Atlas of Crustacean Larvae. Baltimore: Johns Hopkins Univ Press; 2014. p. 284–95.
4. Martin J, Ormsby B. A large brachyuran-like larva of the Hippidae (Crustacea: Decapoda: Anomura) from the Banda Sea, Indonesia: the largest known zoea. P Biol Soc Wash. 1991;104:561–8.
5. Ward P. Shell sculpture as a defensive adaptation in ammonoids. Paleobiol. 1981;7:96–100.
6. Schwartz SS. Predator-induced alterations in *Daphnia* morphology. J Plankton Res. 1991;13:1151–61.
7. Haug C, Haug JT. Defensive enrolment in mantis shrimp larvae (Malacostraca: Stomatopoda). Contrib Zool. 2014;83:185–94.
8. Gaudin TJ, Biewener A. The functional morphology of xenarthrous vertebrae in the armadillo *Dasypus novemcinctus* (Mammalia, Xenarthra). J Morphol. 1992;214:63–81.
9. Chen IH, Kiang JH, Correa V, Lopez MI, Chen PY, McKittrick J, Meyers MA. *Armadillo armor*: Mechanical testing and micro-structural evaluation. J Mech Behav Biomed Mater. 2011;4:713–22.
10. Squires RL, Goedert JL. An extant species of *Leptochiton* (Mollusca: Polyplacophora) in Eocene and Oligocene cold-seep limestones, Olympic Peninsula, Washington. Veliger. 1995;38:47–53.
11. Schwabe E, Wehrtmann IS. Chitons. In: Wehrtmann IS, Cortés J, editors. Marine Biodiversity of Costa Rica, Central America. Springer + Business Media B.V.; 2009. p. 323–31.
12. Speyer SE, Brett CE. Trilobite taphonomy and Middle Devonian taphofacies. Palaios. 1986;1:312–27.
13. Bruton DL, Haas W. Functional morphology of Phacopinae (Trilobita) and the mechanics of enrolment. Palaeontogr Abt A. 1997;245:1–43.
14. Lerosey-Aubril RL, Angiolini L. Permian trilobites from Antalya Province, Turkey, and enrollment in Late Palaeozoic trilobites. Turkish J Earth Sci. 2009; 18:427–48.
15. Feist R, Lerosey-Aubril R, Johnson R. Coaptative devices, enrolment, and life habits in *Paralejurus*, a particular case in scutelluid trilobites. Palaeobiol Palaeoenviron. 2010;90:125–37.
16. Haug C, Mayer G, Kutschera V, Waloszek D, Maas A, Haug JT. Imaging and documenting gammarideans. Int J Zool. 2011;380829.
17. Haug JT, Haug C, Kutschera V, Mayer G, Maas A, Liebau S, Castellani C, Wolfram U, Clarkson ENK, Waloszek D. Autofluorescence imaging, an excellent tool for comparative morphology. J Microsc. 2011;244:259–72.
18. Haug JT, Haug C, Ehrlich M. First fossil stomatopod larva (Arthropoda: Crustacea) and a new way of documenting Solnhofen fossils (Upper Jurassic, Southern Germany). Palaeodiv. 2008;1:103–9.
19. Haug C, Haug JT, Waloszek D, Maas A, Frattigiani R, Liebau S. New methods to document fossils from lithographic limestones of southern Germany and Lebanon. Palaeontol Electron. 2009;12:6T.
20. Haug JT, Briggs DEG, Haug C. Morphology and function in the Cambrian Burgess Shale megacheiran arthropod *Leanchoilia superlata* and the application of a descriptive matrix. BMC Evol Biol. 2012;12:162.
21. Walossek D. The upper Cambrian *Rehbachiella* and the phylogeny of Branchiopoda and Crustacea. Lethaia. 1993;26:1–318.
22. Garm A, Watling L. The crustacean integument: setae, setules, and other ornamentation. Funct Morphol Divers. 2013;1:167–98.
23. Martin JW. Brachyura. In: Martin JW, Olesen J, Høeg JT, editors. Atlas of Crustacean Larvae. Baltimore: Johns Hopkins Univ Press; 2014. p. 295–310.
24. Aikawa H. On larval forms of some Brachyura. Rec Oceanogr Works Japan. 1929;2:17–55.

25. Gurney R. Larvae of decapod Crustacea. London: Ray Society; 1942.
26. Knight M. The larval development of the sand crab *Emerita rathbunae* Schmitt (Decapoda, Hippidae). Pacific Sci. 1967;21:58–76.
27. Siddiqui FA, Ghory FS. Complete larval development of *Emerita holthuisi* Sankolli, 1965 (Crustacea: Decapoda: Hippidae) reared in the laboratory. Turkish J Zool. 2006;30:121–35.
28. Haug JT, Haug C. An unusual fossil larva, the ontogeny of achelatan lobsters, and the evolution of metamorphosis. Bull Geosci. 2013;88:195–206.
29. Diaz H, Costlow JD. Seasonal developmental sequences of *Emerita talpoida* (Say) (Anomura, Hippidae). J Coast Res. 1987;3:169–77.
30. Bracken-Grissom HD, Cannon ME, Cabezas P, Feldmann RM, Schweitzer CE, Ahyong ST, Felder DL, Lemaitre R, Crandall KA. A comprehensive and integrative reconstruction of evolutionary history for Anomura (Crustacea: Decapoda). BMC Evol Biol. 2013;13:128.
31. Stuck KC, Truesdale FM. Larval and early postlarval development of *Lepidopa benedicti* Schmitt, 1935 (Anomura: Albuneidae) reared in the laboratory. J Crustac Biol. 1986;6:89–110.
32. Konishi K. Larval development of the spiny sand crab *Lophomastix japonica* (Duruflé, 1889)(Crustacea, Anomura, Albuneidae) under laboratory conditions. Publ Seto Mar Biol Lab. 1987;32:123–39.
33. Hanson AJ. The larval development of the sand crab *Hippa cubensis* (De Saussure) in the laboratory (Decapoda, Anomura). Crustaceana. 1969;16:143–57.
34. Thatje S, Fuentes V. First record of anomuran and brachyuran larvae (Crustacea: Decapoda) from Antarctic waters. Polar Biol. 2003;26:279–82.
35. Esteve J, Zamora S, Gozalo R, Liñán E. Sphaeroidal enrolment in middle Cambrian solenopleuropsine trilobites. Lethaia. 2010;43:478–93.
36. Esteve J, Hughes NC, Zamora S. Thoracic structure and enrolment style in Middle Cambrian *Eccaparadoxides pradoanus* presages caudalization of the derived trilobite trunk. Palaeontology. 2013;56:589–601.
37. Moore KA, Williams DD. Novel strategies in the complex defense repertoire of a stonefly (*Pteronarcys dorsata*) nymph. Oikos. 1990;57:49–56.
38. Williamson DI. Larval morphology and diversity. In: Abele LG, editor. The Biology of Crustacea, vol. 2. New York: Academic; 1982. p. 43–110.
39. Rice AL. Crab zoeal morphology and its bearing on the classification of the Brachyura. Trans Zool Soc London. 1980;35:271–372.
40. Martin JW. Crabs of the family Homolodromiidae, III. First record of the larvae. J Crustac Biol. 1991;11:156–61.
41. Wear RG. Life-history studies on New Zealand Brachyura: 4. Zoea larvae hatched from crabs of the family Grapsidae. New Zeal J Mar Freshw Res. 1970;4:3–35.
42. Lago RP. Larval development of *Spiroplax spiralis* (Barnard, 1950) (Brachyura: Hexapodidae) in the laboratory; the systematic position of the family on the basis of larval morphology. J Crustac Biol. 1988;8:576–93.

5

Palaeospondylus as a primitive hagfish

Tatsuya Hirasawa[1*], Yasuhiro Oisi[2] and Shigeru Kuratani[1]

Abstract

Background: The taxonomic position of the Middle Devonian fish-like animal *Palaeospondylus* has remained enigmatic, due mainly to the inability to identify homologous cranial elements. This animal has been classified into nearly all of the major vertebrate taxa over a century of heuristic taxonomic research, despite the lack of conclusive morphological evidence.

Results: Here we report the first comparative morphological analysis of hagfish embryos and *Palaeospondylus*, and a hitherto overlooked resemblance in the chondrocranial elements of these animals; *i.e.*, congruence in the arrangement of the nasal capsule, neurocranium and mandibular arch-derived velar bar. The large ventral skeletal complex of *Palaeospondylus* is identified as a cyclostome-specific lingual apparatus. Importantly, the overall morphological pattern of the *Palaeospondylus* cranium coincides well with the cyclostome pattern of craniofacial development, which is not shared with that of crown gnathostomes. Previously, the presence of the vertebral column in *Palaeospondylus* made its assignment problematic, but the recent identification of this vertebral element in hagfish is consistent with an affinity between this group and *Palaeospondylus*.

Conclusion: These lines of evidence support the hagfish affinity of *Palaeospondylus*. Moreover, based on the less specialized features in its cranial morphology, we conclude that *Palaeospondylus* is likely a stem hagfish.

Keywords: Cyclostomes, Devonian, Fossil taxa, Hagfish, Homology, *Palaeospondylus*

Background

Palaeospondylus gunni [1] from the Middle Devonian of Scotland (Fig. 1, Additional file 1: Figure S1, Additional file 2: Figure S2) has long been an enigmatic fossil in vertebrate palaeontology [1–11]. Over the past 125 years, attempts have been made to classify this fossil as a cyclostome [1, 6, 8], frog tadpole [2], lungfish larva [3, 7, 10], holocephalan [4], elasmobranch [5], placoderm [9], and even a secondarily boneless osteichthyan [11]; however, none of these assignments has been supported by conclusive evidence [11–13]. This problem has been attributed to an inability to homologize its skeletal elements [14], and previous hypotheses have never successfully explained its anatomical configuration. Indeed, the arrangement of skeletal elements of *Palaeospondylus* has never yet been integrally compared to any of the actual ontogenetic stages of certain taxa.

Morphological homology [15–17] is not recognized *a priori*, but is adopted when the topographical relationships among a series of comparable body parts are found consistently, through heuristic search, in a certain clade. Logically, cladistic analyses cannot be applied prior to the identification of homologies [15, 18], and it is difficult to undertake a dynamic homology approach [19] when analyzing unintelligible morphological data. Unlike molecular-based phylogenetic analyses utilizing discrete characters (base or amino acid sequences), most morphology-based analyses require anatomical realization for alignments of corresponding units, that is, character codings. For yet-unassigned fossil taxa, such as *Palaeospondylus*, the discovery of the most parsimonious set of homologous body parts is the only solution to the classification problem [14].

Until recently, a lack of morphological data in hagfish, developmental data in particular, has prevented the heuristic search for homologies in entire vertebrate groups. Recent studies of the embryonic development of the hagfish [20–23] have gone some way to addressing this deficiency. Analyses of the embryonic development of the hagfish resolved the disparity in cranial morphology of vertebrates [22, 24], and for the first time, made it possible to compare the cranial skeleton of *Palaeospondylus*

* Correspondence: hirasawa@cdb.riken.jp
[1]Evolutionary Morphology Laboratory, RIKEN, 2-2-3 Minatojima-minami, Chuo-ku, Kobe, Hyogo 650-0047, Japan
Full list of author information is available at the end of the article

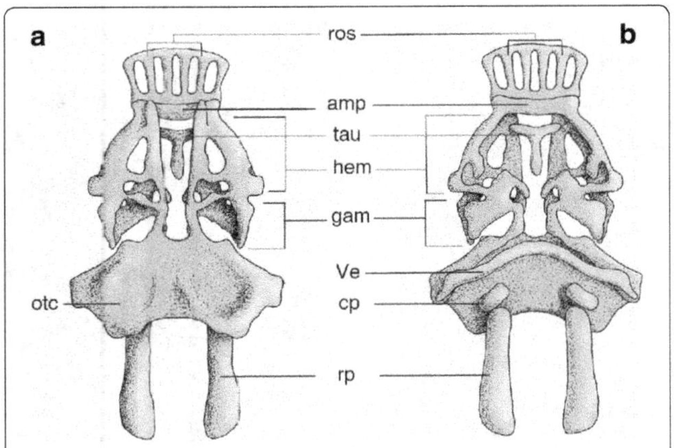

Fig. 1 Cranial skeletons of *Palaeospondylus*. **a**,**b** Restoration of *Palaeospondylus gunni* cranial skeleton in dorsal (**a**) and ventral (**b**) views. amp, ampyx; cp, caudal plate; gam, gammation; hem, hemidome; otc, otic capsule; ros, rostralia; rp, rostral plate; tau, tauidion; Ve, V-shaped element

with those of a wide range of vertebrate groups. Between the cyclostomes and crown gnathostomes (jawed gnathostomes), there is a profound disparity in craniofacial pattern, which originates from the difference in arrangement of craniofacial primordia during embryonic development [22, 24]. Importantly, although the cyclostome pattern likely represents the ancestral condition for the vertebrates as an entire group, crown gnathostome development does not follow the cyclostome pattern [22, 24]. Based on this disparity, which is due to the complete loss of some vertebrate plesiomorphies in the crown gnathostomes, the possibility of the crown gnathostome affinity of a certain taxon can be eliminated by identifying a set of homologous elements seen in the extant cyclostomes.

From the perspective of morphological disparity, classification of the enigmatic taxon *Palaeospondylus* into a higher taxonomic group within the vertebrates is achievable through the identification of congruence in arrangement of a suite of homologous elements. Attempts to identify homologous elements between *Palaeospondylus* and various vertebrate groups have been made, but the hagfish embryo has remained unexamined. In this study, we compared embryos of the inshore hagfish *Eptatretus burgeri* and *Palaeospondylus*, focusing on phylogenetically informative features, and discovered homologous skeletal parts between them.

Methods

Observation of hagfish embryos

Embryos of the inshore hagfish *Eptatretus burgeri* were collected according to the developmental stage table of [25], as described previously [20]. In this study, we reexamined three-dimensional models reconstructed from the serial histological sections of the hagfish embryos previously [22, 23].

Observation of *Palaeospondylus* fossils

Specimens of *Palaeospondylus* were observed using a stereomicroscope (M7A, Wild, Heerbrugg, Switzerland) at the American Museum of Natural History (AMNH), New York, USA. Images were taken using a digital camera (E-PL6, Olympus Corp., Tokyo, Japan) mounted on the stereomicroscope via an adaptor (NY-1S, Microscope Network Co., Ltd., Saitama, Japan). The specimens of *Palaeospondylus* expose either dorsal or ventral aspects of their cranial skeletons on the slabs. Based on positional relationship to vertebral column, which lay dorsally in the live body, we identified dorsal and ventral views of the *Palaeospondylus* cranial skeletons on the slabs.

Results

Key features in the embryonic development of the modern hagfish *Eptatretus*

In *E. burgeri* embryos at stages 53–60 (Fig. 2a–d), the neurocranium consists of two pairs of longitudinal bars with transverse commissures and processes, rostrally attached to the nasal capsule. Caudally, the longitudinal bars are contiguous to the otic capsule, which continues medially to the parachordal that surrounds the rostral part of the notochord (Fig. 2a, b). From the boundary between the neurocranium and otic capsule, a pair of velar bars grows ventromedially, and later, by stage 53, becoming fused at the median plane to form a V-shaped element. The dental plate of the oral apparatus is located below the neurocranium, associated with the rostral part of the lingual plate that supports the muscles of the cyclostome 'tongue' [22]. The topographical relationships between these cranial elements in hagfish embryos continue to change over the course of development [23]. Prior to hatching, the nasal capsule grows cartilages supporting the rostrally extended nasal duct (Fig. 2e). Simultaneously, cartilages in the oronasal septum extend rostrally, and supporting cartilages arise within tentacles, features that are specific to adult individuals in all known hagfish species. These changes result in a caudad shift in the relative position of the nasal capsule in the cranium.

Homologies between *Eptatretus* and *Palaeospondylus*

As in cyclostomes, the cranium of *Palaeospondylus* is generally believed to consist of endoskeletal elements [5, 8, 9]. In many *Palaeospondylus* specimens, the vertebral column is preserved, and it is generally accepted that the massive skeletal element at the level of the first vertebral element represents the otic capsule [3–9] (Figs. 1 and 3a, b, Additional file 1: Figure S1, Additional file 2: Figure S2). We use this feature as a landmark in our identification of other skeletal elements in this species. Our observations (Additional file 1: Figure S1) and a review of the literature [5, 8, 26] (Additional file 2: Figure S2)

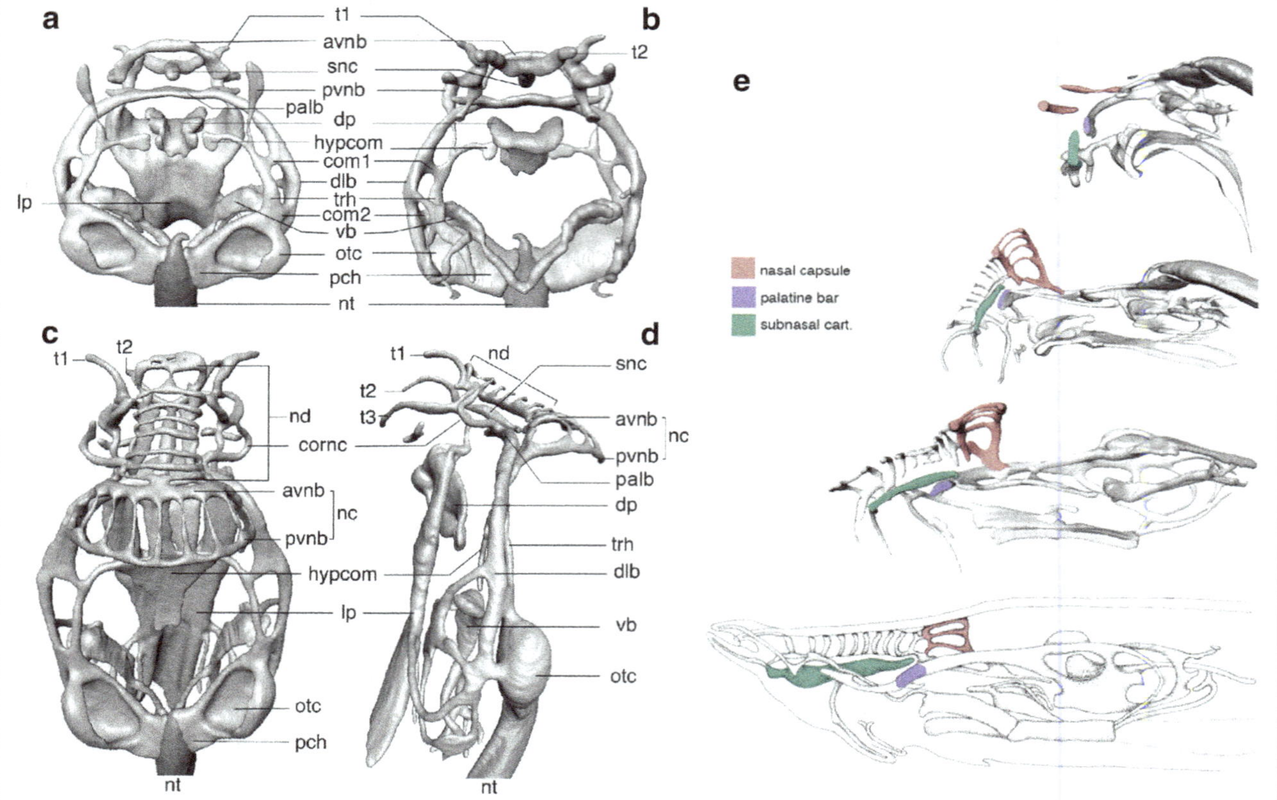

Fig. 2 Cranial skeletons of the *Eptatretus* embryo. **a** *Eptatretus burgeri* embryo at stage 53 in dorsal view. **b** *E. burgeri* embryo at stage 53 in ventral view. **c** *E. burgeri* embryo at stage 60 in dorsal view. **d** *E. burgeri* embryo at stage 60 in left lateral view. **e** Relative growth of the developing hagfish chondrocrania. From top to bottom, stage 53, stage 60, prehatching and adult hagfish crania. Note, due to the rostral growth of the snout by addition of nasal duct cartilage and extension of subnasal cartilage (*green*), that the position of the nasal capsule (*red*) shifts relatively caudal through development. The palatine bar, or the transverse commissure on the rostral tip of the dorsal longitudinal bar, is colored blue for the reference. avnb, anterior vertical nasal bar; com1, 2, commissures of dlb; cornc, cornual cartilage; dlb, dorsal longitudinal bar; dp, dental plate; hypcom, hypophyseal commissure; lp, lingual plate; nc, nasal capsule; nd, nasal duct cartilage; nt, notochord; otc, otic capsule; palb, palatine bar; pch, parachordal; pvnb, posterior vertical nasal bar; rp, rostral plate; snc, subnasal cartilage; t1-3, cartilaginous support for tentacles; trh, trabecula of hagfish; vb, velar bar

indicate that *Palaeospondylus* possessed a neurocranium comparable to that of the hagfish embryo, in terms of the possession of two pairs of longitudinal bars (the dorsal longitudinal bar and hagfish trabecula), the palatine bar, and other commissures connecting these bars (Figs. 2 and 3a, b). Rostrally, several (presumably six) longitudinal bars run in parallel ('rostralia' in the previous description [5]; Fig. 1a, Additional file 1: Figure S1, Additional file 2: Figure S2). These bars are contiguous caudally with a transverse bar and are expanded towards the rostral end to contact with adjacent elements (Fig. 1a, Additional file 1: Figure S1, Additional file 2: Figure S2), forming a cage-shaped structure [5, 9]. The shape and position of this structure, and in particular the presence of both the anterior and posterior transverse bars, show strong morphological similarities with those of the nasal capsule in the hagfish embryo (Fig. 3a). The nasal capsule of *Palaeospondylus* was previously identified at a lateral part of the neurocranium ('hemidome') [10], but in that model, the other neurocranial elements, including the trabeculae (in this case, the trabecula of crown gnathostomes), could not be identified. In contrast, the homologization between *Palaeospondylus* and the hagfish embryo described above succeeds in identifying each skeletal element at a comparable position.

The ventral V-shaped element in *Palaeospondylus* (Fig. 1b, Additional file 1: Figure S1, Additional file 2: Figure S2) resembles the velar bar of the hagfish embryo, as both elements are suspended ventrally at the level of the junction between the neurocranium and otic capsule (Fig. 3b). The midline of the V-shaped element of *Palaeospondylus* is mostly unfused, but this does not contradict the homology as, in the hagfish embryo, the velar bar develops from paired cartilages and later becomes fused at the median plane by stage 53. In some previous studies, this element has been identified as the gnathostome ceratohyal [8, 9]; however, this fails to

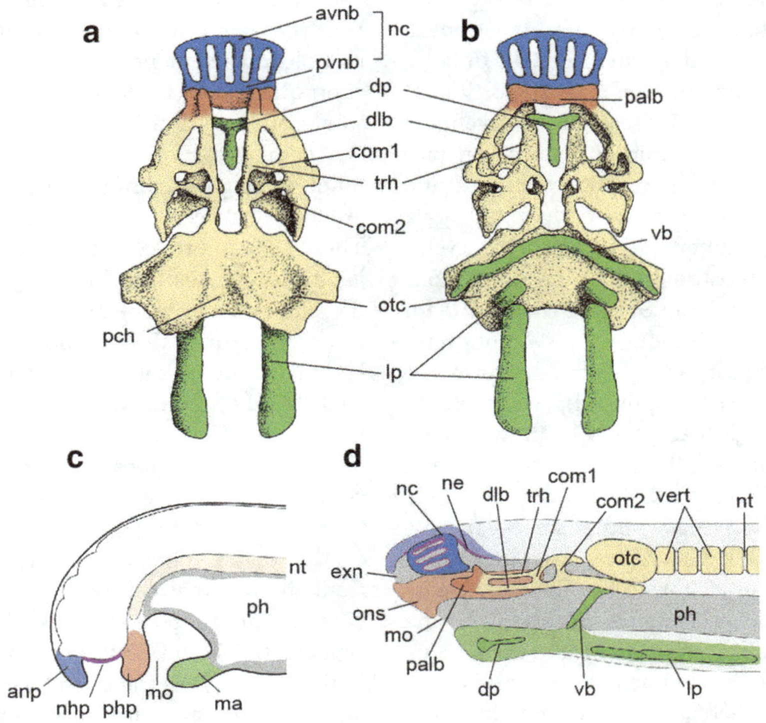

Fig. 3 Developmental origins of *Palaeospondylus* cranial skeleton. **a**,**b** Diagrammatic anatomy of the cranial skeleton in dorsal (**a**) and ventral (**b**) views. **c** Cyclostome pattern in embryonic development of lamprey, hagfish, and also of *Palaeospondylus*. **d** Hypothetical developmental configuration of *Palaeospondylus* derived from the cyclostome craniofacial pattern. anp, anterior nasal process and its derivatives; ma, mandibular arch; mo, mouth; ne, nasal epithelium; nhp, nasohypophyseal plate; ph, pharynx; php, post-hypophyseal process and its derivatives; vert, vertebra. For other abbreviations, see Fig. 2

explain the apparent absence of more posterior branchial arches in *Palaeospondylus*. The ventral, median T-shaped skeletal element in *Palaeospondylus*, often called the 'tauidion' [5] (Fig. 1b), may occur distant from the neurocranium [5, 8, 9, 26] (Additional file 2: Figure S2), and we suggest that it is comparable to the hagfish dental plate (Figs. 2b, d and 3b, Additional file 2: Figure S2). In the previous gnathostome hypothesis, this 'tauidion' has been analogized to the gnathostome vomer [10], but that identification is less plausible, as the 'tauidion' was independent of the other cranial elements. The absence of dermal bones in *Palaeospondylus* [11] also argues against the interpretation of the 'tauidion' as the gnathostome vomer, which is a dermal bone.

Two paired skeletal plates cover the ventral aspect of the caudal half of *Palaeospondylus* cranium (Fig. 1). The rostral pair is located at the level of the otic capsule, and the larger caudal pair covers the ventral aspect at the levels of the caudal half of the otic capsule and the rostral vertebral elements (Fig. 1b). These plates likely represent the lingual apparatus of the hagfish (Fig. 3b), as proposed also by Bulman [8]. These plates were later identified as lungfish-specific cranial ribs (occipital ribs) [10]. This view, however, is not consistent with the absence of the cranial rib in larval lungfishes [27, 28].

Discussion

The history of studies on *Palaeospondylus* [1–10] constitutes a heuristic search for homologous body parts, a task that we suggest may be completed with the present study, in which it was finally compared with the hagfish embryo. Previous studies comparing *Palaeospondylus* with hagfishes relied solely on adult morphology [8]. Importantly, it has recently been shown that adult hagfish species possess vestigial vertebral elements [21]. The presence of a vertebral column composed of cylindrical vertebrae in *Palaeospondylus* (Additional file 1: Figure S1), which simply sheathed the notochord [5, 8], thus does not necessarily preclude its affinity with the modern hagfish. Our comparative analysis of *Palaeospondylus* and the hagfish embryo, revealing previously unrecognized topographical relationships between skeletal elements specific to these taxa, was made possible by the introduction of modern techniques to the study of hagfish embryonic development in recent years [20–23, 29].

According to comparative developmental biology two mutually exclusive systems of craniofacial anatomical configuration are present in vertebrates [22, 24]. In both, during embryonic development, the rostral portion of the cephalic crest-derived ectomesenchyme (trigeminal crest cells) initially follows a common distribution pattern, comprising the pre- and postoptic crest cells and mandibular arch crest cells (up to early pharyngula stage) [30]. However, the difference between the cyclostomes and crown gnathostomes later becomes conspicuous (Additional file 3: Table S1) in association with differences in placode distribution [22, 31]. In cyclostomes, the craniofacial morphology develops from the mandibular arch (MA), post-hypophyseal process (PHP), and anterior nasal process (ANP) [24, 32]. In crown gnathostomes, the embryonic head consists of lateral and medial nasal prominences (LNP, MNP), postoptic ectomesenchyme (PO), maxillary and mandibular processes (MX, MN) [24, 32]. Given that the craniofacial pattern of stem gnathostomes (osteostracans and galeaspids) conforms with the cyclostome pattern [22], the cyclostome pattern (involving MA, PHP, and ANP) thus represents the ancestral developmental pattern for the total-group vertebrates, whereas the crown gnathostome pattern (involving LNP, MNP, PO, MX, and MN) represents a derived condition. However, the craniofacial development of crown gnathostomes does not, even transiently, recapitulate the cyclostome pattern at any developmental stage [22, 24].

The *Palaeospondylus* cranium is congruent with the cyclostome craniofacial pattern described above (Fig. 3). In this comparison, the nasal capsule, subnasal rostral cranium (palatine bar, and rostral part of the dorsal longitudinal bar and trabecula), and dental and lingual plates of the *Palaeospondylus* are reconstructed to develop from the ANP, PHP and MA of the cyclostome pattern, respectively (Fig. 3c,d; Additional file 4: Table S2). The inferred positions of the nasal epithelium and adenohypophysis in *Palaeospondylus* are consistent with this scheme (Fig. 3d). Contrastingly the *Palaeospondylus* cranium cannot be derived from the crown gnathostome pattern, which involves a dorsoventrally bifurcated mandibular arch, separate paired nostrils, and adenohypophysis. A previous hypothesis that suggested a Devonian lungfish affinity of *Palaeospondylus* [10] was flawed in that it did not account for the neurocranial element developed medially to MX.

More detailed homology relationship of skeletal elements can be established between *Palaeospondylus* and the hagfish embryo at specific developmental stages, but not with the lamprey skeletal elements at any developmental stage. The cage-shaped nasal capsule, two separated longitudinal bars (the dorsal longitudinal bar and trabecula), velar bar, and large lingual plates (Fig. 3a, b) are seen exclusively in the hagfish embryo and *Palaeospondylus*, and the topographical arrangement of these homologous elements is consistent. Based on these synapomorphies, it is parsimonious that *Palaeospondylus* was related to hagfishes.

A comparison of the rostral portion of the neurocranium among vertebrates further underscores the congruence between *Palaeospondylus* and the hagfish. There is a morphological disparity among the crown gnathostomes, lamprey and hagfish, which results from the difference in 'trabeculae' within embryonic heads; in fact, the term 'trabecula' does not designate the same body part among the crown gnathostomes, lamprey and hagfish. The neurocranium of the crown gnathostomes is heterogeneous in cell lineage [30, 33]; the rostral part, or trabecula, develops from the cephalic neural crest cells [34–38], while the caudal part, or parachordals, from the mesoderm [33]. In the cyclostomes, a large part of the rostral portion of the neurocranium develops from mesodermal cells [23, 39–43], and only the nasal capsule and oropharyngeal skeleton develop from the cephalic (ANP and PHP, respectively) crest cells [24, 44]. In the lamprey, the trabecula ('lamprey trabecula') consists of a mesodermal neurocranial wall secondarily elongated rostrally [39–43]. In the hagfish, the corresponding prechondrogenic precursor later splits dorsoventrally into two bars: the dorsal longitudinal bar and 'hagfish trabecula' [23, 24]. Of these, the configuration of rostral elements of *Palaeospondylus* neurocranium agrees only with the hagfish-type, as the dorsal longitudinal bar and 'hagfish trabecula' occupy the corresponding position in this taxon.

One previous study argued against the cyclostome affinity of *Palaeospondylus*, citing the presence of a paired fin located caudally distant from the cranium [9]. The extant hagfishes, on the other hand, possess a cartilage at the origin of the lingual muscle (the 'cyclostome tongue' [45]), or the perpendicular muscle cartilage, which is located far caudal to the cranium [46] (Fig. 4). The perpendicular muscle cartilage of the extant hagfish is not decay-prone [47], and thus can be preserved separately from the cranial skeleton in fossils. Here, we suggest that the 'paired fin' of *Palaeospondylus* may in fact represent such a skeletal element at the origin of the lingual muscle.

Unique among known cyclostomes, *Palaeospondylus* possessed distinct features, including the biomineralization of cartilages and the well-developed vertebral column (notochordal centrum). These features, however, may have evolved from cyclostome-like conditions. The calcified cartilage of *Palaeospondylus* lacked perichondral ossification [48], which is specific to the gnathostomes [38], and it is conceivable that the calcified cartilage of *Palaeospondylus* evolved either through loss of perichondral ossification in gnathostomes, or through

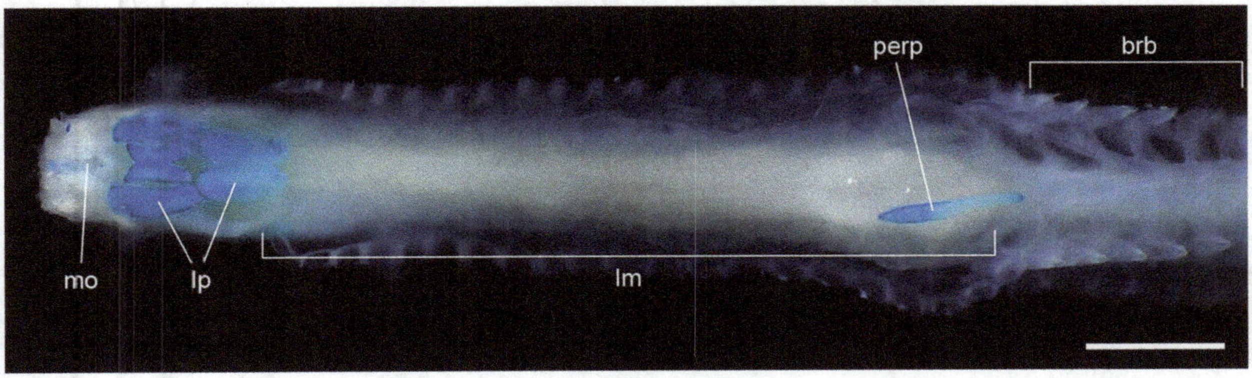

Fig. 4 Alcian blue-stained specimen of the adult hagfish *Eptatretus burgeri* in ventral view. brb, branchial bascket; lm, lingual muscle ('cyclostome tongue'); lp, lingual plate; mo, mouth; perp, perpendicular muscle cartilage (cartilage at the origin of the lingual muscle). Scale bar, 1 cm

acquisition of biomineralization in cyclostomes. The latter possibility appears more plausible, as the hypertrophied cell lacunae in the cartilages of *Palaeospondylus* [11, 48] are reminiscent of cyclostome cartilages in their thin layers of extracellular matrix [21, 49]. A recent study also suggested that the vertebral element is synapomorphic to the vertebrates [21, 50], and the cylindrical vertebral column in *Palaeospondylus* (Additional file 1: Figure S1) may have evolved secondarily from arcualia in basal cyclostomes, through the invasion of cartilaginous cells into the fibrous sheath within the elastica externa [51]. The absence of arcualia in *Palaeospondylus* fossils does not preclude this scenario, since calcification often occurs differentially among vertebral elements, as seen in elasmobranchs [52].

There are nonetheless minor differences between *Palaeospondylus* and the known hagfish species, including the absence of some skeletal elements in *Palaeospondylus*. We suggest that this is likely due to a taphonomic bias. In the extant cyclostomes, there are two types of cartilage, hard and soft, which differ in the amount of extracellular matrix [49, 53, 54]. Soft cartilage is less resistant to decay than hard cartilage [47]. A similar heterogeneity of cartilage composition may have been present in *Palaeospondylus*, given that the branchial skeleton has not been identified in *Palaeospondylus*. The absence of branchial skeleton in *Palaeospondylus* fossils may indicate that cartilages of the branchial basket were not calcified, as in extant cyclostomes, for the functional reason that the branchial basket changes shape during ventilation. Among fossil cyclostomes, the Late Devonian putative stem-lamprey *Euphanerops* is a unique taxon that possessed calcified branchial basket [55]. The gill basket of *Euphanerops* was, however, unique in extending caudally across the half of the trunk, suggesting a functional requirement different from that in other cyclostomes.

In the extant hagfish chondrocranium, the rostralmost skeletal element made of hard cartilage is the subnasal cartilage (Fig. 2e). The subnasal cartilage is expected to be found rostral to the nasal capsule [47], but is apparently missing in *Palaeospondylus*. Since the proportion of the *Palaeospondylus* cranium resembles the embryonic, rather than the adult, cranium in known hagfish species (Fig. 2e), the absence of overt subnasal cartilage in *Palaeospondylus* fossils may reflect a very short snout containing a delicate subnasal cartilage. Cartilages supporting the nasal duct and tentacles in the extant hagfish are composed of soft cartilage [54], thus they are expected to decay rapidly.

Given the potential extensive taphonomic bias and the lack of data about soft tissue anatomy unlike circumstances of other species [56–58], it is difficult to build a character matrix to conduct a cladistic analysis of *Palaeospondylus*. Nevertheless, the above comparison strongly suggests that *Palaeospondylus* and the hagfish share a cranial skeletal configuration that is distinguishable from those of the lamprey and crown gnathostomes. On the other hand, some features seen in adult individuals of the known hagfish species are less conspicuous in *Palaeospondylus*. In particular, in the extant hagfishes, as well as in the Late Carboniferous hagfish *Myxinikela siroka* [59], the position of the nasal capsule, which initially develops at the rostral end of the cranium becomes relatively caudal in the cranium of adult individuals, whereas in *Palaeospondylus*, the nasal capsule remained at the rostral end. Based on this synapomorphy between extant hagfishes and *Myxinikela*, we suggest that the phylogenetic position of *Palaeospondylus* is best explained as a stem hagfish lineage basal to *Myxinikela* (Fig. 5).

The hagfish affinity of *Palaeospondylus* is important for the time of divergence between hagfish and lamprey lineages. The oldest reliable fossil record of the

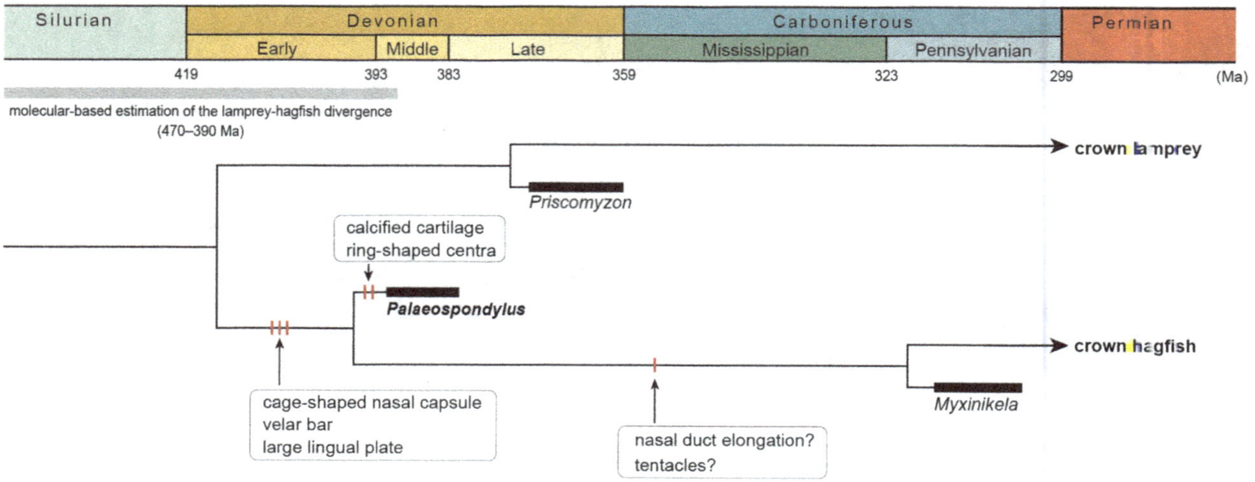

Fig. 5 Proposed phylogenetic position of *Palaeospondylus*. The molecular-based estimation of the lamprey-hagfish divergence was adopted from [63]

divergence time is the fossil lamprey *Priscomyzon riniensis* from the Upper Devonian (Famennian: 372–359 Ma in [60]) [61, 62]. An estimate using nucleotide and amino acid sequences, however, dated the divergence time to 470–390 Ma [63] (by the Middle Devonian), leaving a gap between the molecular-based estimate and the fossil record. Our classification of *Palaeospondylus* into the hagfishes pushes the oldest fossil record of the divergence time back to the Middle Devonian (393–383 Ma, Fig. 5), filling the gap between the molecular estimate and the fossil record, as well as depicting a hagfish species before the end-Devonian mass extinction that wiped out many vertebrate groups, including placoderms [64, 65].

Early cyclostomes may thus have been more morphologically diverse than previously recognized. At present, the early evolution of the cyclostomes has been less clear than that of the gnathostomes [62, 65]. Further analysis of *Palaeospondylus*, as well as other putative cyclostomes in the fossil record [55, 58, 66], may shed new light on the evolution of this lineage.

Conclusions

We identified congruences in the arrangement of a suite of homologous skeletal parts between *Palaeospondylus* and the hagfish embryo, and conclude that *Palaeospondylus* was closely related to hagfishes. The gnathostome affinity of this species is improbable, since the topography of cranial skeletal elements of *Palaeospondylus* is consistent exclusively with the developmental craniofacial pattern of the cyclostomes. *Palaeospondylus* more closely resembles the embryo than the adult of the extant hagfish, suggesting that this species is placed at a stem hagfish position basal to the hitherto known hagfish species in the phylogeny.

Additional files

Additional file 1: Figure S1. Fossils of *Palaeospondylus* from the Middle Devonian of the Scotland. (a) entire specimen of *Palaeospondylus gunni* (AMNH FF 10743) in ventral view. (b) cranial skeleton of *P. gunni* (AMNH FF 7586) in dorsal view. (c) Line drawing of B. (d) cranial skeleton of *P. gunni* (AMNH FF 10742) in ventral view. (e) Line drawing of D. amp, ampyx; cp, caudal plate; gam, gammation; hem, hemidome; otc, otic capsule; ros, rostralia; rp, rostral plate; ve, V-shaped element; vert, vertebra. Scale bar, 1 mm.

Additional file 2: Figure S2. Referred *Palaeospondylus* specimens illustrated in previous studies. (a,b) A specimen in dorsal (a) and ventral (b) views, from [26]. (c,d) three-dimensional model reconstructed from sections in dorsal (c) and ventral (d) views, from [5]. (e) NHM (Natural History Museum, London) P 16123 in dorsal view, from [8]. (f), NHM P 16125 in ventral view, from [8]. amp, ampyx; cp, caudal plate; gam, gammation; hem, hemidome; otc, otic capsule; ros, rostralia; ve, V-shaped element; vert, vertebra.

Additional file 3: Table S1. Comparison of craniofacial primordia in cyclostomes and crown gnathostomes. Based on [22].

Additional file 4: Table S2. Comparison of cranial skeletal elements among extant cyclostomes and *Palaeospondylus*. Names of skeletal elements in *Palaeospondylus* are shown by abbreviations employed in the present paper. For the nomenclature of the lamprey and hagfish chondrocrania, see [23].

Acknowledgements
We thank Kinya G. Ota and Osamu Kakitani for collecting and preparing specimens of hagfish embryos, Satoko Fujimoto for preparing Alcian blue-stained hagfish specimens, John Maisey and Alana Gishlick for fossil specimen access, and Douglas Sipp for useful comments.

Funding
This study was funded by RIKEN.

Authors' contributions
TH and SK designed the project. YO and SK obtained data about the hagfish embryology. TH examined fossils of *P. gunni*. TH and SK prepared figures and wrote the manuscript. All authors read and approved the final manuscript.

Competing interests
The authors declare no competing financial interests. Readers are welcome to comment on the online version of the paper.

Author details
[1]Evolutionary Morphology Laboratory, RIKEN, 2-2-3 Minatojima-minami, Chuo-ku, Kobe, Hyogo 650-0047, Japan. [2]Development and Function of Inhibitory Neural Circuits, Max Planck Florida Institute for Neuroscience, Jupiter, FL 33458, USA.

References

1. Traquair RH. On the fossil fishes at Achanarras Quarry, Caithness. Ann Mag Nat Hist 6th Ser. 1890;6:479–86.
2. Dawson JW. Some salient points in the science of the earth. London: W. Drysdale and Co.; 1893.
3. Kerr JG. The zoological position of *Palaeospondylus* (Traquair). Proc Camb Philol Soc. 1900;10:298–9.
4. Dean B. Still another memoir on *Palaeospondylus*. Science. 1904;19:425–6.
5. Sollas WJ, Sollas IBJ. An account of the Devonian fish *Palaeospondylus gunni* Traquair. Philos Trans R Soc Lond B. 1904;196:267–94.
6. Stensiö EA. The Downtonian and Devonian vertebrates of Spitsbergen. Part I, Family Cephalaspidae. Skr Svalbard Nordishavet. 1927;12:1–391.
7. Miller AE. Note on the tail skeleton of *Lepidosiren paradoxa*, with remarks on the affinities of *Palaeospondylus*. Proc Zool Soc Lond. 1930;100:783–9.
8. Bulman OMB. Note on *Palaeospondylus gunni*, Traquair. Ann Mag Nat Hist 10th Ser. 1931;8:179–90.
9. Moy-Thomas JA. The Devonian fish *Palaeospondylus gunni* Traquair. Philos Trans R Soc Lond B. 1940;230:391–413.
10. Thomson KS, Sutton M, Thomas B. A larval Devonian lungfish. Nature. 2003; 426:833–4.
11. Johanson Z, Kearsley A, den Blaauwen J, Newman M, Smith MM. No bones about it: An enigmatic Devonian fossil reveals a new skeletal framework: A potential role of loss of gene regulation. Semin Cell Dev Biol. 2010;21:414–23.
12. Joss J, Johanson Z. Is *Palaeospondylus gunni* a fossil larval lungfish? Insights from *Neoceratodus forsteri* development. J Exp Zool Part B Mol Dev Evol. 2007;308B:163–71.
13. Newman MJ, Den Blaauwen JL. New information on the enigmatic Devonian vertebrate *Palaeospondylus gunni*. Scot J Geol. 2008;44:89–91.
14. Colbert EH. Evolution of the vertebrates. 2nd ed. New York: Wiley; 1969.
15. Patterson C. Morphological characters and homology. In: Joysey KA, Friday AE, editors. Systematics association special volume 21: problems of phylogenetic reconstruction. London: Academic; 1982. p. 21–74.
16. Hall BK. Homoplasy and homology: Dichotomy or continuum? J Hum Evol. 2007;52:473–9.
17. Wagner GP. Homology, genes, and evolutionary innovation. Princeton: Princeton Univ Press; 2014.
18. Sereno PC. Logical basis for morphological characters in phylogenetics. Cladistics. 2007;23:565–87.
19. Ramirez MJ. Homology as a parsimony problem: a dynamic homology approach for morphological data. Cladistics. 2007;23:588–612.
20. Ota KG, Kuraku S, Kuratani S. Hagfish embryology with reference to the evolution of the neural crest. Nature. 2007;446:672–5.
21. Ota KG, Fujimoto S, Oisi Y, Kuratani S. Identification of vertebra-like elements and their possible differentiation from sclerotomes in the hagfish. Nat Commun. 2011;2:373.
22. Oisi Y, Ota KG, Kuraku S, Fujimoto S, Kuratani S. Craniofacial development of hagfishes and the evolution of vertebrates. Nature. 2013;493:175–80.
23. Oisi Y, Ota KG, Fujimoto S, Kuratani S. Development of the chondrocranium in hagfishes, with special reference to the early evolution of vertebrates. Zool Sci. 2013;30:944–61.
24. Kuratani S, Oisi Y, Ota KG. Evolution of the vertebrate cranium: viewed from the hagfish developmental studies. Zool Sci. 2016;33:229–38.
25. Dean B. On the embryology of *Bdellostoma stouti*. A genera account of myxinoid development from the egg and segmentation to hatching. In: Festschrift zum siebenzigsten Geburtstag von Carl von Kupffer. Jena: Gustav Fischer; 1899. p. 220–76.
26. Traquair RH. A still further contribution to our knowledge of *Palaeospondylus gunni*. Proc R Soc Edinb. 1893;12:312–21.
27. Greil A. Entwickelungsgeschichte des Kopfes und des Blutgefäßsystems von *Ceratodus forsteri*. II. Die epigenetischen Erwerbungen während der Stadien 39-48. Denkschr Med-Naturwiss Ges Jena. 1913;4:935–1492.
28. Johanson Z, Ericsson R, Long J, Evans F, Joss J. Development of the axial skeleton and median fin in the Australian lungfish, *Neoceratodus forsteri*. Open Zool J. 2009;2:91–101.
29. Ota KG, Kuratani S. The history of scientific endeavors towards understanding hagfish embryology. Zool Sci. 2006;23:403–18.
30. Wada N, Nohno T, Kuratani S. Dual origins of the prechordal cranium in the chicken embryo. Dev Biol. 2011;356:529–40.
31. Kuratani S. Evolution of the vertebrate jaw from developmental perspectives. Evol Dev. 2012;14:76–92.
32. Haeckel E. Anthropogenie oder Entwickelungsgeschichte des menschen: Keimes- und stammesgeschichte. Leipzig: W. Engelmann; 1877.
33. Couly GF, Coltey PM, Le Douarin NM. The triple origin of skull in higher vertebrates: a study in quail-chick chimeras. Development. 1993;117:409–29.
34. Le Douarin N. The neural crest. Cambridge: Cambridge Univ Press; 1982.
35. Maderson PFA, editor. Developmental and evolutionary aspects of the neural crest. New York: Wiley; 1987.
36. Hall BK, Hörstadius S. The neural crest. New York: Oxford Univ Press; 1988.
37. Kuratani S. Craniofacial development and the evolution of the vertebrates: the old problems on a new background. Zool Sci. 2005;22:1–19.
38. Hirasawa T, Kuratani S. Evolution of the vertebrate skeleton: morphology, embryology, and development. Zool Lett. 2015;1:2.
39. Johnels AG. On the development and morphology of the skeleton of the head of *Petromyzon*. Act Zool. 1948;29:139–279.
40. Kuratani S, Murakami Y, Nobusada Y, Kusakabe R, Hirano S. Developmental fate of the mandibular mesoderm in the lamprey, *Lethenteron japonicum*: comparative morphology and development of the gnathostome jaw with special reference to the nature of the trabecula cranii. J Exp Zool Part B Mol Dev Evol. 2004;302B:458–68.
41. Newth DR. Experiments on the neural crest of the lamprey embryo. J Exp Biol. 1951;28:247–60.
42. Newth DR. On the neural crest of the lamprey embryo. J Emb Exp Morphol. 1956;4:358–75.
43. Langille RM, Hall BK. Role of the neural crest in development of the trabeculae and branchial arches in embryonic sea lamprey, *Petromyzon marinus* (L). Development. 1988;102:301–10.
44. Dupret V, Sanchez S, Goujet D, Tafforeau P, Ahlberg PE. A primitive placoderm sheds light on the origin of the jawed vertebrate face. Nature. 2014;507:500–3.
45. Oisi Y, Fujimoto S, Ota KG, Kuratani S. On the peculiar morphology and development of the hypoglossal, glossopharyngeal and vagus nerves and hypobranchial muscles in the hagfish. Zool Let. 2015;1:6.
46. Marinelli W, Strenger A. *Vergleichende anatomie und morphologie der wirbeltiere. II.* Myxine glutinosa (*L.*). Wien: Franz Deuticke; 1956.
47. Sansom RS, Gabbott SE, Purnell MA. Decay of vertebrate characters in hagfish and lamprey (Cyclostomata) and the implications for the vertebrate fossil record. Proc R Soc Lond B. 2011;278:1150–7.
48. Johanson Z, Kearsley A, den Blaauwen J, Newman M, Smith MM. Ontogenetic development of an exceptionally preserved Devonian cartilaginous skeleton. J Exp Zool Part B Mol Dev Evol. 2012;318B:50–8.
49. Ota KG, Kuratani S. Expression pattern of two collagen type 2 α1 genes in the Japanese Inshore Hagfish (*Eptatretus burgeri*) with special reference to the evolution of cartilaginous tissue. J Exp Zool Part B Mol Dev Evol. 2010; 314B:157–65.
50. Ota KG, Oisi Y, Fujimoto S, Kuratani S. The origin of developmental mechanisms underlying vertebral elements: implications from hagfish evo-devo. Zoology. 2014;117:77–80.
51. Gadow HF. The evolution of the vertebral column: a contribution to the study of vertebrate phylogeny. Cambridge: Cambridge Univ Press; 1933.
52. Ridewood WG. On the calcification of the vertebral centra in sharks and rays. Philos Trans R Soc Lond B. 1921;210:311–407.
53. Parker WK. On the skeleton of the Marsipobranch fishes. Part I. The Myxinoids (Mixyine, and Bdellostoma). Philos Trans R Soc Lond. 1883;174:373–409.

54. Cole FJ. A monograph on the general morphology of Myxinoid fishes, based on a study of Myxine. Part 1. The anatomy of the skeleton. Trans R Soc Edinb. 1905;41:749–88.
55. Janvier P, Arsenault M. The anatomy of *Euphanerops longaevus* Woodward, 1900, an anaspid-like jawless vertebrate from the Upper Devonian of Miguasha, Quebec, Canada. Geodiversitas. 2007;29:143–216.
56. Shu DG, Conway Morris S, Han J, Zhang ZF, Yasui K, Janvier P, et al. Head and backbone of the Early Cambrian vertebrate *Haikouichthys*. Nature. 2003; 421:526–9.
57. Conway Morris S, Caron JB. A primitive fish from the Cambrian of North America. Nature. 2014;512:419–22.
58. McCoy VE, Saupe EE, Lamsdell JC, Tarhan LG, McMahon S, Kidgard S, et al. The 'Tully monster' is a vertebrate. Nature. 2016;532:496–9.
59. Bardack D. First fossil hagfish (Myxinoidea) - a record from the Pennsylvanian of Illinois. Science. 1991;254:701–3.
60. Walker JD, Geissman JW, Bowring SA, Babcok LE. compilers: Geological Time Scale v. 4.0. Boulder: Geological Society of America; 2012.
61. Gess RW, Coates MI, Rubidge BS. A lamprey from the Devonian period of South Africa. Nature. 2006;443:981–4.
62. Janvier P. Early jawless vertebrates and cyclostome origins. Zool Sci. 2008;25: 1045–56.
63. Kuraku S, Kuratani S. Time scale for cyclostome evolution inferred with a phylogenetic diagnosis of hagfish and lamprey cDNA sequences. Zool Sci. 2006;23:1053–64.
64. Sallan LC, Coates MI. End-Devonian extinction and a bottleneck in the early evolution of modern jawed vertebrates. Proc Natl Acad Sci U S A. 2010; 107:10131–5.
65. Friedman M, Sallan LC. Five handred million years of extinction and recovery: a phanerozoic survey of large-scale diversity patterns in fishes. Palaeontology. 2012;55:707–42.
66. Kuratani S, Hirasawa T. Palaeontology: Getting the measure of a monster. Nature. 2016;532:447–8.

6

Occasional cooperative breeding in birds and the robustness of comparative analyses concerning the evolution of cooperative breeding

Michael Griesser[1,2*] and Toshitaka N. Suzuki[2]

Abstract

Cooperative breeding is a widespread and intense form of cooperation, in which individuals help raise offspring that are not their own. This behaviour is particularly well studied in birds, using both long-term and comparative studies that have provided insights into the evolution of reproductive altruism. In most cooperatively breeding species, helpers are offspring that remain with their parents beyond independency and help in the raising of younger siblings. However, many cooperatively breeding species are poorly studied, and in 152 species, this behaviour only has been observed infrequently (i.e., occasional cooperative breeding). Here we argue that the parental care mode of these 152 species needs to be treated with caution, as factors associated with occasional cooperative breeding may differ from those associated with "regular" cooperative breeding. In most cooperatively breeding species, helpers provide alloparental care at the nests of their parents or close relatives; however, only in one occasionally cooperatively breeding species do offspring remain into the next breeding season with their parents. Accordingly, different factors are likely to be associated with regular and occasional cooperative breeding. The latter behaviour resembles interspecific feeding (i.e., individuals feed offspring of another species), which occurs when birds lose their brood and begin feeding at a nearby nest, or when birds mistakenly feed at another nest. Thus, we advise researchers to exclude occasional cooperative breeders in comparative analyses until their status is clarified, or to categorize them separately or according to the typically observed parental care mode. This approach will increase the robustness of comparative analyses and thereby improve our understanding of factors that drive the evolution of cooperative breeding.

Keywords: Cooperative breeding, Alloparental care, Comparative studies, Interspecific feeding, Incidental observations

Introduction

Cooperation among individuals occurs at all levels of biological organisation [1] and understanding the factors that select for cooperation is a fundamental goal of evolutionary biology. Darwin perceived that reproductive cooperation, such as sterile casts in eusocial insects, represented a challenge to his theory of natural selection [2]. Over the past 50 years, research has shown that cooperation often occurs among related individuals [1, 3], providing opportunities for kin selected fitness benefits [4]. One intensive form of cooperation is alloparental care [1, 5–7], in which individuals help raise the offspring of others while often foregoing their own reproduction [8, 9]. This behaviour has evolved in all major vertebrate lineages and is particularly well investigated in birds, using large-scale comparative studies [10–14].

Earlier comparative studies have investigated the association between eco-climatic and life-history factors and the occurrence of cooperative breeding, yielding sometimes contradictory findings [10–13]. While some studies have suggested that cooperative breeding is associated with stable climatic conditions and saturated habitats [10, 12, 15], other studies have indicated that

* Correspondence: michael.griesser@uzh.ch
[1]Department of Anthropology, University of Zurich, Winterthurerstrasse 190, 8057 Zurich, Switzerland
[2]Department of Evolutionary Studies of Biosystems, SOKENDAI (The Graduate University for Advanced Studies), Kanagawa, Japan

it is associated with unpredictable climatic conditions [11, 16, 17]. These contradictory findings may in part reflect heterogeneity in the quality of data on the parental care mode of birds. To consider this uncertainty, we categorize species in which cooperative breeding has been observed infrequently as "occasional cooperative breeders" ($N = 152$ species, representing 15 % of both cooperatively and occasionally cooperatively breeding species [18, 19]; supplementary material Additional file 1: Table S1). Observe that we categorized the Darwin finches *Geospiza scandens* and *G. fortis* as occasional cooperative breeders, based on a detailed study on helping at the nest in these two species [19] (see below). However, the question of how to categorize the mode of parental care in occasionally cooperatively breeding species remains open, and small numbers of observations may not be sufficient to categorize these species [20].

Here, we review the current understanding of the routes to cooperative breeding, and summarize previously published studies of the family structure of occasional cooperative breeders. We propose that occasional cooperative breeding shows parallels with interspecific feeding (i.e., individuals feed offspring of another species), and thus should not be categorized together with "regular" cooperatively breeding species in comparative studies.

Routes to cooperative breeding

In nearly all cooperatively breeding species (i.e., 93 % [18, 21, 22]), helpers are offspring that remain associated with their parents until the next breeding season and provide alloparental care at the nest of their parents or close relatives [9, 21, 22]. Field studies have demonstrated that helpers can gain both direct and indirect, kin-selected fitness benefits from providing alloparental care [3, 22, 23]. In some cooperatively breeding species, however, helpers are not related to breeders, but these individuals often have a share in reproduction, the species breeds polygynously or polyandrously (such as ani *Crotphasgus sp.*), or helpers queue for a breeding position [5, 22]. Thus, this route to cooperative breeding is most likely facilitated by direct fitness benefits [23]. Both of these phenomena occur regularly within populations, but the number of pairs that receive alloparental care can vary between 0 and 100 % depending on the species and annual conditions [5, 6]. Also, individuals may express a high flexibility in their parental care contributions in some species and frequently switch roles between breeders and helpers even within a breeding season [24]. Finally, three individuals have been observed feeding at the same nest in few instances in 152 species (labelled occasional cooperative breeding [18]), but the factors selecting for this behaviour remain unclear.

Occasional cooperative breeding

By definition, occasional cooperative breeding occurs rarely, and is thus difficult to investigate and remains poorly understood. This behaviour may occur commonly in some species, but be overlooked in species that are poorly investigated. However, while most cooperatively breeding species are not well studied, (i.e., cooperatively breeding species have a mode of two independent Zoological Record entries; data obtained from [25]; Fig. 1), many occasionally cooperatively breeding species in fact are well studied (mode of independent Zoological Record entries of occasional cooperative breeder = 15; $N = 69$ species have more than 200 independent Zoological Record entries; Spearman rank correlation: $P < 0.001$; Fig. 1). For example, occasional cooperative breeding has been

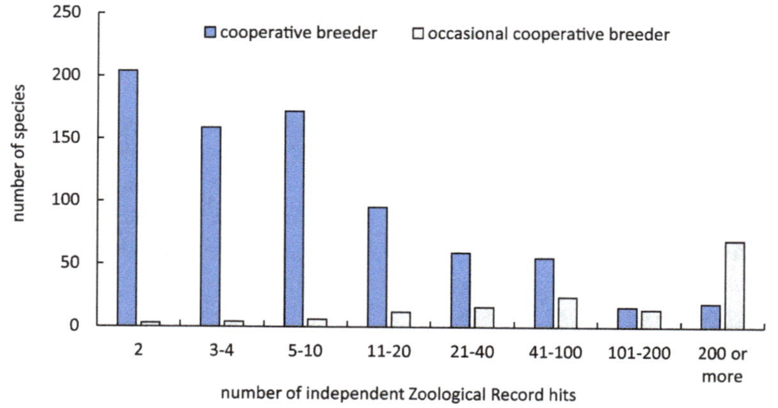

Fig. 1 Distribution of the number of independent Zoological Record entries for cooperatively breeding species, including species that are suspected cooperative breeders (data available for $N = 782$ out of 864 species), and occasionally cooperatively breeding species (data available for $N = 146$ out of 152 species). Most cooperatively breeding species have only two independent Zoological Record entries, while most occasionally cooperatively breeding species are well studied

observed in model species, including mute swan *Cygnus olor*, common guillemot *Uria aalge*, blue tit *Parus caeruleus*, white stork *Ciconia ciconia*, common tern *Sterna hirundo* or house sparrow *Passer domesticus*. In these species, "helpers" are most likely unrelated to the breeders, since the offspring do not remain with their parents into the next breeding season, but disperse earlier (see Additional file 2: Table S2 for details) [21, 26–28]. Only in one occasional cooperative breeder, the Verreaux's eagle owl *Bubo lacteus*, offspring may remain up for to 2 years in the parental territory [29], and thus live in family groups in a manner similar to that of most cooperatively breeding species.

A single study on Darwin finches *Geospiza scandens* and *G. fortis* described occasional cooperative breeding in these species in detail [19]. Eleven unpaired adult males were observed to provide alloparental care during a severe drought. However, the helpers were not related to the breeding pair, and many helpers bred earlier on in their lives, excluding that they helped to gain breeding experience. Moreover, none of the females that received help paired up with a helper later on in their lives, and helpers did not seem to gain access to reproduction. Rather, the authors suggested that the alloparental care reflected "misdirected" parental care [19, 30, 31].

Occasional cooperative breeding and interspecific feeding

Occasional cooperative breeding shows parallels with interspecific feeding (i.e., individuals feed offspring of other species [32]; Figure 2). In both cases, individuals do not feed related offspring, and both behaviours occur rarely. Interspecific feeding has been observed in 51 species (Additional file 2: Table S2; excluding cases where individuals from two species share a nest or egg dumping leads to mixed broods). It has been suggested that the main reason behind interspecific feeding is the loss of the individual's own brood, leading to those individuals to begin feeding at a nearby nest of another species [30, 32, 33]. Alternatively, parents may have their own nest but accidentally feed at another nest due to the close proximity of the other nest, or in response to the begging calls of nestlings [20, 32].

Interspecific feeding has been documented in a wide range of bird species (Table 1, Additional file 2: Table S2; see also [32]), suggesting that individuals may feed the young of conspecifics (i.e., engage in occasional cooperative breeding) for the same reasons as individuals engage in interspecific feeding [19]. Indeed, ten species are known to engage both in occasional cooperative breeding and interspecific feeding (i.e., 19.6 % of all occasional cooperative breeders; Additional file 1: Table S1, Additional file 2: Table S2). Interspecific feeding and occasional cooperative breeding, however, differ in one very critical point: the former is easy to recognize and categorize in the field, but a single observation of three individuals feeding at a nest does not allow categorization. It became quickly evident that the Japanese great tit is not a cooperatively breeding species, but one that may engage in interspecific feeding (Figure 2). If the Japanese great tit male had fed at another nest of Japanese great tits, however, we may have classified this observation as regular or occasional cooperative breeding. The generally used definition of cooperative breeding (an individual helps in the raising of offspring that are not their own, often while foregoing their own reproduction [5, 6, 8, 9]) does not differentiate whether the behaviour occurs within species (i.e. cooperative breeding), or across species (i.e. interspecific feeding). Thus, we propose to adjust the definition of cooperative breeding to: an individual that helps in the raising of *conspecific* offspring that are not their own, often while foregoing their own reproduction.

Occasional cooperative breeding and the evolution of regular cooperative breeding

Historically, observations of occasional cooperative breeding have fuelled a debate on the factors favouring the evolution of cooperative breeding [20, 34]. It has been suggested that the initial evolution of alloparental care may be a non-adaptive response to the begging of nestlings [31]. While this behaviour may provide a first step towards cooperative breeding [34], a number of arguments have been put forward regarding why the behaviour of helpers is modified by natural selection, thus making the behaviour adaptive [34]. Physiological studies showed that helpers in cooperative breeders express higher levels of prolactin than individuals that do not help, and that the prolactin level correlates with the helping effort [35, 36]. Moreover, helpers seem to adjust their feeding effort depending on the need of the nestlings [23, 37–40]. Finally, comparative data have shown that helping behaviour is higher in species in which helping at the nest provides a greater fitness benefit, and helpers in species that exhibit extra-pair mating have a higher degree of kin discrimination [4].

Conclusions

Interspecific feeding and occasional cooperative breeding have intrigued researchers for decades [30] and stimulated a critical assessment of cooperative breeding [20, 34]. Nevertheless, fitness consequences of both interspecific feeding and occasional cooperative breeding remain unstudied due to the rarity of their occurrence [32]. It thus remains difficult to draw firm conclusions as to whether these behaviours are non-adaptive, or provide fitness benefits to the actor. Cooperative breeding has been shown to be adaptive, helpers respond to the actual need of offspring [37], and most helpers provide care at nests of relatives, offering kin selected fitness

Table 1 Taxonomic overview on the family level over the parental care mode of occasional cooperative breeders and of 9659 bird species, the number of species showing interspecific feeding, the proportion of cooperatively breeding species within all families, and the number of species in each taxonomic family; CB = cooperative breeder (including both family living cooperative breeders and species with unrelated helpers), FAM = family living species, NO-FAM = non-family living species, UN = unknown parental care mode

Taxonomic family	Parental care mode of occasional CB species			Inferred parental care mode of all species				N species engaging in interspecific feeding	Proportion CB species	Total number of species
	FAM	NO-FAM	UN	CB	FAM	NO-FAM	UN			
Acanthisittidae	0	0	0	1	0	1	0	0	**0.50**	2
Acanthizidae	0	0	0	31	15	10	8	0	**0.48**	64
Accipitridae	3	11	6	15	30	53	139	1	0.06	237
Aegithalidae	0	0	0	5	0	5	0	1	**0.50**	10
Aegithinidae	0	0	0	0	0	0	4	0	0.00	4
Aegothelidae	0	0	0	0	0	1	7	0	0.00	8
Alaudidae	0	2	2	2	6	35	47	0	0.02	90
Alcedinidae	0	2	1	15	2	13	65	0	0.16	95
Alcidae	0	1	0	0	0	18	4	0	0.00	22
Anatidae	4	0	0	0	32	61	61	0	0.00	154
Anhimidae	0	0	0	0	1	2	0	0	0.00	3
Anhingidae	0	0	0	0	0	4	0	0	0.00	4
Anseranatidae	0	0	0	1	0	0	0	0	**1.00**	1
Apodidae	0	0	0	12	0	24	62	0	0.12	98
Apterygidae	0	0	0	1	0	3	1	0	0.20	5
Aramidae	0	0	0	0	1	0	0	0	0.00	1
Ardeidae	0	0	0	0	5	42	15	0	0.00	62
Artamidae	0	0	0	9	0	2	0	0	**0.82**	11
Atrichornithidae	0	0	0	0	1	1	0	0	0.00	2
Balaenicipitidae	0	0	0	0	0	1	0	0	0.00	1
Bombycillidae	0	0	0	0	0	4	4	0	0.00	8
Brachypteraciidae	0	0	0	0	1	0	4	0	0.00	5
Bucconidae	0	0	0	5	0	4	24	0	0.15	33
Bucerotidae	1	0	0	21	8	12	13	0	**0.39**	54
Bucorvidae	0	0	0	2	0	0	0	0	**1.00**	2
Burhinidae	0	0	0	0	2	4	3	0	0.00	9
Callaeatidae	0	0	0	0	0	2	0	0	0.00	2
Campephagidae	1	0	0	5	5	19	55	0	0.06	84
Caprimulgidae	0	0	1	0	4	10	71	0	0.00	85
Cardinalidae	0	1	0	13	6	21	18	2	0.22	58
Cariamidae	0	0	0	0	2	0	0	0	0.00	2
Casuariidae	0	0	0	0	1	2	0	0	0.00	3
Cathartidae	0	0	0	0	3	1	3	0	0.00	7
Certhiidae	0	0	0	0	2	4	1	0	0.00	7
Charadriidae	0	0	2	1	8	33	23	0	0.02	65
Chionidae	0	0	0	0	1	1	1	0	0.00	3
Chloropseidae	0	0	0	0	0	2	6	0	0.00	8
Ciconiidae	0	0	2	0	2	9	8	0	0.00	19
Cinclidae	0	0	0	0	0	2	3	0	0.00	5

Table 1 Taxonomic overview on the family level over the parental care mode of occasional cooperative breeders and of 9659 bird species, the number of species showing interspecific feeding, the proportion of cooperatively breeding species within all families, and the number of species in each taxonomic family; CB = cooperative breeder (including both family living cooperative breeders and species with unrelated helpers), FAM = family living species, NO-FAM = non-family living species, UN = unknown parental care mode *(Continued)*

Taxonomic family	Parental care mode of occasional CB species			Inferred parental care mode of all species				N species engaging in interspecific feeding	Proportion CB species	Total number of species
	FAM	NO-FAM	UN	CB	FAM	NO-FAM	UN			
Cinclosomatidae	0	0	0	3	0	0	2	0	**0.60**	5
Cisticolidae	0	0	0	14	48	27	29	1	0.12	118
Climacteridae	0	0	0	5	1	1	0	0	**0.71**	7
Cnemophilidae	0	0	0	0	0	3	0	0	0.00	3
Coerebidae	0	0	0	0	0	0	1	0	0.00	1
Coliidae	0	0	0	6	0	0	0	0	**1.00**	6
Colluricinclidae	0	0	0	6	0	0	7	0	**0.46**	13
Columbidae	0	0	0	0	3	276	25	0	0.00	304
Conopophagidae	0	0	0	0	0	8	0	0	0.00	8
Coraciidae	0	0	1	1	0	3	8	0	0.08	12
Corcoracidae	0	0	0	2	0	0	0	0	**1.00**	2
Corvidae	1	0	0	48	35	31	5	0	**0.40**	119
Cotingidae	0	0	0	1	5	60	28	0	0.01	94
Cracidae	1	0	0	0	38	6	6	0	0.00	50
Cracticidae	0	0	0	7	0	1	4	0	**0.58**	12
Cuculidae	0	2	1	4	4	87	47	0	0.03	142
Dasyornithidae	0	0	0	0	0	3	0	0	0.00	3
Dendrocolaptidae	0	0	0	0	5	7	34	0	0.00	46
Dicaeidae	0	0	0	0	4	14	26	0	0.00	44
Dicruridae	0	0	0	1	4	7	12	0	0.04	24
Diomedeidae	0	0	0	1	0	13	0	0	0.07	14
Dromadidae	0	0	0	0	0	1	0	0	0.00	1
Dromaiidae	0	0	0	0	1	0	0	0	0.00	1
Dulidae	0	0	0	1	0	0	0	0	**1.00**	1
Emberizidae	0	5	0	7	40	200	85	6	0.02	332
Estrildidae	1	0	0	0	26	89	23	0	0.00	138
Eupetidae	0	0	0	0	1	1	7	0	0.00	9
Eurylaimidae	0	0	0	3	0	3	9	0	0.20	15
Eurypygidae	0	0	0	0	0	1	0	0	0.00	1
Falconidae	0	5	1	15	3	14	31	0	0.24	63
Falcunculidae	0	0	0	1	0	0	1	0	**0.50**	2
Formicariidae	0	0	0	0	1	1	59	0	0.00	61
Fregatidae	0	0	0	0	2	3	0	0	0.00	5
Fringillidae	1	6	1	3	30	89	37	3	0.02	159
Furnariidae	0	0	0	8	2	16	209	0	0.03	235
Galbulidae	0	0	0	18	0	0	0	0	**1.00**	18
Gaviidae	0	0	0	0	0	5	0	1	0.00	5
Glareolidae	0	0	0	0	3	7	7	0	0.00	17

Table 1 Taxonomic overview on the family level over the parental care mode of occasional cooperative breeders and of 9659 bird species, the number of species showing interspecific feeding, the proportion of cooperatively breeding species within all families, and the number of species in each taxonomic family; CB = cooperative breeder (including both family living cooperative breeders and species with unrelated helpers), FAM = family living species, NO-FAM = non-family living species, UN = unknown parental care mode *(Continued)*

Taxonomic family	Parental care mode of occasional CB species			Inferred parental care mode of all species				N species engaging in interspecific feeding	Proportion CB species	Total number of species
	FAM	NO-FAM	UN	CB	FAM	NO-FAM	UN			
Gruidae	0	0	0	0	13	1	1	0	0.00	15
Haematopodidae	1	0	0	1	5	3	1	0	0.10	10
Heliornithidae	0	0	0	0	0	2	1	0	0.00	3
Hemiprocnidae	0	0	0	0	0	0	4	0	0.00	4
Hirundinidae	0	6	0	0	3	65	19	1	0.00	87
Hydrobatidae	0	0	0	0	0	10	10	0	0.00	20
Ibidorhynchidae	0	0	0	0	0	0	1	0	0.00	1
Icteridae	0	1	0	11	26	62	12	0	0.10	111
Indicatoridae	0	0	0	0	0	17	0	0	0.00	17
Irenidae	0	0	0	0	0	1	1	0	0.00	2
Jacanidae	0	0	0	0	2	2	4	0	0.00	8
Laniidae	0	1	0	6	8	16	1	0	0.19	31
Laridae	3	0	1	0	18	28	51	1	0.00	97
Leptosomidae	0	0	0	0	0	1	0	0	0.00	1
Machaerirhynchidae	0	0	0	0	0	0	2	0	0.00	2
Malaconotidae	1	0	0	7	14	27	6	0	0.13	54
Maluridae	0	0	0	28	0	0	0	0	**1.00**	28
Megapodiidae	0	0	0	0	0	4	15	0	0.00	19
Melanocharitidae	0	0	0	0	1	4	7	0	0.00	12
Meliphagidae	1	5	0	21	13	89	54	0	0.12	177
Menuridae	0	0	0	0	2	0	0	0	0.00	2
Meropidae	0	0	0	20	0	0	6	0	**0.77**	26
Mesitornithidae	0	0	0	2	0	1	0	0	**0.67**	3
Mimidae	0	0	0	7	8	4	16	2	0.20	35
Momotidae	0	0	0	0	0	0	9	0	0.00	9
Monarchidae	1	0	0	6	16	23	51	0	0.06	96
Motacillidae	0	4	1	0	10	32	23	1	0.00	65
Muscicapidae	0	7	0	26	24	55	179	3	0.09	284
Musophagidae	0	0	0	5	1	13	4	0	0.22	23
Nectariniidae	0	0	1	4	5	48	67	1	0.03	124
Neosittidae	0	0	0	2	0	0	0	0	**1.00**	2
Numididae	0	0	0	0	5	1	0	0	0.00	6
Nyctibiidae	0	0	0	0	0	0	7	0	0.00	7
Odontophoridae	0	0	0	0	21	10	0	0	0.00	31
Opisthocomidae	0	0	0	1	0	0	0	0	**1.00**	1
Oriolidae	0	0	1	2	1	15	12	0	0.07	30
Orthonychidae	0	0	0	0	1	1	3	0	0.00	5
Otididae	0	0	0	0	10	5	10	0	0.00	25

Table 1 Taxonomic overview on the family level over the parental care mode of occasional cooperative breeders and of 9659 bird species, the number of species showing interspecific feeding, the proportion of cooperatively breeding species within all families, and the number of species in each taxonomic family; CB = cooperative breeder (including both family living cooperative breeders and species with unrelated helpers), FAM = family living species, NO-FAM = non-family living species, UN = unknown parental care mode *(Continued)*

Taxonomic family	Parental care mode of occasional CB species			Inferred parental care mode of all species				N species engaging in interspecific feeding	Proportion CB species	Total number of species
	FAM	NO-FAM	UN	CB	FAM	NO-FAM	UN			
Pachycephalidae	0	0	0	1	1	0	36	0	0.03	38
Paradisaeidae	0	0	0	0	0	6	34	0	0.00	40
Pardalotidae	0	0	0	1	0	3	0	0	0.25	4
Paridae	1	2	0	14	10	29	4	3	0.25	57
Parulidae	0	4	1	0	7	60	45	2	0.00	112
Passeridae	0	0	0	11	2	24	10	1	0.23	47
Pedionomidae	0	0	0	0	1	0	0	0	0.00	1
Pelecanidae	0	0	0	0	1	3	4	0	0.00	8
Pelecanoididae	0	0	0	0	0	4	0	0	0.00	4
Petroicidae	0	1	0	12	0	6	25	1	**0.28**	43
Peucedramidae	0	0	0	0	1	0	0	0	0.00	1
Phaethontidae	0	0	0	0	0	3	0	0	0.00	3
Phalacrocoracidae	0	1	0	0	1	34	1	0	0.00	36
Phasianidae	0	0	0	2	70	77	26	0	0.01	175
Philepittidae	0	0	0	0	0	0	4	0	0.00	4
Phoenicopteridae	0	0	0	0	0	5	0	0	0.00	5
Phoeniculidae	0	0	0	5	1	2	0	0	**0.63**	8
Picathartidae	0	0	0	0	0	2	0	0	0.00	2
Picidae	2	2	3	18	44	31	123	3	0.08	216
Pipridae	0	0	0	0	2	24	27	0	0.00	53
Pittidae	0	0	0	0	0	2	29	0	0.00	31
Pityriaseidae	0	0	0	1	0	0	0	0	**1.00**	1
Platysteiridae	0	0	0	13	12	5	0	0	**0.43**	30
Ploceidae	0	1	1	9	9	57	31	0	0.08	106
Podargidae	0	0	0	0	2	1	11	0	0.00	14
Podicipedidae	5	4	0	0	5	14	0	0	0.00	19
Polioptilidae	0	1	0	0	1	4	9	1	0.00	14
Pomatostomidae	0	0	0	5	0	0	0	0	**1.00**	5
Procellariidae	0	0	0	0	0	64	12	0	0.00	76
Promeropidae	0	0	0	0	1	1	0	0	0.00	2
Prunellidae	0	0	0	13	0	0	0	0	**1.00**	13
Psittacidae	[illegible]	1	0	22	55	63	206	0	0.06	346
Psophiidae	0	0	0	3	0	0	0	0	**1.00**	3
Pteroclididae	0	0	0	0	0	16	0	0	0.00	16
Ptilonorhynchidae	0	0	0	0	1	11	8	0	0.00	20
Pycnonotidae	0	0	0	21	38	39	35	0	0.16	133
Rallidae	0	1	0	19	6	14	93	0	0.14	132

Table 1 Taxonomic overview on the family level over the parental care mode of occasional cooperative breeders and of 9659 bird species, the number of species showing interspecific feeding, the proportion of cooperatively breeding species within all families, and the number of species in each taxonomic family; CB = cooperative breeder (including both family living cooperative breeders and species with unrelated helpers), FAM = family living species, NO-FAM = non-family living species, UN = unknown parental care mode *(Continued)*

Taxonomic family	Parental care mode of occasional CB species			Inferred parental care mode of all species				N species engaging in interspecific feeding	Proportion CB species	Total number of species
	FAM	NO-FAM	UN	CB	FAM	NO-FAM	UN			
Ramphastidae	0	0	0	38	12	21	52	0	**0.31**	123
Recurvirostridae	0	0	0	0	4	1	4	0	0.00	9
Reguliidae	0	0	0	0	0	5	1	0	0.00	6
Remizidae	0	0	0	7	0	4	1	0	**0.58**	12
Rhabdornithidae	0	0	0	0	0	3	0	0	0.00	3
Rheidae	0	0	0	1	1	0	0	0	**0.50**	2
Rhinocryptidae	0	0	0	0	0	1	28	0	0.00	29
Rhipiduridae	0	0	1	0	1	2	39	0	0.00	42
Rhynochetidae	0	0	0	1	0	0	0	0	**1.00**	1
Rostratulidae	0	0	0	0	1	1	0	0	0.00	2
Sapayoaidae	0	0	0	0	0	1	0	0	0.00	1
Scolopacidae	0	0	0	0	1	33	53	0	0.00	87
Scopidae	0	0	0	1	0	0	0	0	**1.00**	1
Sittidae	0	0	0	2	5	11	7	2	0.08	25
Spheniscidae	0	1	0	0	0	17	0	1	0.00	17
Steatornithidae	0	0	0	0	0	1	0	0	0.00	1
Stercorariidae	0	1	0	2	0	4	2	0	0.25	8
Strigidae	2	3	0	1	14	27	111	1	0.01	153
Struthionidae	0	0	0	0	1	0	0	0	0.00	1
Sturnidae	1	1	0	18	8	77	6	1	0.17	109
Sulidae	0	0	0	0	1	8	1	0	0.00	10
Sylviidae	0	1	1	21	16	50	177	0	0.08	264
Thamnophilidae	0	0	0	2	143	5	46	0	0.01	196
Thinocoridae	0	0	0	0	0	2	2	0	0.00	4
Thraupidae	0	0	0	32	74	110	51	0	0.12	267
Threskiornithidae	0	0	0	0	3	14	15	0	0.00	32
Timaliidae	0	0	0	81	18	130	54	0	**0.29**	283
Tinamidae	0	0	0	0	0	7	39	0	0.00	46
Todidae	0	0	0	5	0	0	0	0	**1.00**	5
Trochilidae	0	0	0	0	0	310	8	0	0.00	318
Troglodytidae	3	0	0	17	15	9	35	3	0.22	76
Trogonidae	0	0	0	0	1	0	38	0	0.00	39
Turdidae	1	1	0	5	13	115	32	5	0.03	165
Turnicidae	0	0	0	0	3	10	3	0	0.00	16
Tyrannidae	1	0	0	6	37	56	293	3	0.02	392
Tytonidae	0	0	0	0	0	6	10	0	0.00	16

Table 1 Taxonomic overview on the family level over the parental care mode of occasional cooperative breeders and of 9659 bird species, the number of species showing interspecific feeding, the proportion of cooperatively breeding species within all families, and the number of species in each taxonomic family; CB = cooperative breeder (including both family living cooperative breeders and species with unrelated helpers), FAM = family living species, NO-FAM = non-family living species, UN = unknown parental care mode *(Continued)*

Taxonomic family	Parental care mode of occasional CB species			Inferred parental care mode of all species				N species engaging in interspecific feeding	Proportion CB species	Total number of species
	FAM	NO-FAM	UN	CB	FAM	NO-FAM	UN			
Upupidae	0	0	0	1	0	0	0	0	**1.00**	1
Urocynchramidae	0	0	0	0	0	0	1	0	0.00	1
Vangidae	0	0	0	4	3	5	8	0	0.20	20
Viduidae	0	0	0	0	0	17	0	0	0.00	17
Vireonidae	0	0	0	0	5	15	31	0	0.00	51
Zosteropidae	0	1	0	6	5	49	34	0	0.06	94
Total	37	86	29	864	1257	3654	3884	51	0.09	9659

Families with more than 25 % cooperative breeders are highlighted in bold. Taxonomy follows Jetz et al. [44]. See main text and [21] for definitions of the parental care mode. The inferred parental care mode follows Cockburn [18], updated based on the Handbook of the Birds of the World [45]. Observe that we categorized the Darwin finches *Geospiza scandens* and *G. fortis* as occasional cooperative breeders, based on a detailed study on helping at the nest in these two species [19]

benefits [22]. Moreover, family living is a stepping stone for the evolution of cooperative breeding [21, 26, 41, 42], but most occasionally cooperatively breeding species do not live in family groups (Additional file 1: Table S1).

Further studies are needed to assess whether "helpers" in occasional cooperative breeders gain direct fitness benefits from doing so (e.g. by having a share in reproduction through extra pair mating or egg dumping). Until the factors that facilitate cooperative breeding in these species are better understood, it may be misleading to categorize such species as cooperative breeders in comparative studies. Thus, we advise excluding these 152 occasional cooperative breeders from analyses of cooperative breeding until their mode of parental care is critically assessed, or to categorize

Fig. 2 An example of interspecific feeding. During fieldwork on Japanese great tits *Parus minor* in Karuizawa, Nagano Prefecture, Japan (36.33'138" N, 138.56'260" E) 15-18 June 2015, we observed a nest of a great spotted woodpecker *Dendrocopos major* where in addition to the parents (**a**), a Japanese great tit male fed the young woodpeckers (**b**). Based on video recordings, the feeding rate of the parents was much lower than the feeding rate by the great tit (feeding rates: male parent: 5.2/hr, female parent: 4.1/hr, great tit male: 17.2/hr, total time assessed: 14.5 hrs). Inspection with an infrared camera confirmed that the cavity contained four great spotted woodpecker nestlings but no great tit nestlings (**c-d**). Moreover, the woodpeckers often displaced the male Japanese great tit from the cavity

them separately or according to the typically observed parental care mode. This approach will increase the robustness of comparative analyses and thereby improve our understanding of factors that drive the evolution of cooperative breeding.

Ethical statements

Our field work was approved by the Animal Care and Use Committees at SOKENDAI (The Graduate University for Advanced Studies) and the Forestry Agency of Japan. Our work adhered to the Guidelines for the Use of Animals in Research of the Animal Behavior Society/ Association for the Study of Animal Behaviour.

Competing interests

The authors declare that they have no competing interests.

Authors' contributions

Data collection: MG, TS; data analyses: MG, TS; writing manuscript: MG, TS. Both authors read and approved the final manuscript.

Acknowledgements

We thank Katharine Bowgen, Kaat Brulez and Xiaoyan Long for helping with the collection of the comparative data, and Sjouke Kingma and an anonymous reviewer for helpful comments. Fuchiko Mito supported us during fieldwork and writing.

Data sharing

We adhere to the data sharing policies outlined in Mills et al. [43].

References

1. Bourke AFG. Principles of social evolution. Oxford series in ecology and evolution. Oxford: Oxford University Press; 2011.
2. Darwin C. On the origin of species. London: John Murray; 1859.
3. Hamilton WD. Genetical evolution of social behaviour I. J Theor Biol. 1964;7(1):1–16.
4. Cornwallis CK, West SA, Griffin AS. Routes to indirect fitness in cooperatively breeding vertebrates: kin discrimination and limited dispersal. J Evol Biol. 2009;22(12):2445–57.
5. Cockburn A. Evolution of helping behavior in cooperatively breeding birds. Annu Rev Ecol Syst. 1998;29:141–77.
6. Stacey PB, Koenig WD. Cooperative breeding in birds. Cambridge: Cambridge University Press; 1990.
7. Solomon NG, French JA. Cooperative breeding in mammals. Cambridge: Cambridge University Press; 1997.
8. Koenig WD, Dickinson JL. Ecology and evolution of cooperative breeding in birds. Cambridge: Cambridge University Press; 2004.
9. Hatchwell BJ. The evolution of cooperative breeding in birds: kinship, dispersal and life history. Philos Trans R Soc Lond B Biol Sci. 2009;364(1533): 3217–27. doi:10.1098/rstb.2009.0109.
10. Arnold KE, Owens IPF. Cooperative breeding in birds: a comparative test of the life history hypothesis. Proc R Soc B-Biol Sci. 1998;265(1398):739–45.
11. Jetz W, Rubenstein DR. Environmental uncertainty and the global biogeography of cooperative breeding in birds. Curr Biol. 2011;21(1):72–8. doi:10.1016/j.cub.2010.11.075.
12. Gonzalez J-CT, Sheldon BC, Tobias JA. Environmental stability and the evolution of cooperative breeding in hornbills. Proc R Soc B Biol Sci. 2013;280(1768):20131297.
13. Cornwallis CK, West SA, Davis KE, Griffin AS. Promiscuity and the evolutionary transition to complex societies. Nature. 2010;466(7309):969–72.
14. Downing PA, Cornwallis CK, Griffin AS. Sex, long life and the evolutionary transition to cooperative breeding in birds. Proc R Soc B. 2015;282(1816): 20151663.
15. Emlen ST. The Evolution of helping.1. an ecological constraints model. Am Nat. 1982;119(1):29–39.
16. Rubenstein DR, Lovette IJ. Temporal environmental variability drives the evolution of cooperative breeding in birds. Curr Biol. 2007;17(16):1414–9.
17. Koenig WD, Mumme RL. Population ecology of the cooperative breeding acorn woodpecker. Princeton: Princeton University Press; 1987.
18. Cockburn A. Prevalence of different modes of parental care in birds. Proc R Soc B Biol Sci. 2006;273(1592):1375–83. doi:10.1098/rspb.2005.3458.
19. Price T, Millington S, Grant P. Helping at the nest in Darwin's finches as misdirected parental care. Auk. 1983;100:192–4.
20. Jamieson IG. Behavioral heterochrony and the evolution of birds' helping at the nest: An unselected consequences of communal breeding? Am Nat. 1989;133:394–406.
21. Drobniak SM, Wagner G, Mourocq E, Griesser M. Family living: an overlooked but pivotal social system to understand the evolution of cooperative breeding. Behav Ecol. 2015;26(3):805–11. doi:10.1093/beheco/arv015.
22. Riehl C. Evolutionary routes to non-kin cooperative breeding in birds. Proc R Soc B Biol Sci. 2013;280(1772):20132245.
23. Dickinson JL, Hatchwell BJ. Fitness consequnces of helping. In: Koenig WD, Dickinson J, editors. Ecology and evolution of cooperative breeding in birds. Cambridge: Cambridge University Press; 2004. p. 48–66.
24. Preston SA, Briskie JV, Burke T, Hatchwell BJ. Genetic analysis reveals diverse kin-directed routes to helping in the rifleman Acanthisitta chloris. Mol Ecol. 2013;22(19):5027–39.
25. Valcu M, Dale J, Griesser M, Nakagawa S, Kempenaers B. Global gradients of avian longevity support the classic evolutionary theory of ageing. Ecography. 2014;37(10):930–8. doi:10.1111/ecog.00929.
26. Covas R, Griesser M. Life history and the evolution of family living in birds. Proc R Soc B Biol Sci. 2007;274(1616):1349–57.
27. Griesser M, Nystrand M, Ekman J. Reduced mortality selects for family cohesion in a social species. Proc R Soc B Biol Sci. 2006;273(1596):1881–6.
28. Griesser M, Barnaby J. Families, a place of loving care and violent conflicts. The role of nepotism, cooperation and competition in the evolution of avian families. In: Columbus F, editor. Behavioral and chemical ecology: new research. New York: Nova; 2010. p. 47–90.
29. Hockey PAR, Dean WJJ, Ryan PG. Roberts birds of Southern Africa. 7th ed. Trustees of the John Voelcker Bird Book Fund: Cape Town; 2005.
30. Skutch AF. Helpers among birds. Condor. 1961;63:198–226.
31. Jamieson I. The unselected hypothesis for the evolution of helping-behavior - too much or too little emphasis on natural-selection. Am Nat. 1991;138(1):271–82.
32. Shy MM. Interspecific feeding among birds: a review. J Field Ornithology. 1982;53:370–93.
33. Jacot A, Reers H, Forstmeier W. Individual recognition and potential recognition errors in parent-offspring communication. Behav Ecol Sociobiol 2010;64(10): 1515–25.
34. Emlen ST, Reeve HK, Sherman PW, Wrege PH, Ratnieks FLW, Shellmanreeve J. Adaptive versus nonadaptive explanations of behavior - the case of alloparental helping. Am Nat. 1991;138(1):259–70.
35. Carlson AA, Russell AF, Young AJ, Jordan NR, McNeilly AS, Parlow AF, et al. Elevated prolactin levels immediately precede decisions to babysit by male meerkat helpers. Horm Behav. 2006;50(1):94–100. doi:10.1016/j.yhbeh.2006.01.009.
36. Schoech SJ, Mumme RL, Wingfield JC. Prolactin and helping behaviour in the cooperatively breeding Florida scrub-jay, Aphelocoma coerulescens. Anim Behav. 1996;52:445–56.
37. Canestrari D, Chiarati E, Marcos JM, Ekman J, Baglione V. Helpers but not breeders adjust provisioning effort to year-round territory resource availability in carrion crows. Anim Behav. 2008;76:943–9. doi:10.1016/j.anbehav.2008.05.013.
38. Kingma SA, Hall ML, Peters A. Multiple benefits drive helping behavior in a cooperatively breeding bird: an integrated analysis. Am Nat. 2011;177(4): 486–95. doi:10.1086/658989.
39. Nam K-B, Simeoni M, Sharp SP, Hatchwell BJ. Kinship affects investment by helpers in a cooperatively breeding bird. Proc R Soc B Biol Sci. 2010; 277(1698):3299–306. doi:10.1098/rspb.2010.0737.

40. Heinsohn R, Legge S. The cost of helping. Trends Ecol Evol. 1999;14(2):53–7.
41. Ekman J, Dickinson JL, Hatchwell BJ, Griesser M. Delayed dispersal. In: Koenig WD, Dickinson JL, editors. Ecology and evolution of cooperative breeding in birds. Cambridge: Cambridge University Press; 2004.
42. Ekman J, Griesser M. Siberian jays: delayed dispersal in absence of cooperative breeding. In: Koenig WD, Dickinson J, editors. Cooperative breeding in vertebrates: studies of ecology, evolution, and behavior. Cambridge: Cambridge University Press; 2016. p. 6–18.
43. Mills JA, Teplitsky C, Arroyo B, Charmantier A, Becker PH, Birkhead TR, et al. Archiving primary data: solutions for long-term studies. Trends Ecol Evol. 2015;30(10):581–9. doi:10.1016/j.tree.2015.07.006.
44. Jetz W, Thomas GH, Joy JB, Hartmann K, Mooers AO. The global diversity of birds in space and time. Nature. 2012;491(7424):444–8. doi:10.1038/nature11631.
45. Del Hoyo J, Elliot A, Sargatal J, Christie DA. Handbook of the birds of the world. Barcelona: Lynx Editions; 2011.

Fine selection of up-regulated genes during ovulation by *in vivo* induction of oocyte maturation and ovulation in zebrafish

Wanlada Klangnurak[1] and Toshinobu Tokumoto[1,2*]

Abstract

Background: Two essential processes, oocyte maturation and ovulation, are independently induced, but proceed cooperatively as the final step in oogenesis before oocytes become fertilizable. Although these two processes are induced by the same maturation-inducing steroid, 17α, 20β-dihydroxy-4-pregnen-3-one (17, 20β-DHP), in zebrafish, it has been suggested that the receptor, and thus the signal transduction pathway is different for each process. Although much progress has been made in understanding the molecular mechanisms underlying the induction of oocyte maturation, the mechanisms for inducing ovulation remain under investigation. In the present study, *in vivo* induction techniques that permit the induction of oocyte maturation and ovulation in living zebrafish (*in vivo* assays) were used to select highly up-regulated genes (genes associated with ovulation). Using an *in vivo* assay, ovarian tissues that induced only oocyte maturation could be obtained. This made it possible for the first time to distinguish maturation-inducing genes from ovulation-inducing genes. Using a genome-wide microarray of zebrafish sequences, the gene expression levels were compared among an ethanol (EtOH)-treated group (non-activated group), a diethylstilbestrol (DES)- or testosterone (Tes)-treated group (maturation-induced group), and a 17, 20β-DHP-treated group (maturation- and ovulation-induced group). Ovulation-specific up-regulated genes were selected. The mRNA expression levels of the selected genes were measured by quantitative polymerase chain reaction (qPCR).

Results: Among 34 genes identified, three that showed ovulation-specific increases were selected as candidates potentially associated with ovulation. The ovulation-specific up-regulation of three candidates, slc37a4a, zgc:65811 and zgc:92184 was confirmed by qPCR.

Conclusion: Our *in vivo* assay provides a new approach to precisely select genes associated with ovulation.

Keywords: Ovulation, Steroids, Oocyte maturation, Zebrafish

Background

Oocyte maturation in zebrafish can be induced by the maturation-inducing hormone (MIH) 17α, 20β-dihydroxy-4-pregnen-3-one (17, 20β-DHP) [1, 2]. 17, 20β-DHP binds to a membrane-bound progestin receptor (mPR) and acts via non-genomic pathways to induce oocyte maturation [3, 4]. By contrast, the signals for ovulation are activated by 17, 20β-DHP through a nuclear isoform of the progesterone receptor (nPR) [5]. Thus, it has been suggested that the actual pathways for inducing oocyte maturation and ovulation involve a coordinated process between the genomic and non-genomic pathways induced by 17, 20β-DHP [6]. In zebrafish, oocyte maturation can be induced within three hours and ovulation within four hours *in vivo* [7]. Although considerable progress has been made in understanding the mechanism of oocyte maturation, additional studies are needed to develop a clearer understanding of the mechanisms underlying ovulation.

The gene expression profile observed during ovarian development provides basic knowledge for understanding the fish reproductive system. The transcription profile of the fathead minnow (*Pimephales promelas*) ovary in different ovarian stages (atretic, previtellogenic, vitellogenic,

* Correspondence: tokumoto.toshinobu@shizuoka.ac.jp
[1]Integrated Bioscience Section, Graduate School of Science and Technology, National University Corporation Shizuoka University, Ohya 836, Suruga-ku, Shizuoka 422-8529, Japan
[2]Department of Biology, Faculty of Science, National University Corporation Shizuoka University, Shizuoka 422-8529, Japan

and post-ovulatory follicles) has been described [8]. The authors of that study provided a list of genes that might shift their expression levels during several ovarian stages. A study has reported on the changes in the expression of steroidogenic enzymes, including steroidogenic acute regulatory protein (StAR), during the primary growth of follicles and on the effect of human chorionic gonadotrophin (hCG) in zebrafish [9]. Analysis of transcriptional levels in pre-ovulatory rainbow trout (*Oncorhynchus mykiss*) ovaries revealed a number of genes linked to maturational competence [10, 11]. It is well known that maturation and ovulation are overlapping events [12, 13]. Thus far, it has not been possible to distinguish the changes in gene expression for the induction of ovulation from those for the induction of oocyte maturation. Studies of the expression profile of genes associated with ovulation in the ovulated eggs and follicles may be perturbed by genes associated with oocyte maturation.

Previously, we established a procedure that enabled the preparation of ovarian tissue containing oocyte maturation-induced oocytes *in vivo*. Similarly, ovulation can be induced in live zebrafish [7]. This technique makes it possible to select the up-regulated genes that induce ovulation by comparing the gene expression between genes associated with oocyte maturation in matured oocytes and both genes associated with oocyte maturation and ovulation in ovulated eggs. Recently, the role of eicosanoids in ovulation and spawning was revealed via an *in vivo* bioassay [14].

In the present study, we used an *in vivo* bioassay to prepare matured and ovulated ovarian samples. Specifically, the up-regulated genes that induce ovulation were selected by microarray analysis. The mRNA abundance of highly up-regulated genes was confirmed by qPCR analysis. The results indicate that this procedure is practical and can be used to select genes associated with ovulation.

Methods

Experimental animals

Zebrafish were raised in a proper chamber with a recirculating water system maintained at 28.5 °C under a 14 h light:10 h dark cycle [15]; the fish were fed a diet of brine shrimp in the morning and fish feed pellets (Croma, Kobe, Japan) in the evening. Female fish were raised until they possessed full-grown immature oocytes and were then used in this experiment. All experiments were conducted in accordance with procedures approved by the Shizuoka University Animal Care Committee.

Experimental design for collecting tissue

To investigate genes associated with ovulation, artificial inductions of maturation and ovulation were conducted by an *in vivo* bioassay following the method reported by Tokumoto *et al.* [7]. Briefly, female fish possessing full-grown immature oocytes were selected from a mixed group of 10–50 males and females that had been held in a 20 cm × 25 cm × 25 cm square acrylic case with the standard water system used in our aqualab. Females were transferred into a glass case containing 100 ml of water per fish. The fish were exposed to agents *in vivo* by adding each agent, at a 10,000-fold stock in EtOH (final concentration; 0.01% EtOH, 0.1 μM 17, 20β-DHP, 5 μM diethylstilbestrol (DES), 10 μM testosterone (Tes)), to water and were then incubated at 28.5 °C. More than five fish were used for each treatment. Female zebrafish were treated for three hours under these conditions. After incubation, the female zebrafish were killed by spinal severance followed by dissection. Ovarian samples were cut from the body cavity under sterile conditions. One side of the ovary was placed in a 1.5 ml micro tube and immediately frozen in liquid nitrogen to preserve the RNA until extraction. To assess the oocyte developmental stage, the other side of the ovary was placed in fresh zebrafish Ringer's solution (116 mM NaCl, 2.9 mM KCl, 1.8 mM $CaCl_2$, and 5 mM HEPES, pH 7.2) and observed using a stereomicroscope. The specimen showing the best ovarian status based on observation after collection in each treatment was selected for RNA sample preparation. Three replicates were separately prepared from three different batches (more than 20 individuals each) of zebrafish.

RNA extraction and reverse transcription for microarray

The total RNA for microarray analysis was extracted from the ovarian tissue using ISOGEN (Nippon Gene, Tokyo, Japan) in accordance with the manufacturer's protocol. ISOGEN is a phenol-based pre-made reagent for RNA extraction.

Complementary DNA was prepared from 500 ng of total RNA from each replicate, as described in the manual for Agilent Low RNA Input Linear Amplification kit (Agilent Technologies, Palo Alto, CA, USA). Double-stranded cDNA was synthesized using the kit, and Cy3-labeled cRNA was prepared by cDNA *in vitro* transcription in the presence of cyanine 3-CTP dyes. Fluorescently labeled RNA was then purified with Qiagen RNeasy spin columns in accordance with the manufacturer's protocol (Qiagen, Hilden, Germany). After purification, the cRNA was stored at −80 °C until use.

Microarray

cRNA was fragmented and used to hybridize to the zebrafish G2519F 4X44K microarray containing 43,803 sequences (Agilent Technologies, Palo Alto, CA, USA). Hybridization, washing, and scanning were performed in accordance with the manufacturer's protocol. Microarrays were scanned on a DNA microarray scanner (Agilent

Technologies, Palo Alto, CA, USA) at a resolution of 5 µm. Raw digitized expression values from each probe set were extracted using feature extraction software, and the features were flagged manually for poor quality. The microarray data were then analyzed using Gene Spring ver. 11.0 (Agilent Technologies, Palo Alto, CA, USA). Three sets of samples from 3 biological replicates (as described above in Experimental design for collecting tissue) were analyzed by three separate arrays.

Gene Selection

The median signal of the samples treated with EtOH, DES, Tes and 17, 20β-DHP from the microarray analyses were analyzed by Subio Platform ver. 1.18.4667 (Subio Inc., Amami, Japan). Two types of analyses were performed to find the candidate genes: statistical selection and non-statistical selection. For statistical selection, two separate statistical analyses were performed. The 17, 20β-DHP-treated group was separately compared with the groups treated with DES, Tes and EtOH. Any genes exhibiting at least a 1.8-fold greater expression in the 17, 20β-DHP-treated group were selected by the Venn diagram analysis. In addition, analysis of variance (ANOVA) was performed across all groups by using the basic plug-in to find genes that were expressed at significantly different levels among the four treated groups. Finally, overlapping genes that were up-regulated in the 17, 20β-DHP-treated group and the significantly different genes from ANOVA were selected. For non-statistical selection, any genes showing a 10-fold greater expression in the 17, 20β-DHP-treated group compared with the EtOH-, DES- or Tes-treated groups were selected. A Venn diagram analysis was performed with these groups. Central overlapped genes were selected as candidates for genes associated with ovulation (Fig. 2).

Quantitative RT-PCR

The mRNA abundance of the first 20 genes from the 33 genes selected (Table 1) and a gene from the non-statistical gene selection (Table 2) was assessed by qPCR to confirm their expression levels in the EtOH-, DES-, Tes- and 17, 20β-DHP-treated samples. Specific primers were designed via Primer 3 [16] (Table 3). The annealing temperature gradient was investigated for all primers to find the most appropriate temperature for qPCR amplification (data not shown). A widely used reference gene, elongation factor 1 alpha (*ef1α*, GenBank accession number L47669), was tested using the same sample set to validate the normalization procedure.

RNA from four treated samples was extracted from ovarian tissue, again using ISOGEN. Total RNA (1 µg) was reverse transcribed using illustra Ready-To-Go RT-PCR Beads (GE Healthcare Life Sciences, Buckinghamshire, UK), following the manufacturer's instructions. The qPCR reactions were performed in a 20 µl volume that contained 5 µl of 10 times-diluted cDNA, 1 µl of each primer (10 µM), and 10 µl of SYBR green PCR Master Mix (Roche Applied Science, Mannheim, Germany). Real-time qPCR was conducted by LightCycler® Nano System (Roche Applied Science, Mannheim, Germany). The thermal cycle began with an initial denaturation step at 95 °C for 5 min, followed by 45 cycles of denaturation at 95 °C for 10 sec, annealing at Ta°C (Table 3) for 10 sec, and extension at 72 °C for 15 sec. The final melting curve analysis was observed at 65 °C for 20 sec, followed by 95 °C for 20 sec. The mRNA abundance of each target gene was calculated from a serially diluted pooled cDNA, and each sample was normalized against its reference gene expression level. Biological triplicates were observed for each treatment. Triplicate reactions were performed for each individual sample. The normalized mRNA abundance was calculated as the mean ± SE.

Statistical analyses

The mRNA abundance of each gene among the different treated samples was computed using the Kruskal-Wallis one-way ANOVA nonparametric test because not all of the data set met the ANOVA assumptions. The Mann-Whitney *U* test was calculated to compare whether any significant difference existed between the treated sample pairs. Significance was set at $P \leq 0.05$.

Results

As reported in 2011 by Tokumoto et al., oocyte maturation and ovulation can be induced by adding agents to the water in which the fish are maintained [7]. Using this *in vivo* bioassay method, we succeeded in preparing maturation-induced ovarian samples by DES or Tes treatment and ovulation-induced samples with 17, 20β-DHP treatment. Oocytes showed dramatic morphological changes in the samples treated with DES and Tes compared with the control treatment (EtOH) (Fig. 1). Oocyte maturation was induced within three hours after adding DES or Tes to the water, but ovulation did not progress. Oocytes underwent germinal vesicle breakdown and became transparent, which is a morphological characteristic of matured oocytes [7, 12]. In this manner, an ovarian sample that exhibited only oocyte maturation could be prepared *in vivo*.

17, 20β-DHP induced both oocyte maturation and ovulation, and fertilizable eggs were obtained at four hours. Oocytes in 17, 20β-DHP-treated fish expelled their follicles, and a fertilization membrane developed after contact with water, which is a morphological characteristic of ovulated eggs [7, 12]. We compared the mRNA expression levels in ovaries of fish treated with the following: EtOH (control group); DES or Tes (oocyte-maturation-

Table 1 Top 20 candidates for genes associated with ovulation selected by the statistical selection method. The fold changes among treated samples are shown

Gene name	Accession #	Description	Fold change relative to EtOH			Fold change relative to DES			Fold change relative to Tes			Fold change relative to 17, 20β-DHP		
			DES	Tes	17, 20β-DHP	EtOH	Tes	17, 20β-DHP	EtOH	DES	17, 20β-DHP	EtOH	DES	Tes
cyp11a1	NM_152953	Danio rerio cytochrome P450, subfamily XIA, polypeptide 1), mRNA	31.7	1.3	433.7	<0.01	<0.01	13.7	0.8	24.9	340.8	<0.01	0.1	<0.01
zgc:136308	NM_001045247	Danio rerio zgc:136308, mRNA	1.2	4.5	24.5	0.8	3.6	19.7	0.2	0.3	5.4	<0.01	0.1	0.2
efna1	NM_200783	Danio rerio ephrin A1, mRNA	1.7	5.5	16.5	0.6	3.2	9.6	0.2	0.3	3.0	0.1	0.1	0.3
slc37a4a	NM_214738	Danio rerio solute carrier family 37 (glucose-6-phosphate transporter), member 4a, mRNA	7.9	6.4	15.3	0.1	0.8	1.9	0.2	1.2	2.4	0.1	0.5	0.4
slc25a10	NM_201172	Danio rerio solute carrier family 25 (mitochondrial carrier; dicarboxylate transporter), member 10, mRNA]	2.5	1.7	12.1	0.4	0.7	4.8	0.6	1.5	7.1	0.1	0.2	0.1
spata18	NM_001018678	Danio rerio spermatogenesis associated 18 , mRNA	4.1	5.2	11.6	0.2	1.2	2.8	0.2	0.8	2.2	0.1	0.4	0.4
sox21a	NM_131286	Danio rerio SRY-box containing gene 21a , mRNA	3.1	3.5	7.9	0.3	1.1	2.6	0.3	0.9	2.3	0.1	0.4	0.4
api5	NM_199540	Danio rerio apoptosis inhibitor 5, mRNA	3.2	3.1	7.0	0.3	1.0	2.2	0.3	1.0	2.2	0.1	0.5	0.4
nup85	NM_001003625	Danio rerio nucleoporin 85	2.3	3.6	7.0	0.4	1.6	3.0	0.3	0.6	1.9	0.1	0.3	0.5
gbx2	NM_152964	Danio rerio gastrulation brain homeo box 2	1.4	2.3	6.9	0.7	1.7	5.0	0.4	0.6	3.0	0.1	0.2	0.3
zgc:65811	NM_200552	Danio rerio zgc:65811 , mRNA	2.7	2.8	6.4	0.4	1.0	2.4	0.4	1.0	2.3	0.2	0.4	0.4
cnn2	NM_213349	Danio rerio calponin 2 , mRNA	2.5	1.5	5.6	0.4	0.6	2.2	0.7	1.7	3.6	0.2	0.5	0.3
CK139976	CK139976	AGENCOURT_16876226 NCI_CGAP_ZEmb3 Danio rerio cDNA clone IMAGE:7059552 5', mRNA sequence]	1.4	1.4	5.4	0.7	1.0	3.9	0.7	1.0	3.7	0.2	0.3	0.3
bcl3	XM_688922	PREDICTED: Danio rerio B-cell CLL/lymphoma 3, mRNA	1.3	1.6	4.5	0.8	1.2	3.5	0.6	0.8	2.9	0.2	0.3	0.3
asic2	NM_214788	Danio rerio acid-sensing (proton-gated) ion channel 2, mRNA	2.2	2.1	4.4	0.4	0.9	2.0	0.5	1.1	2.1	0.2	0.5	0.5
zgc:56525	NM_200279	Danio rerio zgc:56525, mRNA	1.9	2.3	4.2	0.5	1.2	2.2	0.4	0.8	1.8	0.2	0.5	0.5
asic4b	NM_214786	Danio rerio acid-sensing (proton-gated) ion channel family member 4b, mRNA	0.8	1.7	4.2	1.3	2.2	5.5	0.6	0.4	2.5	0.2	0.2	0.4
rhbdl3	NM_001017556	Danio rerio rhomboid, veinlet-like 3 (Drosophila), mRNA [NM_001017556]	0.6	1.3	3.7	1.7	2.2	6.3	0.8	0.5	2.9	0.3	0.2	0.3
LOC100003798	XM_001343224	PREDICTED: Danio rerio hypothetical protein LOC100003798, mRNA.	1.1	0.5	3.1	0.9	0.4	2.9	2.2	2.4	6.8	0.3	0.3	0.1
nptna	NM_001160156	Danio rerio neuroplastin a, mRNA	1.5	1.6	3.1	0.7	1.1	2.1	0.6	0.9	1.9	0.3	0.5	0.5

Table 2 Two genes associated with ovulation selected by the non-statistical selection method. The fold changes among treated samples are shown

Gene name	Accession #	Description	Fold change relative to EtOH			Fold change relative to DES			Fold change relative to Tes			Fold change relative to 17, 20β-DHP		
			DES	Tes	17, 20β-DHP	EtOH	Tes	17, 20β-DHP	EtOH	DES	17, 20β-DHP	EtOH	DES	Tes
cyp11a1	NM_152953	Danio rerio cytochrome P450, family 11, subfamily A, polypeptide 1	31.7	1.3	433.7	0.0	0.0	13.7	0.8	24.9	340.8	<0.01	0.1	<0.01
Zgc:92184	NM_001002344	Danio rerio zgc:92184	1.4	0.9	20.7	0.7	0.6	14.9	1.1	1.6	23.6	<0.01	0.1	<0.01

induced group); or 17, 20β-DHP (an oocyte-maturation- and ovulation-induced group). To select the genes responsible for ovulation, we set the time period for preparing the mRNA at three hours after adding the agents to the water. Because we planned to analyze the changes in gene expression levels in the whole ovary, it was necessary to excise the whole ovary. To avoid the destruction of the ovary due to ovulation, the samples had to be obtained before ovulation. We speculated that the expression of genes associated with ovulation would have already started one hour before ovulation.

Transcriptome analysis was conducted using microarrays to select genes that were up-regulated during ovulation. We compared the expression levels of mRNA from ovaries in fish treated with EtOH, DES, Tes or 17, 20β-DHP using a conventional microarray containing probes of genome-wide variety (Agilent 4x44k). The signals from samples treated with 17, 20β-DHP, DES or Tes were significantly higher for nearly all of the genes than were the signals from samples treated with EtOH (data not shown). Although it was difficult to identify alterations in gene expression during the reproductive course, we identified the transcriptional levels of genes associated with ovulation by comparing signals between matured and ovulated oocytes.

Table 3 Primers used for qPCR analyses of candidate genes.

Target gene	Accession #	Forward Primer 5' to 3'	Reverse Primer 5' to 3'	product size, bp	Ta, °C
cyp11a1	NM_152953	AAAGCCTGAAGACGGTGCTA	AGCAGGACGCCATATTTTTG	117	60
zgc:136308	NM_001111227	GCAAACACGACACAACTCCTGC	TGTGTCCTCCATCAGGTCTGTTTAC	130	56
efna1	NM_200783	AGCAGTTGGCGAAGGTGATG	CGGTAATGGAGGAGGCGTTC	99	56
slc37a4a	NM_214738	CTCCAGCAAAAATGAAAGCA	CACCCCAAACACCACCAG	91	56
Slc25a10	NM_201172	TAATATACTCACACACTTCCTG	CTGTATTCTCCTTTAGAGTTC	114	56
Spata	NM_001018678	AGAAATAACACTCAAAGAGG	AGACTACAAGGAGAAACACT	174	55
sox21a	NM_131286	GTTCCCTCATCTTATGTACT	TTAAACTCCACTCATATCGT	92	56
api5	NM_199540	GAGTCAAATCTTACCTTTCA	ACATACATCAGGGCATAATA	86	-
nup85	NM_001003625	CACTCTTACAGACCATGCCCATATT	CCAGTGTCTCCATTTCACATCAAAC	80	60
gbx2	AF288762	CATTAACACAACCATAATCC	GTCACTAACACAGTCTCACAT	106	-
zgc65811	NM_200552	TCCTCATGTTAGTTTAAGGTCACGG	ACAGAAATGAAAGAGAAGCAGAAGT	90	60
cnn2	NM_213349	GGACTACAGATGGGAACAAATAAAT	ATGTGTGACTTGGGATAATACAGAT	92	60
CK139976	CK139976	CAGTATCTGCGATGTTTAATGTCAG	GCTGTCTGAGTCTTCCATTTGA	102	61
bcl3	XM_688922	TGAAGAAAGAGGTGTGAGTTGATAG	TTAAAGAGACACAATGCTGAACGAA	115	61
asic2	NM_214788	AATGTTCTTTGAGGATGGATGGTT	GTGACCTTGTATTTAGATTGAGAGC	80	60
zgc:56525	NM_200279	CAACATTTCCTCCAGTGCTAAAG	CTTCAGTTCACTCTGCATCTTATTC	94	55
asic4b	NM_214786	GAGGAGTACATCAGAGACAACTTTC	CTTTCTTCTGCTCAATCGTTTCATA	82	57
rhbdl3	NM_001017556	TCACTATAACAGAGGTTGTTGTCTT	TTCAGGAAGTATGGCGATGATAC	88	-
LOC100003798	XM_001343224.1	GTACCTCAGTCAATCTCTAATCCTC	GAACCTTTGTTCATCTCTTCTGTTT	85	60
nptna	NM_001160156	ATACAGGCATTTCCAGGCTTTATTT	AAAGAACTGTCCAACCAGAATCAT	110	60
zgc:92184	NM_001002344	CCCGTGGCGGCGATATGCTT	TCCCCCGCAGCGTCTGATGA	544	56
ptgs2a	NM_153657	ATGTTTGCTTTCTTCGCCCA	AGATCCACTCCATGACCCAG	101	55
ef1α	L47669	CTTCTCAGGCTGACTGTGC	CCGCTAGCATTACCCTCC	358	60

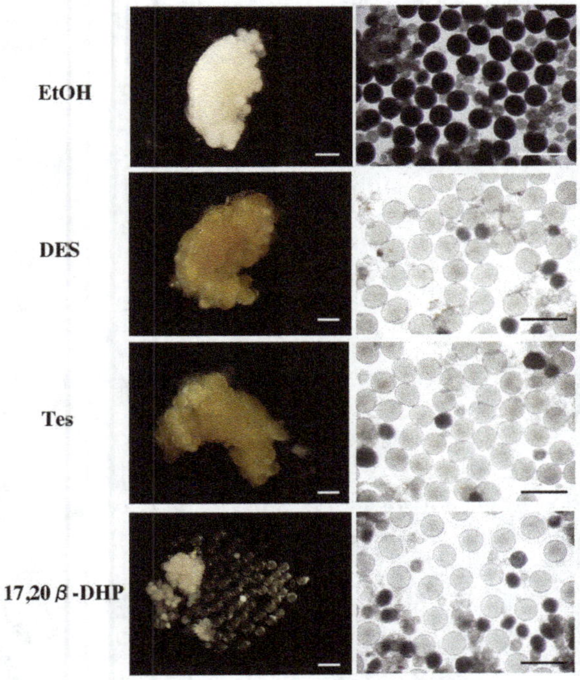

Fig. 1 The *in vivo* bioassay was performed at a final concentration of 5 μM of DES, 1 μM of Tes or 0.01 μM of 17, 20β-DHP. One side of the ovary was observed by stereomicroscopy. The morphologies of the ovarian samples after three hours of treatment by EtOH, DES, Tes and 17, 20β-DHP were photographed. Ovaries before (*left panels*) and after (*right panels*) splitting are shown. After treatment with EtOH, the oocytes remained opaque and showed no morphological change after exposure to water. Oocytes after treatment with DES or Tes became transparent. A fertilization membrane developed in oocytes ovulated by 17, 20β-DHP treatment after exposure to water. Scale bars indicate 1 cm

The genes that were up-regulated to induce oocyte maturation (genes associated with oocyte maturation) were excluded from the 17, 20β-DHP-treated group that contained both genes associated with maturation- and ovulation. Genes that were significantly up-regulated only in the 17, 20β-DHP treated group were selected. In the standard statistical selection, ANOVA was performed across all groups to detect significantly different up- and down-regulated genes among all treated groups ($P \leq 0.05$) (Fig. 2). A total of 5448 genes were identified (category a in Fig. 2). Subsequently, genes were selected that represented expression levels that were up-regulated more than 1.8-fold in the 17, 20β-DHP-treated group. By comparing the EtOH-, DES- and Tes-treated groups, 8237 (1), 443 (2) and 768 (3) genes were selected, respectively. Overlapping genes among these three groups were selected by Venn diagram analysis (category b in Fig. 2). Finally, overlapping genes in categories A and B were selected. Thirty-three genes were selected as candidates for genes associated with ovulation (Table 1). For non-statistical gene selection, 53, 5 and 14 genes that were 10 times higher in 17, 20β-DHP compared with the EtOH- (4) DES- (5) and Tes- (6) treated groups, respectively, were analyzed by Venn diagram. Only two genes were selected as candidates for genes associated with ovulation by this method (Table 2).

The expression levels of the first 20 genes from the statistical selection and 1 gene from the non-statistical selection were analyzed by qPCR with ptgs2a, well-known up-regulated gene before ovulation, as a positive control [14]. Although 18 of them (with the exception of api5, gbx2 and rhbdl3) showed excellent amplification when detected by absolute quantification analysis and melting temperature profiles (data not shown), only three genes, slc37a4a, zgc:65811 and zgc:92184 showed significantly greater expression in the 17, 20β-DHP-treated group (Fig. 3, $P < 0.05$ with all other groups). An inconsistent expression trend was observed between the microarray detection and qPCR analysis in the other genes. As a result, we selected only three genes, slc37a4a, zgc:65811 and zgc:92184, as candidates potentially associated with ovulation by the microarray analysis platform. We then conducted time course analysis of expression levels of selected genes and ptgs2a during DHP treatment. As expected, levels of all four genes were elevated at three hours, but slc37a4a and ptgs2a were more strongly up--regulated at five hours (Fig. 4). This suggests zgc:65811 and zgc:92184 as the genes most likely to be associated with ovulation.

Discussion

The *in vivo* bioassay established by Tokumoto et al. (2011) provided a novel means of distinguishing the pathways that induce ovulation and are known to be induced by genomic actions from the pathways involved in oocyte maturation, the non-genomic action-dependent pathways [7]. This *in vivo* assay is a practical technique for preparing matured and ovulated ovarian samples that was created to understand the influence of endocrine disrupting chemicals (EDCs) and steroid hormones on oocyte maturation and ovulation in live zebrafish. Both diethylstilbestrol (DES) and a naturally occurring steroid hormone, 17, 20β-DHP, were effective when these compounds were externally applied to a water column containing live fish. The results showed that fish kept in water containing 17, 20β-DHP induced both maturation and ovulation, whereas fish treated with DES underwent oocyte maturation and stopped at this stage. A high expression of genes associated with oocyte maturation was observed in mRNA extracted from matured oocytes. Similarly, both genes associated with maturation and ovulation are highly expressed in ovulated eggs. Genes that are highly expressed in ovulated eggs, but exhibit a low expression in matured oocytes could be selected as candidates for genes associated with ovulation.

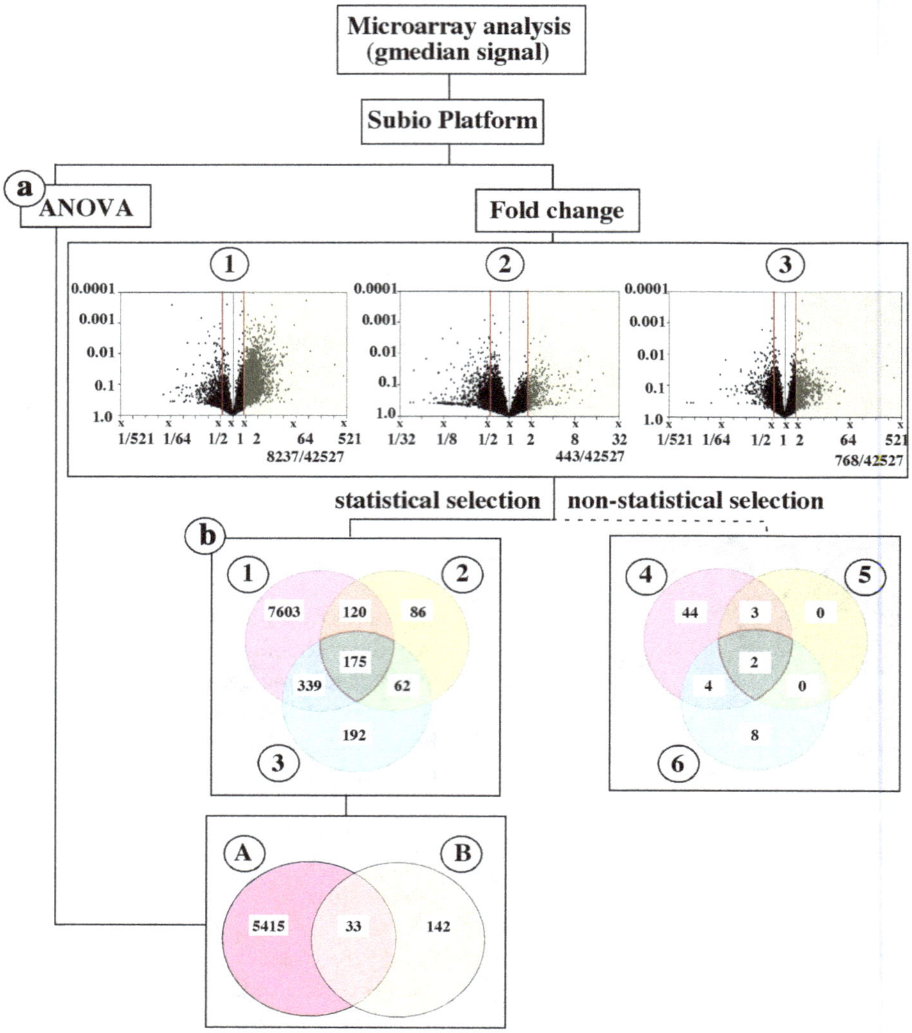

Fig. 2 A diagram showing the process for selecting candidates of genes associated with ovulation by the Subio platform (Subio Inc., Amami, Japan). In standard statistical selection, ANOVA was performed across all groups to detect significant differences in both up- and down-regulation among all treated groups ($P \leq 0.05$) (**a**). More than 1.8-fold greater expressions in 17, 20β-DHP than in EtOH (1) DES (2) or Tes (3) have been reported. Overlapping genes among these three groups were selected by Venn diagrams (**b**). Finally, overlapping genes in categories A and B were selected. Thirty-three genes were selected as candidates for genes associated with ovulation. Twenty highly elevated genes analyzed by qPCR were listed in Table 1. For non-statistical gene selection, more than 10-fold greater expressions in the 17, 20β-DHP-treated sample compared with the EtOH- (4) DES- (5) or Tes- (6) treated samples were analyzed by Venn diagrams. Two genes were selected as candidates for genes associated with ovulation (Table 2)

The ovulation-inducing pathway has been studied for more than two decades [17], which has provided a deeper understanding of the reproductive system. Metalloproteinases responsible for the rupture of follicle cells were identified in Medaka [18]. The first transcriptome analysis was conducted in rainbow trout (*Oncorhynchus mykiss*) [11]. The gene sets that were up- or down-regulated during spawning were identified by a microarray analysis. Many biological activities, such as proteolysis, inflammation, coagulation, vasodilatation, and angiogenesis, have been reported as essential mechanisms in the acquisition of maturational competence and ovulation occurring in the pre-ovulatory ovarian follicle [19]. Recently, specifically up-regulated genes in the pre-ovulatory, peri-ovulatory, and peri-spawning intervals have been identified in zebrafish [14]. However, none of these studies eliminated the maturation-inducing genes that are expressed in ovulated ovarian tissue, although it is well known that oocyte maturation and ovulation are overlapping events [12, 13]. This current study is the first to eliminate maturation-inducing genes from ovulation-inducing genes.

The microarray analysis identified a number of up-regulated genes by the statistical selection method. Of 20 genes, only solute carrier family 37 (the glucose-6-

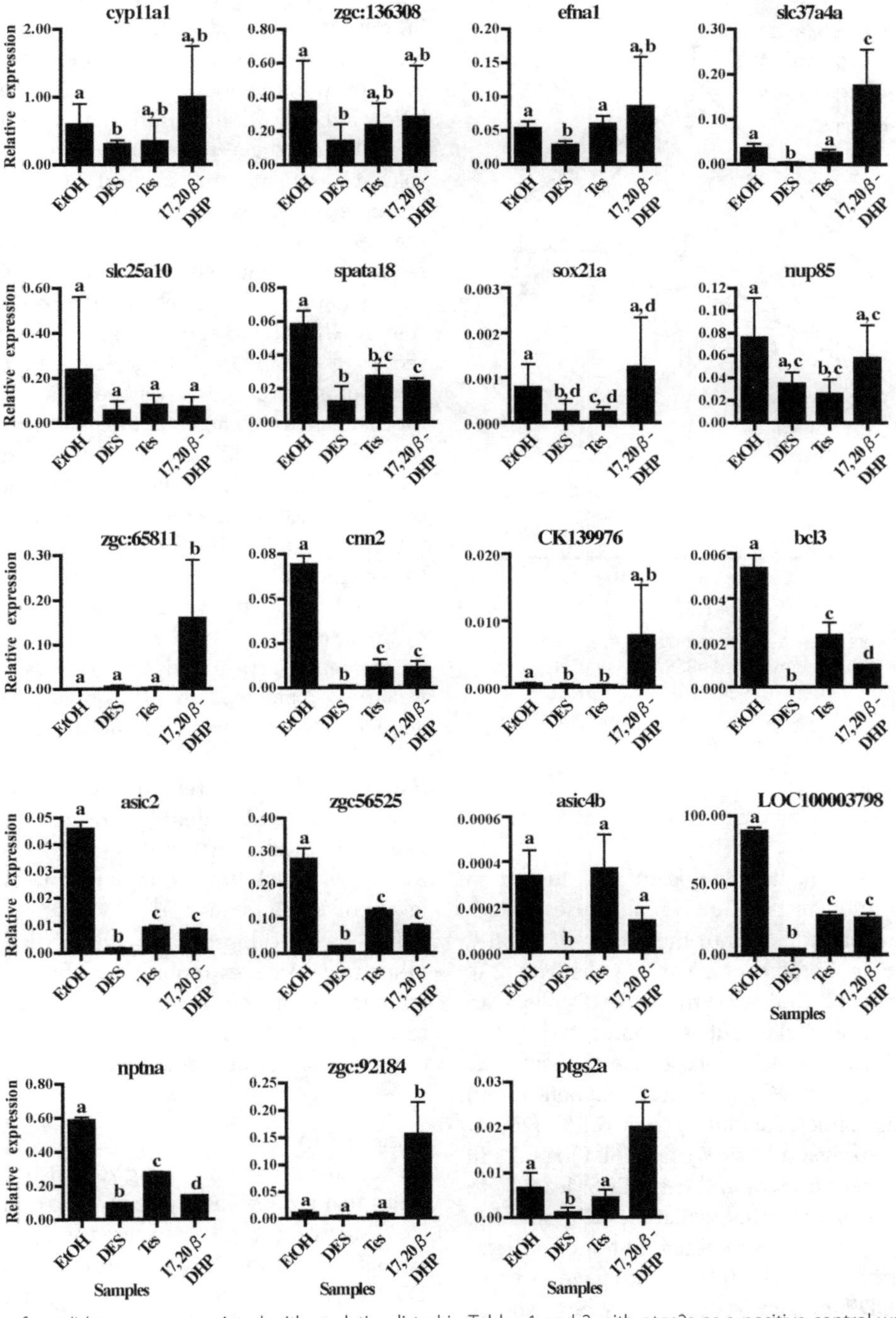

Fig. 3 qPCR analysis of candidate genes associated with ovulation listed in Tables 1 and 2 with ptgs2a as a positive control was conducted using cDNAs prepared from ovaries from fish treated by EtOH, DES, Tes or 17, 20β-DHP for three hours. The mRNA abundance was observed in triplicate for each sample and all data were normalized with the number of transcripts of elongation factor 1α (EF1α) in each sample. Expression values are represented as the mean ± SE of three independent samples. Different letters represent significant differences among the data ($P \leq 0.05$)

phosphate transporter) member 4a (slc37a4a) and zgc:65811 were confirmed as specifically up-regulated during ovulation by qPCR analysis. The mRNA of slc37a4a was reported to be expressed ubiquitously in human [20], whereas its variant is primarily expressed in the brain, heart, and skeletal muscle [21]. This well-characterized protein is one of four protein families that play important roles as sugar transporters [22]. The slc37a4a is located on the endoplasmic reticulum (ER) via 10 transmembrane domains [23]. It encodes the glucose-6-phosphate transporter (G6PT) protein, which plays a central role in the translocation of glucose-6-

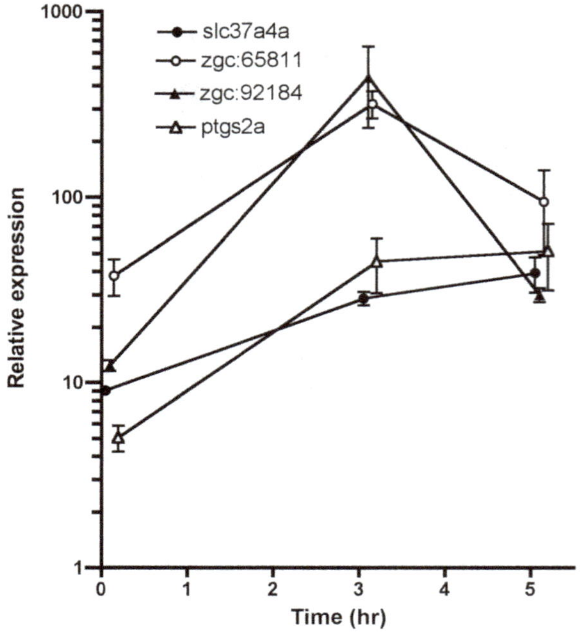

Fig. 4 Total mRNAs were prepared from samples at 0, 3, 5 hour treated with 17, 20β-DHP. We conducted qPCR analysis of slc37a4a, zgc:65811, zgc:92184 and ptgs2a. The mRNA abundance was observed in triplicate for each sample and all data were normalized with the number of transcripts of elongation factor 1α (EF1α) in each sample. Expression values are represented as the mean ± SE three independent samples

phosphate (G6P) between the cytoplasm and the lumen of ER [24]. A mutation in the G6PT gene causes a deficiency in the transport of G6P into the lumen of the ER, which leads to type Ib glycogen storage disease [25]. This finding supports the idea that G6PT plays an important role in intracellular glucose homeostasis.

In dysregulated myocardial glucose metabolism, the intracellular accumulation of glucose 6-phosphate (G6P) can activate endoplasmic reticulum stress (ERS) [26]. It may be that high expression of slc37a4a (G6PT) results in excess G6P in the cell, thereby inducing ERS, leading to apoptosis which in turn induces ovulation [27]. Another candidate zgc:65811 was not annotated, but showed relatively high similarity with CD9 antigen-like gene of common carp (64%). CD9 antigen has previously been shown to be involved in fertilization in knockout mice [28]. Thus it can be speculated that elevation of expression of zgc:65811 before ovulation was to prepare for fertilization.

Another candidate for gene associated with ovulation, zgc:92184, is also not annotated. Blast analysis showed the highest similarity with the GTPase IMAP family member 7-like (LOC100005907) (Gimap7), with 82% identity. Based on this high sequence similarity, we speculate that the biological function of the unknown zgc:92184 gene may be analogous to that of Gimap7.

The Gimap7 gene corresponds to GTPase of the immunity-associated protein family (GIMAPs). GIMAPs are related to immunological functions, such as lymphocyte survival, thymocyte development and apoptosis regulation in cells of the mammalian immune system [29, 30]. It has been suggested that GIMAPs can play opposite roles to regulate the survival of lymphocyte cells; for example, GIMAP4 and GIMAP5 function differently in the regulation of apoptosis. GIMAP4-deficient rats are resistant to apoptosis [31], whereas GIMAP5-deficient mice showed higher rates of apoptosis [32]. Although the role of GIMAP7 in stimulating apoptosis is still uncertain, there are possibility that zgc:92184 in zebrafish is related to apoptosis. Apoptosis has been shown to be an essential mechanism for follicle rupture and leads to the completion of ovulation [33–35]. Two highly up-regulated genes found in this study might have the roles in induction of ovulation through the regulation of apoptosis.

Unfortunately, inconsistent expression levels between the microarray and qPCR analyses have been detected in many genes. The cross hybridization of mRNA resulting in an inaccurate detection of signals in the microarray analysis must be considered. This weakness can lead to the misinterpretation of the gene expression profile [36]. RNA sequencing, which is a modern approach to detect transcriptome profiling, should be adopted to obtain a precise quantity of transcripts and identify other ovulation-inducing genes. Naturally matured and ovulated samples need to be observed further to confirm the transcriptional levels of these genes.

The biological processes underlying the ovulation of other related genes reported in this study should be examined using the recently developed gene knockout technique. This study provides a new system for discovering genes that play essential roles in ovulation.

Conclusion

The *in vivo* assay using living zebrafish allowed us to select genes that are specifically up regulated to induce ovulation. Candidates for genes associated with ovulation, slc37a4a, zgc:65811 and zgc:92184, represented high expression in 17, 20β-DHP-exposed samples (ovulated sample). Precise selection by *in vivo* assay provides new insights into the molecular mechanisms regulating ovulation.

Acknowledgments

We are grateful to Dr. N. Sakai, Dr. T. Saito and Dr. Umeda for valuable advice on the microarray analysis.

Funding

This work was supported by the Grants-in-Aid for Scientific Research on Priority Areas from the Ministry of Education, Culture, Sports, Science and Technology of Japan and JSPS KAKENHI Grant Number 25440154 and 16 K07419 (to TT).

Authors' contributions

TT performed the microarray analyses. WK carried out the gene selection and qPCR analyses. TT and WK participated in the discussions and preparation of the manuscript. Both authors read and approved the final manuscript.

Competing interests

The authors declare that they have no competing interests.

References

1. Kondo T, Yanagawa T, Yoshida N, Yamashita M. Introduction of cyclin B induces activation of the maturation-promoting factor and breakdown of germinal vesicle in growing zebrafish oocytes unresponsive to the maturation-inducing hormone. Dev Biol. 1997;190:142–52.
2. Tokumoto T, Tokumoto M, Horiguchi R, Ishikawa K, Nagahama Y. Diethylstilbestrol induces fish oocyte maturation. Proc Natl Acad Sci U S A. 2004;101:3686–90.
3. Zhu Y, Rice CD, Pang Y, Pace M, Thomas P. Cloning, expression, and characterization of a membrane progestin receptor and evidence it is an intermediary in meiotic maturation of fish oocytes. Proc Natl Acad Sci U S A. 2003;100:2231–6.
4. Tokumoto M, Nagahama Y, Thomas P, Tokumoto T. Cloning and identification of a membrane progestin receptor in goldfish ovaries and evidence it is an intermediary in oocyte meiotic maturation. Gen Comp Endocrinol. 2006;145:101–8.
5. Zhu Y, Liu D, Shaner ZC, Chen S, Hong W, Stellwag EJ. uclear progestin receptor (pgr) knockouts in zebrafish demonstrate role for pgr in ovulation but not in rapid non-genomic steroid mediated meiosis resumption. Front Endocrinol (Lausanne). 2015;6:37.
6. Tokumoto T. Zebrafish as a model for reproductive biology and environmental screening. In: Charles A, Lessman CA, Ethan CA, vol., editor. Zebrafish topics in reproductive toxicology and development. City: Nova science publishers, Inc; 2014. p. 107–22.
7. Tokumoto T, Yamaguchi T, Ii S, Tokumoto M. In vivo induction of oocyte maturation and ovulation in zebrafish. PLoS One. 2011;6:e25206.
8. Villeneuve DL, Garcia-Reyero N, Martinovic D, Cavallin JE, Mueller ND, Wehmas LC, Kahl MD, Linnum AL, Perkins EJ, Ankley GT. Influence of ovarian stage on transcript profiles in fathead minnow (Pimephales promelas) ovary tissue. Aquat Toxicol. 2010;98:354–66.
9. Ings JS, Van Der Kraak GJ. Characterization of the mRNA expression of StAR and steroidogenic enzymes in zebrafish ovarian follicles. Mol Reprod Dev. 2006;73:943–54.
10. Bobe J, Maugars G, Nguyen T, Jalabert B. Specific gene expression profiles are associated with follicular maturational competence acquisition in rainbow trout (Oncorhynchus mykiss). Fish Physiol Biochem. 2003;28:309–11.
11. Bobe J, Montfort J, Nguyen T, Fostier A. Identification of new participants in the rainbow trout (Oncorhynchus mykiss) oocyte maturation and ovulation processes using cDNA microarrays. Reprod Biol Endocrinol. 2006;4:39.
12. Patiño R, Thomas P, Yoshizaki G. Ovarian follicle maturation and ovulation: an integrated perspective. Fish Physiol Biochem. 2003;28:305–8.
13. Lubzens E, Young G, Bobe J, Cerda J. Oogenesis in teleosts: how eggs are formed. Gen Comp Endocrinol. 2010;165:367–89.
14. Knight OM, Van Der Kraak G. The role of eicosanoids in 17alpha, 20beta-dihydroxy-4-pregnen-3-one-induced ovulation and spawning in Danio rerio. Gen Comp Endocrinol. 2015;213C:50–8.
15. Westerfield M. The Zebrafish Book: a Guide for the laboratory Use of Zebrafish (Danio rerio). City: Univ. of Oregon Press, Eugene, OR; 1995.
16. Rozen S, Skaletsky H. Primer3 on the WWW for general users and for biologist programmers. Methods Mol Biol. 2000;132:365–86.
17. Goetz FW, Garczynski M. The ovarian regulation of ovulation in teleost fish. Fish Physiol Biochem. 1997;17:33–8.
18. Ogiwara K, Takano N, Shinohara M, Murakami M, Takahashi T. Gelatinase A and membrane-type matrix metalloproteinases 1 and 2 are responsible for follicle rupture during ovulation in the medaka. Proc Natl Acad Sci U S A. 2005;102:8442–7.
19. Bobe J, Nguyen T, Fostier A. Ovarian function of the trout preovulatory ovary: new insights from recent gene expression studies. Comp Biochem Physiol A Mol Integr Physiol. 2009;153:63–8.
20. Lin B, Annabi B, Hiraiwa H, Pan CJ, Chou JY. Cloning and characterization of cDNAs encoding a candidate glycogen storage disease type 1b protein in rodents. J Biol Chem. 1998;273:31656–60.
21. Lin B, Pan CJ, Chou JY. Human variant glucose-6-phosphate transporter is active in microsomal transport. Hum Genet. 2000;107:526–9.
22. He L, Vasiliou K, Nebert DW. Analysis and update of the human solute carrier (SLC) gene superfamily. Hum Genomics. 2009;3:195–206.
23. Pan CJ, Chen SY, Lee S, Chou JY. Structure-function study of the glucose-6-phosphate transporter, an eukaryotic antiporter deficient in glycogen storage disease type Ib. Mol Genet Metab. 2009;96:32–7.
24. Chen SY, Pan CJ, Nandigama K, Mansfield BC, Ambudkar SV, Chou JY. The glucose-6-phosphate transporter is a phosphate-linked antiporter deficient in glycogen storage disease type Ib and Ic. Faseb J. 2008;22:2206–13.
25. Hiraiwa H, Pan CJ, Lin B, Moses SW, Chou JY. Inactivation of the glucose 6-phosphate transporter causes glycogen storage disease type 1b. J Biol Chem. 1999;274:5532–6.
26. Kundu BK, Zhong M, Sen S, Davogustto G, Keller SR, Taegtmeyer H. Remodeling of glucose metabolism precedes pressure overload-induced left ventricular hypertrophy: review of a hypothesis. Cardiology. 2015;130:211–20.
27. Tabas I, Ron D. Integrating the mechanisms of apoptosis induced by endoplasmic reticulum stress. Nat Cell Biol. 2011;13:184–90.
28. Kaji K, Oda S, Shikano T, Ohnuki T, Uematsu Y, Sakagami J, Tada N, Miyazaki S, Kudo A. The gamete fusion process is defective in eggs of Cd9-deficient mice. Nat Genet. 2000;24:279–82.
29. Filen S, Lahesmaa R. GIMAP Proteins in T-Lymphocytes. J Signal Transduct. 2010;2010:268589.
30. Krucken J, Schroetel RM, Muller IU, Saidani N, Marinovski P, Benten WP, Stamm O, Wunderlich F. Comparative analysis of the human gimap gene cluster encoding a novel GTPase family. Gene. 2004;341:291–304.
31. Carter C, Dion C, Schnell S, Coadwell WJ, Graham M, Hepburn L, Morgan G, Hutchings A, Pascall JC, Jacobs H, et al. A natural hypomorphic variant of the apoptosis regulator Gimap4/IAN1. J Immunol. 2007;179:1784–95.
32. Chen Y, Yu M, Dai X, Zogg M, Wen R, Weiler H, Wang D. Critical role for Gimap5 in the survival of mouse hematopoietic stem and progenitor cells. J Exp Med. 2011;208:923–35.
33. Crespo D, Goetz FW, Planas JV. Luteinizing hormone induces ovulation via tumor necrosis factor alpha-dependent increases in prostaglandin F2alpha in a nonmammalian vertebrate. Sci Rep. 2015;5:14210.
34. Crespo D, Bonnet E, Roher N, MacKenzie SA, Krasnov A, Goetz FW, Bobe J, Planas JV. Cellular and molecular evidence for a role of tumor necrosis factor alpha in the ovulatory mechanism of trout. Reprod Biol Endocrinol. 2010;8:34.
35. Murdoch WJ, McDonnel AC. Roles of the ovarian surface epithelium in ovulation and carcinogenesis. Reproduction. 2002;123:743–50.
36. Gunnarsson L, Kristiansson E, Forlin L, Nerman O, Larsson DG. Sensitive and robust gene expression changes in fish exposed to estrogen–a microarray approach. BMC Genomics. 2007;8:149.

8

Neuromuscular study of early branching *Diuronotus aspetos* (Paucitubulatina) yields insights into the evolution of organs systems in Gastrotricha

Nicolas Bekkouche and Katrine Worsaae*

Abstract

Background: *Diuronotus* is one of the most recently described genera of Paucitubulatina, one of the three major clades in Gastrotricha. Its morphology suggests that *Diuronotus* is an early branch of Paucitubulatina, making it a key taxon for understanding the evolution of this morphologically understudied group. Here we test its phylogenetic position employing molecular data, and provide detailed descriptions of the muscular, nervous, and ciliary systems of *Diuronotus aspetos*, using immunohistochemistry and confocal laser scanning microscopy.

Results: We confirm the proposed position of *D. aspetos* within Muselliferidae, and find this family to be the sister group to Xenotrichulidae. The muscular system, revealed by F-actin staining, shows a simple, but unique organization of the trunk musculature with a reduction to three pairs of longitudinal muscles and addition of up to five pairs of dorso-ventral muscles, versus the six longitudinal and two dorso-ventral pairs found in most Paucitubulatina. Using acetylated α-tubulin immunoreactivity, we describe the pharynx in detail, including new nervous structures, two pairs of sensory cilia, and a unique canal system. The central nervous system, as revealed by immunohistochemistry, shows the general pattern of Gastrotricha having a bilobed brain and a pair of ventro-longitudinal nerve cords. However, in addition are found an anterior nerve ring, several anterior longitudinal nerves, and four ventral commissures (pharyngeal, trunk, pre-anal, and terminal). Two pairs of protonephridia are documented, while other Paucitubulatina have one. Moreover, the precise arrangement of multiciliated cells is unraveled, yielding a pattern of possibly systematic importance.

Conclusion: Several neural structures of *Diuronotus* resemble those found in *Xenotrichula* (Xenotrichulidae) and may constitute new apomorphies of Paucitubulatina, or even Gastrotricha. In order to test these new evolutionary hypotheses, comparable morphological data from other understudied gastrotrich branches and a better resolution of the basal nodes of the gastrotrich phylogeny are warranted. Nonetheless, the present study offers new insights into the evolution of organ systems and systematic importance of so-far neglected characters in Gastrotricha.

Keywords: Neurobiology, Nervous system, Musculature, Ciliary structures, Meiofauna, Chaetonotida, Spiralia, Molecular phylogeny

Background

Gastrotricha are small, often sub-millimetre, interstitial worms, ubiquitously found in most aquatic environments with a long debated phylogenetic position [1–3]. They were first considered closely related to various meiofaunal, protostome groups, such as rotifers (Trochelminthes [4]), kinorhynchs (Nematorhyncha [5]) or Gnathostomulida (Neotrichozoa [6, 7]). Later, molecular phylogenies placed them within Spiralia with uncertain affinities; within the debated group Platyzoa, comprising Gastrotricha, Platyhelminthes and Gnathifera [1, 8, 9]. Recent phylogenomic studies propose a sister group relationship between Platyhelminthes and Gastrotricha [3, 10]. However, the controversy over the phylogenetic position of Gastrotricha masks other problems within the group. Indeed,

* Correspondence: kworsaae@bio.ku.dk
Marine Biological Section, Department of Biology, University of Copenhagen, Universitetsparken 4, 2100 Copenhagen Ø, Denmark

compared to the diversity and omnipresence of these animals, relatively few phylogenetic and detailed morphological studies have been conducted on this group and the evolution of, e.g., nervous system, muscular system and nephridia is unresolved [11–13]. Its diversity also remains largely unexplored, as exemplified by the recent erection of the family Hummondasyidae (Macrodasyida) in 2014 [14] and the genera *Thaidasys* in 2015 [15] and *Bifidochaetus* in 2016 [16].

Diuronotus [17] is another recently described gastrotrich genus (2005), comprising two described species: *Diuronotus aspetos* Todaro et al. [17] and, *Diuronotus rupperti* Todaro et al. [17], and one undescribed species *Diuronotus* sp. [18–20], transferred from *Halichaetonotus* [17]. These are all found in marine interstitial environments of the North Atlantic; *D. aspetos* from Greenland [17] and Germany [2, 21], *D. rupperti* from Denmark [17] and *Diuronotus* sp. from North Carolina, USA [18]. *Diuronotus* was placed in Muselliferidae (Paucitubulatina, Chaetonitida) next to *Musellifer* [22] with which it shares the presence of a ciliated so-called 'muzzle' (or snout) and specific ultrastructural traits of scales and sperm [23].

Gastrotricha are divided into two main taxa: the supposedly monophyletic Macrodasyida and the possibly paraphyletic Chaetonotida, divided further into the Multitubulatina, (consisting of one genus, *Neodasys*, and possessing multiple adhesive glands) and the diverse Paucitubulatina (possessing generally only two adhesive tubes) [24]. Muselliferidae, belonging to Paucitubulatina, is the possible sister group to all remaining Paucitubulatina according to morphological [22, 25] and molecular [26, 27] studies. However, Paps and Riutorts [28] propose an alternative topology in which Xenotrichulidae is positioned as sister group of the remaining Paucitubulatina, and Muselliferidae being the sister group to Chaetonotidae. Kieneke et al. [29] find Proichthydiidae as sister group to the remaining Paucitubulatina, with Muselliferidae forming a clade together with Xenotrichulidae sister group to other Paucitubulatina. These different topologies overall suggest a key position of Muselliferidae within Gastrotricha, emphasizing the importance of this family for understanding the evolution of Gastrotricha. Indeed, some features of Muselliferidae, namely its marine habitat and well-developed hermaphroditism, are thought to be plesiomorphic character traits of Chaetonotida. However, detailed morphological studies on this family are lacking, most likely due to the paucity of these animals and their late discovery [26, 30].

Recently, a series of papers employing confocal laser scanning microscopy (CLSM) described the detailed muscular arrangement of several Paucitubulatina, namely *Musellifer* [22], Xenotrichulidae [22, 31], Chaetonotidae [22, 32], and Dasydytidae [11, 33], and notably, the helicoidal musculature, proposed to be a gastrotrich synapomorphy [34]. These recent reports were used to infer the plesiomorphic arrangement of the musculature of Gastrotricha as constituted by two ventro-lateral longitudinal muscles surrounded by outer circular muscles, and longitudinal splanchnic muscles surrounded by helicoidal and intestinal circular muscles [2]. In Paucitubulatina, the longitudinal muscles appear to be more numerous, and the outer circular muscles, if present, are incomplete and consist of dorso-ventral muscles [2]. These dorso-ventral or semi-circular muscles are found in marine chaetonotids [22], but are often missing or highly reduced in freshwater chaetonotids [11, 22, 33] emphasizing the importance of studying the marine *Diuronotus* in order to resolve their evolution and contribute to the broader understanding of muscular evolution within Gastrotricha.

To date, only one confocal study on *Xenotrichula* has described the nervous system of a member of Paucitubulatina in detail [12], while it has been extensively described for Multitubulatina (*Neodasys*) [13] and in several Macrodasyida with combined immunohistochemistry and CLSM (e.g., [35, 36]), or transmission electron microscopy (TEM) [37]. One of the conclusions of the *Xenotrichula* study [13] is the low structural variation of the nervous system within Gastrotricha, which always comprises a bilobed brain with a ventral commissure, a pair of anteriorly projecting longitudinal nerves, a pair of ventro-lateral nerve cords along the trunk, and a terminal commissure. These features were also interpreted as ancestral conditions of Gastrotricha in Kieneke and Schmidt-Rhaesa (2015) [2, 12]. Yet only a single Paucitubulatina, *Xenotrichula*, was considered for this state reconstruction. Moreover, substantial variation exists, such as the presence of an additional ventral nerve in *Oregodasys cirratus* Rothe & Schmidt-Rhaesa [38] and dorsal nerves in *Xenodasys riedli* (Schöpfer-Sterrer, 1969) [39, 40], or additional trunk commissures in *Dactylopodola* and *Oregodasys cirratus.* These studies highlight the unexplored diversity of gastrotrich nervous systems, which may be especially relevant in the diverse group of Paucitubulatina, in which only one study of the nervous system has been reported so far [13] and given the total lack of data on Muselliferidae.

Several studies have described the ultrastructure and repartition of protonephridia in Gastrotricha, with a few of them addressing species of Paucitubulatina [41, 42]. It has been suggested that members of Paucitubulatina always possess one pair of trunk protonephridia [41], although again, data on Muselliferidae are lacking. Each nephridium was found to encompass two monociliated terminal cells with coaxial cilia, a long canal cell, and a nephridopore cell [2].

In order to enhance our understanding of the evolution of major organ systems within Gastrotricha we acquired new morphological data on *Diuronotus aspetos*, using CLSM techniques and immunohistochemistry to describe the detailed arrangement of the musculature, nervous system, and ciliation. To assess the previously proposed relationship of *D. aspetos* within Muselliferidae, we analyzed the phylogenetic position within Chaetonotida, using molecular data. In this phylogenetic context, the morphology of *Diuronotus* is compared and discussed relative to other Chaetonotida, and Gastrotricha in general, leading to the proposition of several new homologies.

Methods

Collection

For *Diuronotus aspetos*, the samples were taken with a mini van Veen grab from shallow water (3–6 m water depth) of Flakkerhuk (69°38.63'N 51°51.13'W), Disko Island, West Greenland. All specimens were collected during the Arctic summer in August 2013. Sediment was well-sorted sand of fine to medium grain size. The specimens have been extracted with $MgCl_2$ narcotization and decantation.

For DNA, specimens of *Xenotrichula* sp. have been sampled in Ystad, Sweden (55°26.28'N 13°55.44'E) in subterranean environments on a beach with fine to medium-sized sand, and extracted with $MgCl_2$ narcotization and decantation. Marine *Aspidiophorus* sp. were sampled from cultures of *Dinophilus gyrociliatus* from Copenhagen University, where they are contaminants and unfortunately of unknown origin.

Sequence acquisition

Total genomic DNA was obtained from whole specimens using the Qiagen DNeasy Blood & Tissue Kit (Qiagen Inc., Valencia, CA, USA) following the manufacturer's protocol, except for performing the DNA elution in 160 µL of AE buffer in order to increase the final DNA concentration.

Polymerase chain reactions (PCR) using Sanger based markers were prepared to a final volume of 25 µL with 12.5 µL of GoTaq® Green Master Mix (Promega Corporation, Madison, WI, USA), 1 µL of each primer (10 µM concentration), 10–8.5 µL of Milli-Q water (adjusted to amount of DNA template), and 0.5–2 µL of DNA template. Reaction mixtures were heated in a Bio-Rad G1000 Thermal Cycler at 94 °C for three minutes, followed by 35–40 cycles (primer specific) of 94 °C for 30 s, specific primer pair annealing temperatures for 30 s, and an extension at 68 °C for 45 s (unless indicated otherwise), and a final extension phase of 5 min at 72 °C. The COI primer set dgLCO1490/dgHCO2198 [43] was run with two cycling steps, both variable in temperature. COI annealing temperatures were 45 °C for 45 s and 51 °C for 45 s, respectively with extensions of 30s. Overlapping fragments of the small 18S rDNA (ca. 1800 bp) were obtained using paired primers corresponding to fragments 1 and 3 of the 18S rDNA [44]: (1) 18S1f/18S5R (ca. 900 bp) and (3) 18Sa2.0/18S9r (ca. 800 bp) both overlapping. Both primer sets (1) and (3) had annealing temperatures set to 49 °C. The 28S primer set used was 28SD3/28SG758 [45, 46] with an annealing temperature of 53 °C.

All newly generated and integrated sequences were deposited in the GenBank ® database with the following accession numbers KX531001 to KX531007 (Table 1).

Phylogenetic analysis

Sequences were cleaned on BioEdit [47], and a consensus has been realized from the reverse and forward sequences. Sequences were blasted on NCBI [48]. In parallel, COI, 18S and 28S of *Diuronotus aspetos* were found from its transcriptome [3], using Blastall from NCBI. Sangers and transcriptome acquired COI, 18S and 28S genes were aligned and compared, showing low quality and short length of Sangers sequences. Consequently, COI and 28S of the transcriptome were kept, while a consensus of 18S from the transcriptome and the Sangers sequencing was done having an identical overlapping segment. This hybrid approach was possible since the specimens used for the transcriptome and the Sanger sequences came from the same sample. Sequences of *Aspidiophorus* sp. and *Xenotrichula* sp. were added to the dataset. Sequences of other gastrotrichs acquired from GenBank, based on the tree proposed by Kånneby et al. 2012 [49] were added, selecting sequences from each genus (except *Bifidochaetus* [16]) for which sequences were not available at the time of the analysis), and representing the shortest and deepest branches possible. Sequences from Kånneby et al. [26] for *Musellifer*, Kånneby and Todaro [50] for Neogosseidae, and Todaro et al. [51] for the macrodasyidans, outgroups have additionally been collected from GenBank. Subsequently, the sequences were aligned gene per gene with Muscle in Seaview [47], checked by hand, and the three genes were concatenated with Sequence Matrix [52]. Finally, this dataset was analyzed with Bayesian inference in MrBayes 3.2.6 [53] under the model GTR + I + Γ. The gamma shape parameter, the substitution rates, the proportion of invariable sites, and the character state frequencies were all unlinked. The dataset was partitioned according to each gene and by codon position for COI and analyzed with four MCMC chains for each run, for 30,000,000 generations. Chains were sampled every 1000th generations and the burn-in was set to 25 %. Convergence of the two runs as well as analysis quality was ascertained by checking the log likelihood graphs, the average standard deviation of split frequencies, and the model fit with Tracer [54].

Immunohistochemistry and CLSM

Specimens were anesthetized with isotonic magnesium chloride and fixed in 3.7 % paraformaldehyde in phosphate buffered saline (PBS) for 1 to 2 h at room temperature (RT), followed by six rinses in PBS and

Table 1 Sequences used for the phylogenetic reconstruction with their corresponding GenBank accession number. See material and methods for further information on the acquisitions of the sequences of *Diuronotus aspetos*

Species name	18S	28S	COI
Arenotus strixinoi	JQ798537.1	JQ798608.1	JQ798677.1
Aspidiophorus kw654	KX531002	KX531001	No
Aspidiophorus ophiodermus	JN185463.1	JN185510.1	JN185544.1
Aspidiophorus paramediterraneus	JQ798538.1	JQ798609.1	JQ798678.1
Aspidiophorus polystictos TK76	JQ798598.1	JQ798665.1	JQ798727.1
Aspidiophorus polystictos TK75	JQ798597.1	JQ798664.1	JQ798726.1
Aspidiophorus sp.3	JQ798559.1	JQ798629.1	JQ798694.1
Aspidiophorus tentaculatus TK120	JQ798553.1	JQ798625.1	JQ798690.1
Aspidiophorus tentaculatus TK228	JQ798591.1	JQ798659.1	JQ798721.1
Aspidiophorus tetrachaetus	JN185505.1	JN185540.1	JN185576.1
Chaetonotus laroides	JQ798580.1	No	JQ798712.1
Chaetonotus cf. *sphagnophilus*	JQ798604.1	JQ798671.1	JQ798733.1
Chaetonotus cf. *dispar*	JQ798561.1	JQ798631.1	JQ798696.1
Chaetonotus cf. *hystrix*	JQ798603.1	Q798670.1	JQ798732.1
Chaetonotus cf. *laroides* TK86	JQ798602.1	JQ798669.1	JQ798731.1
Chaetonotus cf. *maximus* TK186	JQ798574.1	JQ798646.1	JQ798706.1
Chaetonotus heterocanthus TK100	JQ798543.1	JQ798615.1	JQ798681.1
Chaetonotus mariae	JQ798558.1	JQ798628.1	No
Chaetonotus neptuni MT61	JQ798539.1	JQ798610.1	JQ798679.1
Chaetonotus uncinus	JQ798540.1	JQ798611.1	No
Chaetonotus cf. *novenarius*	JQ798566.1	JQ798636.1	JQ798699.1
Dactylopodola mesotyphle	JF357651.1	JF357699.1	JF432036.1
Dasydytes papaveroi TK157	JQ798571.1	JQ798640.1	JQ798703.1
Diuronotus aspetos	KX531005 and SRX1121926	KX531006 or SRX1121926	KX531007 or SRX1121926
Draculiciteria tesellata MT63	JN185457.1	JN185506.1	JN185541.1
Draculiciteria tesellata TK142	JN185470.1	JN185516.1	JN185549.1
Halichaetonotus aculifer	JQ798550.1	JQ798622.1	JQ798688.1
Halichaetonotus euromarinus	JQ798551.1	JQ798623.1	No
Haltidytes squamosus	JQ798567.1	JQ798637.1	No
Heterolepidoderma macrops	JN185469.1	JN185515.1	JN185548.1
Heterolepidoderma sp.2	JN185485.1	JQ798644.1	JN185563.1
Heteroxenotrichula squamosa	JQ798542.1	JQ798613.1	No
Ichthydium skandicum TK182	JQ798573.1	JQ798645.1	JQ798705.1
Ichthydium squamigerum	JQ798607.1	JQ798674.1	JQ798736.1
Kijanebalola devestiva TK240	KR822112.1	KR822117.1	KR822120.1
Lepidochaetus brasilense TK223	JN185495.1	JQ798658.1	JN185568.1
Lepidochaetus zelinkai TK94	JN185503.1	JN185538.1	JN185574.1
Lepidodermella squamata TK97	JN185504.1	JN185539.1	JN185575.1
Macrodasys sp.1	JF357654.1	JF357702.1	JF432040.1
Megadasys sp.1	JF357656.1	JF357704.1	JF432042.1
Musellifer delamarei	AM231775.1	No	No
Musellifer reichardti	KF578503.1	No	No
Neodasys chaetonotoideus	JQ798535.1	No	JQ798675.1

Table 1 Sequences used for the phylogenetic reconstruction with their corresponding GenBank accession number. See material and methods for further information on the acquisitions of the sequences of *Diuronotus aspetos* *(Continued)*

Neodasys uchidai	JQ798536.1	No	JQ798676.1
Neogossea acanthocolla	KR822114.1	KR822119.1	KR822121
Neogossea antennigera TK232	KR822110.1	KR822115.1	No
Ornamentula paraensis TK147	JQ798562.1	JQ798632.1	JQ798697.1
Polymerurus nodicaudus TK165	JN185502.1	JN185537.1	JN185573.1
Polymerurus rhomboides TK217	JN185493.1	JN185533.1	JN185567.1
Stylochaeta fusiformis	JN185471.1	JN185517.1	JN185550.1
Xenotrichula cf. intermedia	JN185461.1	No	No
Xenotrichula intermedia MT71	JF357664.1	JF357712.1	No
Xenotrichula sp. kw655	KX531003	KX531004	No
Xenotrichula punctata	JN185464.1	JN185511.1	No
Xenotrichula sp. TK121	JF970234.1	No	No
Xenotrichula sp.1	JN185466.1	No	JN185545.1
Xenotrichula velox TK202	JN185488.1	JQ798652	No
Xenotrichula velox TK43	JN185499.1	No	No

storage in PBS with 0.05 % NaN_3. Triple or quadruple stainings were performed for the investigation of muscular, nervous, glandular and ciliary systems, including F-actin staining (Alexa Fluor 488-labelled phalloidin, INVITROGEN, Carlsbad, USA), DNA-staining (405 nm fluorescent DAPI) and antibodies against neurotransmitters and tubulinergic elements (monoclonal mouse anti-acetylated α-tubulin (SIGMA T6793, St. Louis, USA), polyclonal chicken anti acetylated α-tubulin (SAB3500023-100UG), polyclonal rabbit anti-serotonin (5-HT, SIGMA S5545) and anti-FMRF-amide (IMMUNOSTAR 20091, Hudson, USA)). Prior to adding the primary antibody-mix, the samples were pre-incubated with 1 % PTA (PBS + 1 % Triton-X, 0.05 % NaN3, 0.25 % BSA, and 5 % sucrose) for 1 h. Samples were incubated over night at RT in primary antibodies mixed 1:1 with glycerol (in a final 1:400 concentration). Subsequently, specimens were rinsed in PBS six times and incubated with the secondary antibodies conjugated with fluorochromes over night at RT (mixed 1:1 with glycerol; 1:400 goat anti-mouse labeled with CY5 (JACKSON IMMUNO-RESEARCH, West Grove, USA, 115-175-062), 1:400 goat anti-mouse labeled with TRITC (JACKSON IMMUNO-RESEARCH, West Grove, USA, 115-175-062), 1:400 goat anti-rabbit labeled with TRITC (SIGMA T5268), and 1:200 goat anti-chicken labeled with Dylight (JACKSON IMMUNO-RESEARCH, West Grove, USA, 103-495-1550)). They were rinsed in PBS five times and one time in 1 % PTA and pre-incubated for 60 min in Alexa Fluor 488-labeled phalloidin (0.33 M in 1 % PTA). Thereafter, specimens were rinsed in PBS (without NaN_3) and mounted in Fluoromount-G with DAPI (SOUTHERN BIOTECHNOLOGY ASSOCIATES, Inc., Alabama, USA) or Vectashield with DAPI (VECTOR LABORATORIES, Burlingame, USA). The specificity of the antibodies was tested by examining specimens, where either the primary or secondary antibodies were omitted. Chicken anti acetylated α-tubulin staining did not produce satisfying results and is therefore not shown in this study (Sigma SAB3500023-100UG).

The mounted specimens were scanned using an Olympus Fluoview FV-1000 confocal laser scanning microscope (of K. Worsaae, University of Copenhagen, Denmark), with the acquired z-stacks of scans being either projected into 2D-images or analyzed three-dimensionally using IMARIS 7.1 (BITPLANE SCIENTIFIC SOFTWARE, Zürich, Switzerland). This software package was also used to conduct the measurements presented in the following text according to the conventions introduced by Hummon et al. [55], i.e. position in the body is given in units (U) as a relative measurements to total body length, measured from anterior to posterior. Schematic hand drawings and plate setup were done with Adobe Illustrator CS6 and image adjustments conducted in Adobe Photoshop CS6.

Results

Phylogeny

The tree (Fig. 1) shows a fully supported sister group relationship (100 % posterior probability (PP)) between the monophyletic Muselliferidae (100 % PP) and Xenotrichulidae (100 % PP), herein collectively called "group A". Within Muselliferidae the genus *Musellifer* (100 % PP) (represented only by *Musellifer delamarei* (Renaud-Mornant [56]) and *Musellifer reichardti* Kånneby et al. [26]) is found to be the sister group to *Diuronotus aspetos*. Group A is found next to "group B", together constituting the Paucitubulatina

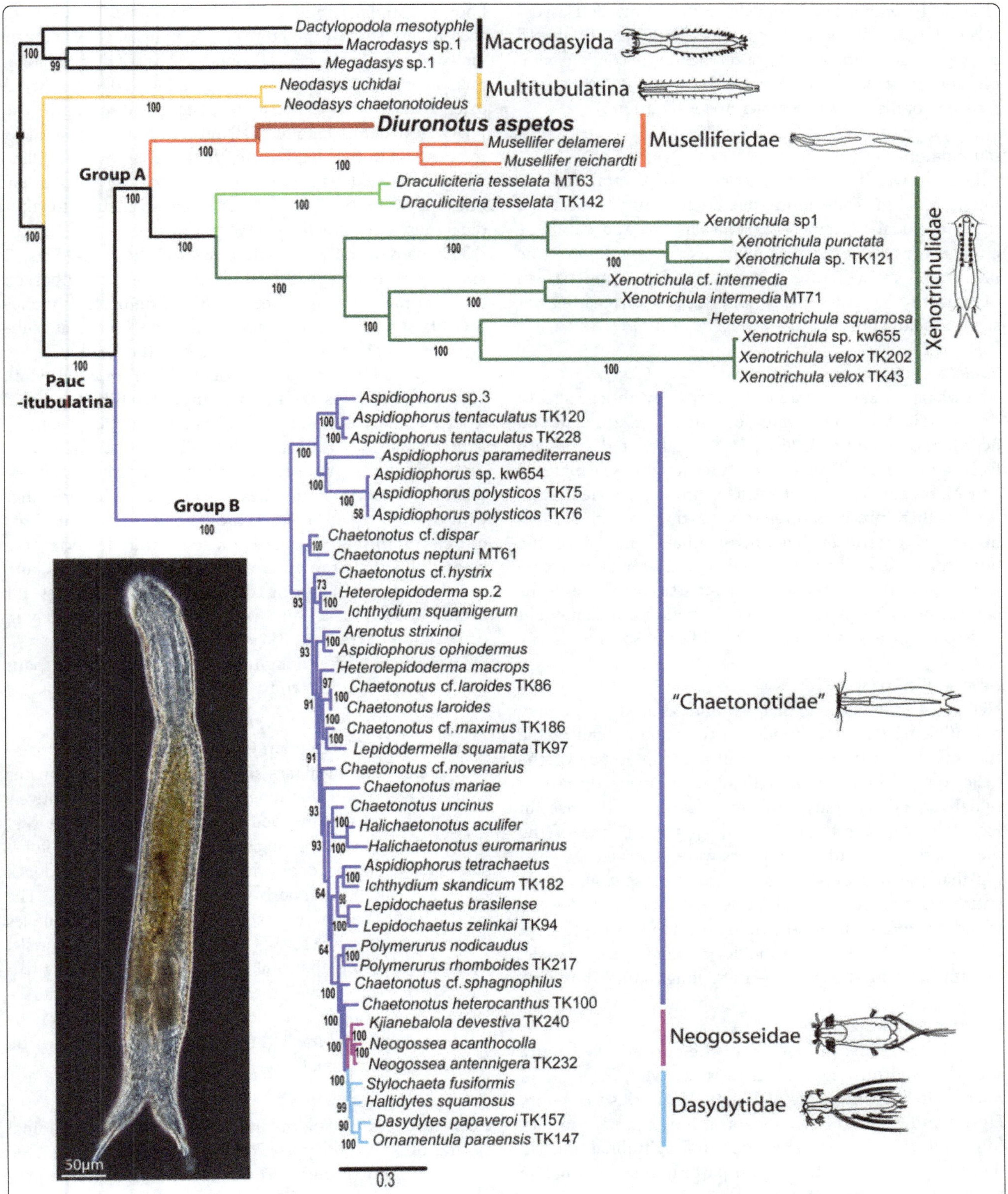

Fig. 1 Phylogenetic position of *Diuronotus aspetos* inferred from Bayesian analysis of 18S, 28S, and COI. This tree is commented in the results section. The analysis includes 58 taxa representing all available genera of Chaetonotida for molecular data on NCBI, and three Macrodasyida as outgroups. Numbers at the nodes represent posterior probabilities in percentages. The picture on the lower left corner is a light micrograph of a live specimen of *Diuronotus aspetos*

(100 % PP), with group B showing a monophyletic Dasydytidae (100 % PP) and Neogosseidae (100 % PP) nested within "Chaetonotidae", the latter hereby becoming paraphyletic. Supports are high in all nodes of the tree, except for some of the numerous inner nodes of group B.

Musculature

The body wall musculature consists of several pairs of longitudinal muscles, numerous dorso-ventral muscles, a thin helicoidal musculature, semi-circular and complete circular muscles, as well as pharyngeal musculature. The pharyngeal musculature is especially dense and has an organization typical of chaetonotid gastrotrichs, as described below in more detail (Figs. 2 and 3).

Radial muscles

The pharynx, *sensu stricto*, is formed by three rows of dense radial pharyngeal muscles (rpm, Figs. 2d, e, m and 3c, d), and extend to U26 (units are calculated as length from anterior end, relative to total length, see material and methods). The radial muscles are cross-striated and each of them presents three to six Z-discs, which are less numerous anteriorly. The myoepithelial nuclei of the pharynx have a distinctive folded and elongated shape (mn, Fig. 2m). The pattern and repartition of these nuclei seems to be specific and corresponding nuclei could be found in the same position in different specimens.

Helicoidal muscles

Helicoidal muscles (hm, Figs. 2n and 3a, b, d, e) are very thin (0.5–1.2 μm) and limited to the anterior half of the specimen. It is difficult to confirm the presence of the helicoidal muscles closest to the pharynx due to the strong signal of other pharyngeal muscles (dashed lines in Fig. 3c-d). In a few locations along the midline of the pharynx very faint diagonal fibers were observed, suggesting that the helicoidal musculature is present along the entire pharyngeal region. Distinct helicoidal muscles are found extending from the midgut/pharynx junction at U26 until U42, enveloping the dorsal longitudinal muscle, but not the ventral or ventro-lateral longitudinal muscles.

Longitudinal musculature

Three longitudinal muscles span the entire body length: a pair of ventral longitudinal muscles (vlm, Figs. 2 and 3), a pair of ventro-lateral longitudinal muscles (vllm, Figs. 2 and 3), and a pair of dorsal longitudinal muscles (dlm, Figs. 2 and 3). The ventral longitudinal muscle bundle splits several times in a pattern described below for the different body regions.

Pharyngeal region Several longitudinal muscles are present along the pharynx. Some are limited to the pharyngeal region, while others are the continuity of the body longitudinal muscles mentioned above. Two sets of muscles are strictly limited to the pharyngeal region: a pair of lateral and a pair of dorsal muscles. The lateral pharyngeal longitudinal muscle (lplm, Figs. 2d, e and 3c, d) extends adjacent to the pharynx along its entire length. The pharyngeal dorsal longitudinal muscles (pdlm, Figs. 2d, e, m, o and 3b-d) extend close to the pharyngeal midline along its total length, ventral to the dorsal longitudinal muscles. Moreover, several somatic and splanchnic longitudinal muscles supply the pharyngeal region.

The paired ventral longitudinal muscle (vlm, Figs. 2 and 3) originating in the head splits along the pharynx into a complex pattern (see Fig. 3b). One of its branches extends more laterally and splits into several sub-branches, supplying the lateral sides of the head.

The paired ventro-lateral longitudinal muscle (vllm, Figs. 2 and 3) lines the pharynx until reaching the head, where it bifurcates at U7, one branch extending ventro-laterally and the other dorso-laterally. Each of them subsequently splits into several minor branches, supplying the lateral sides of the head. These muscles together with the antero-lateral branch of the ventral longitudinal muscle (vlm), and the head diagonal muscle (hdm, see below) all supply the antero-lateral part of the head, and in addition to anchoring the longitudinal muscles for overall body contraction, they may function separately in contraction of the head (Figs. 2j and 3a, c).

The paired dorsal longitudinal muscle (pdlm) spans the anterior-most extremity of the pharynx.

Trunk region Three main longitudinal muscles, i.e. ventral, lateral and dorsal, are supplying the trunk. The paths of the lateral and dorsal longitudinal muscles are relatively straight throughout the body. However, just posterior to the pharynx, the dorsal longitudinal muscle lines the intestine (Figs. 2f and 3c, d, e), while it runs closer to the dorsal body wall more posteriorly (Figs. 2g, h and 3f, g). The ventral longitudinal muscle splits into three muscle bundles at the anterior trunk. Two of these branches run in parallel mid-ventrally along the trunk (lvlm, mvlm, Figs. 2a, c, f-h, p and 3a, b, f-h), whereas the third branch extends dorso-laterally and supplies the dorso-lateral sides of the body until meeting the median-most branch at U86 (dvlm, Figs. 2a, g, l and 3a, b, f).

Posterior region The median branch of the longitudinal ventral muscle splits posteriorly into two bundles at U86. One very short (8 μm) portion medially supplies the posterior part of the adhesive gland of the posterior tube, while another longer branch extends into the primary tube, supplying it for approximately two-thirds of its length to U97 (Fig. 3a). Additionally, the lateral branches of the ventral and lateral longitudinal muscles also extend

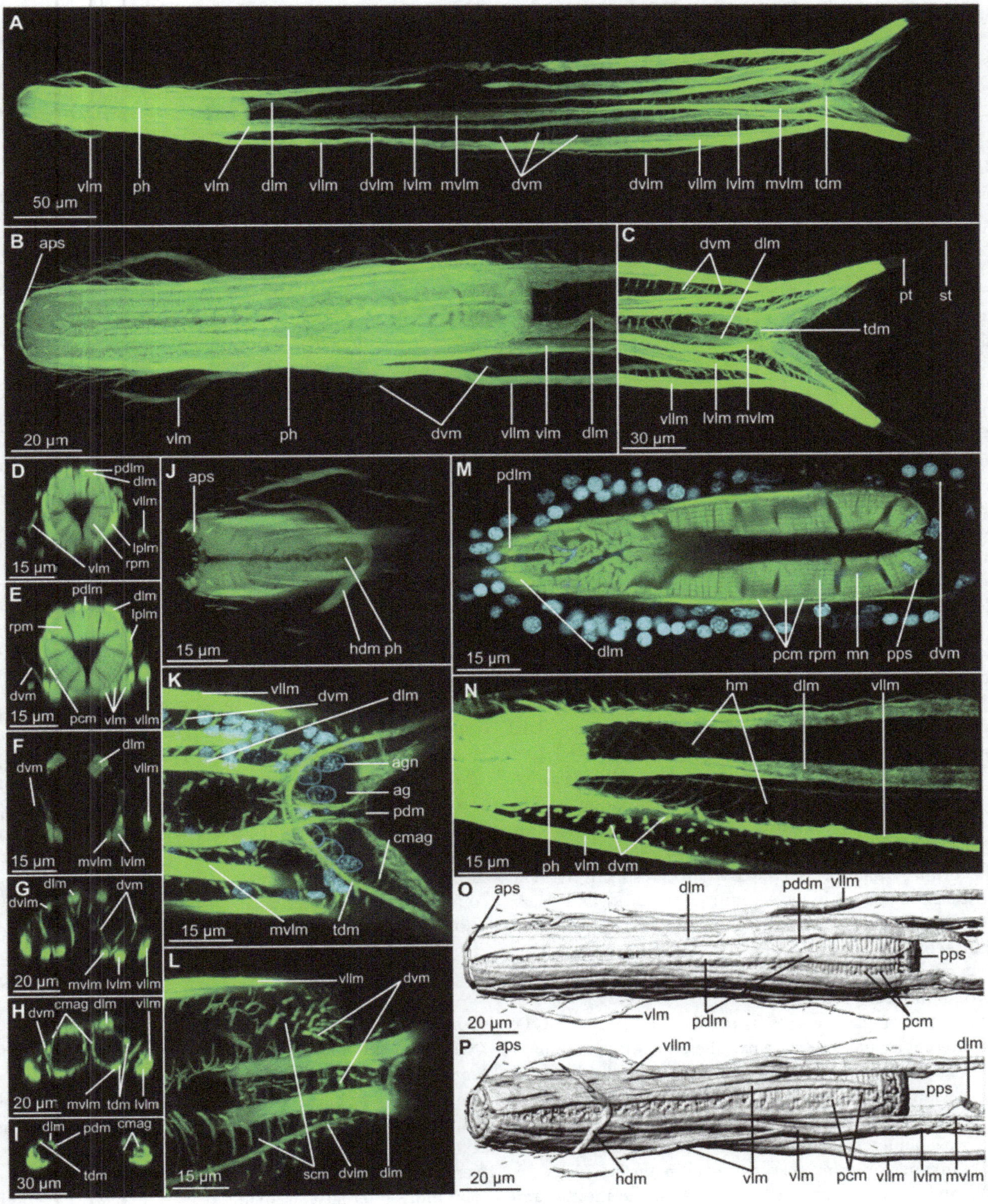

Fig. 2 (See legend on next page.)

(See figure on previous page.)
Fig. 2 CLSM of phalloidin stained muscle of *Diuronotus aspetos*. Anterior of the specimen is pointing *left* for **a**, **b** and **j–p**, and dorsal is pointing at the *top* for **d–i**. **a–n** Muscles in *green*, nuclei in *cyan* **a** Ventral view of the maximum intensity projection (MIP) of the whole specimen. **b** Dorsal MIP of the pharynx. **c** Dorsal MIP of the posterior specimen. **d–i** CLSM virtual transverse section of various parts of the specimen: **d** head, **e** posterior part of the pharynx, **f** anterior of the trunk, **g** posterior of the trunk, **h** post-anal region of the trunk, **i** and furca before bifurcation of the tubes. **j** Dorsal MIP of a sub-stack showing details on the head musculature. **k** Dorsal MIP of a sub-stack showing details of the furca separation. **l** Ventral MIP of a sub-stack showing details of the semicircular musculature. **m** Single section showing details of the inner pharynx. **n** Dorsal MIP of a substack showing details of the helicoidal musculature. **o** and **p**, isosurface reconstruction of the pharynx. **o** Dorsal view, **p** ventral view. *ag* adhesive gland, *agn* adhesive gland nucleus, *aps* anterior pharyngeal sphincter, *cmag* circular muscle of the adhesive gland, *dlm* dorsal longitudinal muscle, *dvlm* dorsal projection of the ventral longitudinal muscle, *dvm* dorso-ventral muscle, *hdm* head diagonal muscle, *hm* helicoidal muscles, *lplm* lateral pharyngeal longitudinal muscle, *lvlm* Lateral extension of the ventral longitudinal muscle, *mn* myocyte nuclei, *mvlm* medial projection of the ventral longitudinal muscle, *pcm* pharyngeal circular muscle, *pddm* pharyngeal dorsal diagonal muscle, *pdm* posterior diagonal muscle, *pdlm* pharyngeal dorsal longitudinal muscle, *ph* pharynx, *pps* posterior pharyngeal sphincter, *pt* primary tube, *rpm* radial pharyngeal muscles, *scm* semi-circular muscle, *st* secondary tube, *tdm* tube diagonal muscle, *vlm* ventral longitudinal muscle, *vllm* ventro-lateral longitudinal muscle, *vlm* ventral longitudinal muscle

along the primary tube. The dorsal longitudinal muscle supplies the anterior third of the secondary tube.

Diagonal muscles

The head diagonal muscle (hdm, Figs. 2j, p and 3a, b, c) forms a V-shape with two medially joined branches. The median part of the muscle is situated in the midline of the body in the posterior region of the head while the two extremities extend to the antero-lateral region of the head.

At the dorso-posterior pharynx, a pair of pharyngeal dorsal diagonal muscles decussate (pddm, Figs. 2p and 3b, d), ventral to the pharyngeal dorsal longitudinal muscle and the dorsal longitudinal muscle. Although their orientation is similar to that of the helicoidal muscles, their exclusively dorsal extension and their greater width differ significantly from those of the helicoidal muscles.

Three diagonal muscles are found in the furca, extending from one side to the contralateral one. The two anterior muscles (tmd, Figs. 2a, b, h, i, k and 3a, b, g, h) extend from the midline at U86, halfway to the primary and the secondary tube, respectively. The posterior diagonal muscle (pdm, Figs. 2i, k, and 3a, b, h) extends from U89 laterally into the first third of the secondary tube.

Circular muscles

Pharyngeal circular muscles (pcm 2E,M,O,P and 3A,B,C,D) are present around the pharynx. These muscles are numerous (ca. 110 in one specimen), positioned proximal to each other, and 1–1.50 μm thick with increasing diameter towards the posterior region.

Two sphincters are present at each extremity of the pharynx: one anterior pharyngeal sphincter (aps, Figs. 2b, o, p and 3a, b) located just posterior to the mouth, and one posterior pharyngeal sphincter (pps, Figs. 2m, o, p and 3a, b) marking the transition between the intestine and the pharynx. The anterior sphincter is smaller, being 1 μm thick and has a diameter of 17 μm, while the posterior sphincter is more prominent with 8 μm thickness and a diameter of 23 μm.

Supplementary circular muscles of the adhesive glands (cmag, Figs. 2h, i, k and 3a, b, g, h) are present in the tubes, forming a muscular layer around the large adhesive glands (ag, Fig. 2k). They thereby supply two cavities, one for each tube. The muscular layer surrounding the primary tube is smaller than the one surrounding the secondary tube, and both structures are connected. This suggests that both tubes are supplied by a single set of glands controlled by muscles. Three adjacent nuclei are found within the layer of circular muscles at the base of the gland (agn, Fig. 2k), near the anus, around U88.

Semi-circular muscles

Ventrally opened semicircular muscles (scm, Figs. 2l and 3a, b, f, g) are present in the posterior part of the specimen, but do not extend into the tubes. They originate ventrally, from each side of the body, and extend to the dorsal side. From there, they project to the contralateral side, external to the longitudinal musculature. They are more numerous in the posterior region, anterior to the furca, where they are separated by 2–5 μm. Semicircular muscles are less numerous and spaced further apart (5–8 μm) in the anterior region of the ovary. They appear to supply only the ovary region; their contraction probably reduces the body diameter and may be involved in the movement/release of eggs.

Dorso-ventral muscles

Numerous thin muscles (1–2 μm in diameter) traverse the entire trunk dorso-ventrally (dvm, Figs. 2a–c, e-h, k, l, n and 3a, b, e-g). These dorso-ventral muscles are spaced approximately 5 μm apart in the region between U18 and U95. In this region, two pairs are found laterally in transverse sections of the pharyngeal region (one external and one more internal pair, dvm, Figs. 2d, e and 3a, b). This number increases more posteriorly in the trunk, where up to five pairs of dorso-ventral muscles can be detected (dvm, Figs. 2g, k, l, and 3a, b, f). The dorso-ventral muscles extend dorso-ventrally between the different

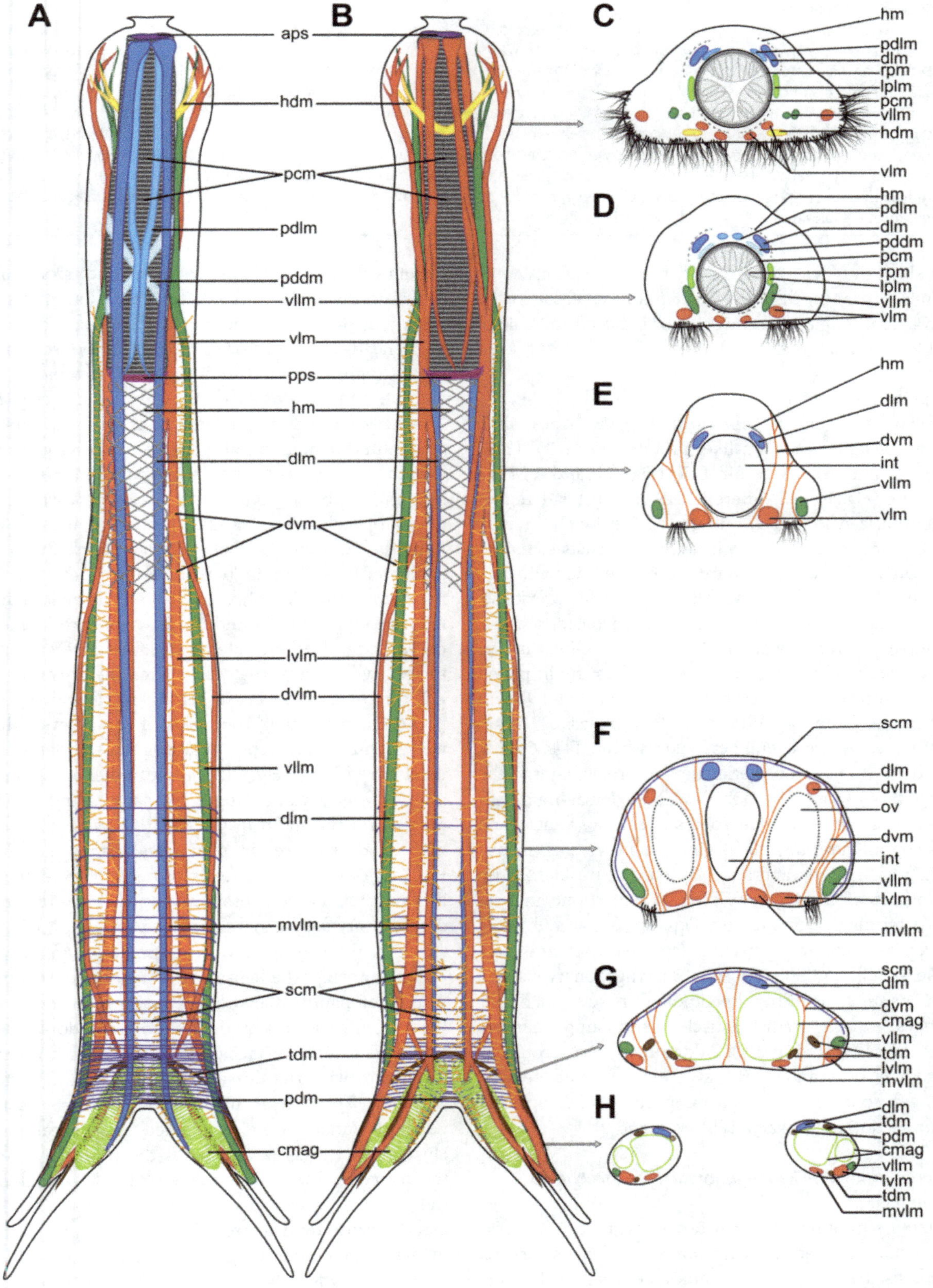

Fig. 3 (See legend on next page.)

(See figure on previous page.)
Fig. 3 Schematic drawings of the musculature of *Diuronotus aspetos*. Anterior end is pointing at the *top* for **a** and **b**, dorsal side is pointing at the *top* for **c**–**h**. **a** Ventral view of the musculature, **b** dorsal view of the musculature, **c**–**h** cross section of the specimen, **c** in the head, **d** posterior part of the pharynx, **e** anterior of the trunk, **f** posterior of the trunk, **g** post-anal region of the trunk, **h** and in the furca before bifurcation of the tubes. Note that in **c** and **d**, the helicoidal pharyngeal musculature is represented in *dash lines* due to the uncertainty of its presence, and it is not drawn in (**a** and **b**). *aps* anterior pharyngeal sphincter, *cmag* circular muscle of the adhesive gland, *dlm* dorsal longitudinal muscle, *dvm* dorsal projection of the ventral longitudinal muscle, *dvm* dorso-ventral muscle, *hdm* head diagonal muscle, *hm* helicoidal muscle, *int* intestine, *lplm* lateral pharyngeal longitudinal muscle, *lvlm* Lateral extension of the ventral longitudinal muscle, *mvlm* medial projection of the ventral longitudinal muscle, *ov* ovary, *pcm* pharyngeal circular muscle, *pdm* posterior diagonal muscle, *pddm* pharyngeal dorsal diagonal muscle, *pdlm* pharyngeal dorsal longitudinal muscle; *pps* posterior pharyngeal sphincter, *rpm* radial pharyngeal muscles, *scm* semi-circular muscle, *tdm* tube diagonal muscle, *vllm* ventro lateral longitudinal muscle, *vlm* ventral longitudinal muscle

longitudinal muscles and the ciliary bands in various combinations. However, they are never found external to the ventro-lateral longitudinal muscles or between the pair of dorsal longitudinal muscles.

Nervous system

The nervous system of *Diuronotus aspetos* is described from acetylated α-tubulin-like immunoreactivity (LIR, Figs. 4, 5 and 6), serotonin-LIR (Fig. 7) and FMRF-amide-LIR (Fig. 8) (all different LIR of the head region are summarized in Fig. 9). Similar to previously investigated Gastrotricha, the nervous system consists of paired nerve cords, which originate from a bilobed dorsal brain, and extend posteriorly. In the following section, previously described and undescribed structures are detailed, such as: i) multiple pairs of longitudinal nerve projections in the head (danp, dlpn, hln, Figs. 6a, b, e, j and 9a, b); ii) paired anterior ventro-median nerves (avmn, Fig. 6a, d and 9b); iii) dorsal nerves posterior to the brain (hdpn, Figs. 6a, h and 9a); iv) paired ventro-lateral nerve cords (vlnc, Figs. 6b-d, f, g, k and 9b); v) paired posterior nerves, projecting into the adhesive tubes (nppt, Fig. 6k), vi) bilobed, dorsal brain with three commissures (main neuropil and anterior ventral and dorsal commissure, together forming a nerve ring encircling the pharynx) (np and anr, Figs. 6a-c, e, h-j and 9a, b); vii) two pairs of ganglia along the nerve cord: one anterior and one terminal (pgg and ang, Figs. 6b, f, k and 9b); viii) four ventral trunk commissures (spc, tvc, pac and pco, Figs. 6b, d, g, k and 9b); and ix) a pharyngeal nervous system, consisting of three longitudinal nerves (one per pharyngeal row of radial muscles) and supplementary minor nerves (Figs. 4 and 5). Additionally, the serotonin-LIR and FMRF-amide-LIR gave very detailed results, allowing us to collect precise data on the number, position and connection of perikarya (Figs. 7, 8 and 9).

Acetylated α-tubulin-like immunoreactivity (acetylated α-tubulin-LIR)

Acetylated α-tubulin-LIR provides information on most neurites, cilia as well as other portions of cytoskeletons of the cells. However, not all minor neurites of the nervous system are traced and the description focuses on the central nervous system and sensory structures.

Stomatogastric nervous system The stomatogastric nervous system, confined to the pharynx, consists mainly of three main longitudinal nerves: a dorsal (dpn, Figs. 4a, d–h, j–n and 5a, b, e) and two ventro-lateral nerves (vpn, Figs. 4b, f–h, l–n and 5e), extending basally along the midline of each row of radial muscles (Fig. 4). The nerves are closely related to three structures: kinocilia, anterior pharyngeal glands and a pharyngeal canal system. The dorsal pharyngeal nerve (dpn, Figs. 4a, d–h, k–o and 5a, b, e) originates at the mouth, where it supplies a buccal nerve ring (bnr, Figs. 4a, b and 5a), encircling the mouth (probably innervating the anterior sphincter (aps, Figs. 2b, o, p and 3a, b) opening and closing the mouth). At U4, two anterior diagonal pharyngeal nerves (adpn Figs. 4a and 5a) originate from the dorsal nerve, extend antero-laterally to the anterior edges of the pharynx and medially join back the dorsal nerve. At U3, one pharyngeal dorso-ventral nerve (pdvn, Figs. 4a, b and 5a) originates from each of the anterior diagonal nerve, and supply ventrally a pharyngeal gland longitudinal nerve (plgn, Figs. 4b and 5c) innervating an anterior pharyngeal gland (apg, Figs. 4a, c, i and 5c) (which opens into the mouth). On the right side of the specimen, a dorso-anterior pharyngeal canal nerve (dpcn, Figs. 4a, d, j and 5a) extends posteriorly from the anterior diagonal nerve, possibly supplying the asymmetric dorsal pharyngeal canal (dpc, Figs. 4a, f, g, l, m and 5e). At U9 a pharyngeal nerve ring (pnr, Figs. 4a, b, e, k and 5b, d) supplying the ventral and dorsal nerves is present. A pair of paramedian dorsal pharyngeal nerves (dpnp, Figs. 4a, f, l and 5e) originates from the dorsal nerve at U19 and extends in a parallel fashion on each side, to fuse again with the dorsal nerve at U26 (Figs. 4a, g, m and 5e). The dorsal nerve extends more posteriorly where it innervates a two-celled pharyngeal posterior cluster (ppc, Figs. 4a and 5i) at the posterior margin of the pharynx. Two nerves extend the terminal part of the ventro-lateral pharyngeal sections: the lateral gland longitudinal nerve (plgn, Figs. 4a and 5c), and a median kinocilium longitudinal nerve (plkn, Figs. 4b, c and 5g), with the latter supplying a mouth and a pharyngeal kinocilia, respectively, at U1 and U6 (mk and pk, Figs. 4b-d, j, i and 5c, g). The gland and

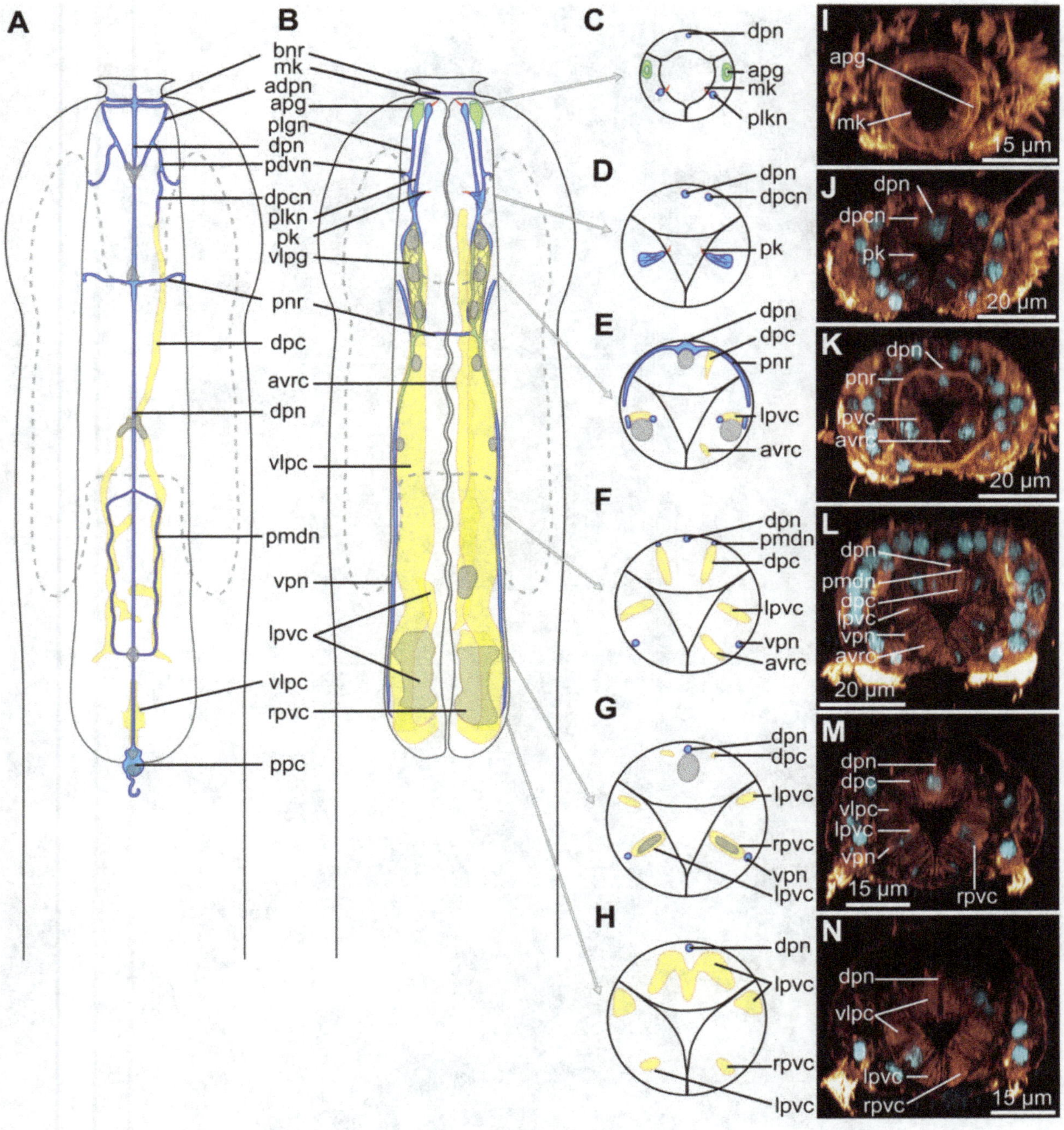

Fig. 4 Pharyngeal nervous system and canal system of *Diuronotus aspetos*. **a**, **b** Anterior is pointing at the *top*; **c**–**n** dorsal is pointing at the *top*. **a**–**h** Schematic drawings with nerves in *blue* and pharyngeal system in *yellow*, nuclei in *grey*, glands in *green* and cilia in *red*. **a** Dorsal section of the pharynx. **b** Ventral section of the pharynx. **c**–**h** Successive transverse sections of the pharynx from anterior to posterior. **i**–**n** CLSM virtual transverse sections at the same levels as (**c**–**h**). Acetylated α-tubulin-LIR in *glow* and DAPI in *cyan*. *adpn*, anterior diagonal pharyngeal nerve, *apg* anterior pharyngeal gland, *avrc* anterior ventro-median right pharyngeal canal, *bnr* buccal nerve ring, *dpc* dorsal pharyngeal canal, *dpcn* dorso-anterior pharyngeal canal nerve, *dpn* dorsal pharyngeal nerve, *lpvc* left posterior ventro-median canal, *mk* mouth kinocilium, *pdvn* pharyngeal dorso-ventral nerve, *pk* posterior pharyngeal kinocilium, *plgn* pharyngeal longitudinal gland nerve, *plkn* pharyngeal longitudinal kinocilium nerve, *pmdn* paramedian dorsal pharyngeal nerves, *pnr* pharyngeal nerve ring, *ppc* posterior pharyngeal cluster, *rpvc* right posterior ventro-median canal, *vlpc* ventro-lateral pharyngeal canal, *vlpg* ventro-lateral pharyngeal ganglion, *vpn* ventral pharyngeal nerve

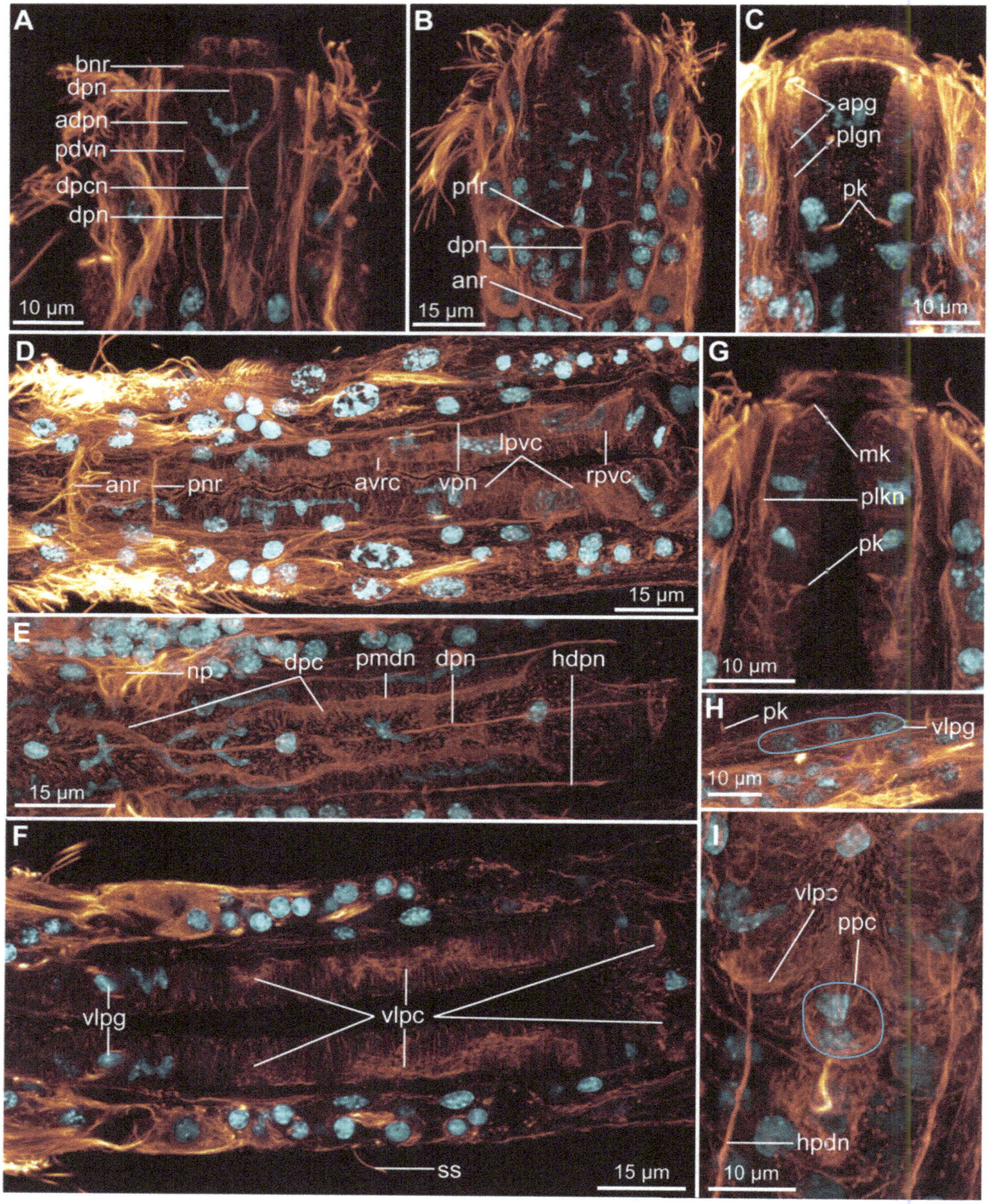

Fig. 5 (See legend on next page.)

(See figure on previous page.)
Fig. 5 CLSM of the pharyngeal nervous system and canal system of *Diuronotus aspetos*. **a–c**, **g**, **i** anterior is pointing at the *top*. **d–f**, **h** anterior pointing *left*. CLSM maximum intensity projection of sub-stacks. Acetylated α-tubulin-LIR in *glow*, DAPI in *cyan*. **a** Dorso-anterior section of the pharynx. **b** Dorso-anterior section of the pharynx, more ventral than **b**). **c** Ventro-anterior section of the pharynx. **d** Ventro-posterior section of the pharynx. **e** Dorso-posterior section of the pharynx. **f** Medio-posterior portion of the pharynx. **g** Medio-anterior section of the pharynx. **h** Details of the ventro-lateral pharyngeal ganglion. **i** Details of the posterior pharyngeal ganglion. *adpn* anterior diagonal pharyngeal nerve, *anr* anterior nerve ring, *apg* anterior pharyngeal gland, *avrc* anterior ventro-median right pharyngeal canal, *bnr* buccal nerve ring, *dpc* dorsal pharyngeal canal, *dpcn* dorso-anterior pharyngeal canal nerve, *dpn* dorsal pharyngeal nerve, *hdpn* head dorso-posterior nerve, *lpvc* left posterior ventro-median canal, *mk* mouth kinocilium, *np* neuropile, *pdvn* pharyngeal dorso-ventral nerve, *pk* posterior pharyngeal kinocilium, *plgn* pharyngeal longitudinal gland nerve, *plkn* pharyngeal longitudinal kinocilium nerve, *pmdn* paramedian dorsal pharyngeal nerves, *pnr* pharyngeal nerve ring, *ppc* posterior pharyngeal cluster, *rpvc* right posterior ventro-median canal, *ss* sensoria, *vlpg* ventro-lateral pharyngeal ganglion, *vlpc* ventro-lateral pharyngeal canal, *vpn* ventral pharyngeal nerve

kinocilium nerves originate at U7 from an elongated ventro-lateral pharyngeal ganglion (vlpg, Figs. 4 and 5h, f), consisting of three nuclei and extending from U7 to U12 (probably integrating the signal collected by the two kinocilia and responsible for the putative terminal gland secrete release). Moreover, the ganglion seems to be related to the ventro-lateral pharyngeal canal (vlpc, Figs. 4a, b, e-g, k-n and 5f, i) described below. The ventral pharyngeal nerves (vpn, Figs. 4b, f, g, l, m and 5d) (supplied by perikarya at U15 and U19) extend from the ganglion, until U28.

Due to the unknown nature of the canal system and its main acetylated α-tubulin-LIR (as well as a weak FMRF-amide-LIR), it is described in this nervous system section. It consists of radially flattened cavities, sometimes asymmetrical, extending longitudinally within the pharynx (Figs. 4a, b, e, f, k, l and 5d, e). Six pharyngeal canals extend the pharynx: i) the unpaired right ventro-anterior pharyngeal canal (avrc, Figs. 4b, e, f, k, l and 5d) extending from U6 to U26; ii–iii) The paired ventro-lateral pharyngeal canals (vlpc, Figs. 4a, b, e-h, k-n and 5f, i), extending from U7 (at the level of the pharyngeal ganglia (vlpg, Figs. 4a and 5h, f)) to U30, and merging dorso-posteriorly; iv-v) the paired ventro-posterior pharyngeal canals extending from U24 for the left one (lpvc, Figs. 4b, g, h, m, n and 5d) to U28 and from U26 for the right one (rpvc, Figs. 4b, g, h, m, n and 5d) to U28; vi) the dorsal pharyngeal canal (dpc, Figs. 4a, f, g, l, m and 5e) extending along the right side from U6 to U15, then reaching the midline and bifurcating in two symmetrical branches, following the paramedian dorsal nerves of the pharynx between U19 and U27. Few nuclei are embedded in the pharyngeal canal system (Fig. 4a, b, g).

Central nervous system The neuropil (np, Figs. 6a, c, e, h, j and 9a) is 14 μm thick and its center is positioned at U15. One 3 μm broad nerve extends antero-medially from the neuropil and branches laterally to form a dorsal and a ventral commissure at U12 and U9, respectively, together constituting an anterior nerve ring (anr, Figs. 5d, 6a-c, i, j and 9a, b). At the dorsal section of the anterior nerve ring (anr, Fig. 9), the acetylated α-tubulin-LIR is relatively weak, and the commissure consists of two transverse (anterior FMRF-amide-like-immunoreactive (LI-reactive) and posterior serotonin-LI-reactive) nerves, which eventually fuse dorso-laterally, forming the lateral sections of the anterior nerve ring. One anterior and one posterior short longitudinal nerves (cpn, Figs. 6b, i and 9b) extend from the ventral portion of the anterior nerve ring, innervating two median ciliary patches (described below). The neuropil supplies ventrally a pair of anterior ventro-median nerves (avmn, Figs. 6b, d, and 9b) extending between the anterior nerve ring and the post-pharyngeal ganglion posteriorly (pgg, Figs. 6b, f and 9b). It extends parallel to the pharyngeal median ciliated cell (pmcc, Fig. 10b, g), probably innervating it. Two pairs of dorso-median anterior nerves projections (danp, Figs. 6a, j and 9a) and two pairs of dorso-lateral anterior nerve projections (dlnp, Figs. 6a, j and 9a) extend from the anterior nerve ring and the lateral sides of the neuropil, respectively, projecting anteriorly. One pair of head lateral nerves (hln, Figs. 6a, b, e, j and 9a, b) extends from the lateral sides of the neuropil and bifurcates, posteriorly supplying a cell with a large and diffuse nucleus (possibly a gland cell (lgcb, Figs. 6a, e and 9a)), and anteriorly forming a nerve projection. Each of these anterior nerve projections probably innervates head sensory organs. Dorso-laterally, the posterior sides of the neuropil supply the ventro-lateral nerve cord (vlnc, Figs. 6b, c, d, f, g, j, k and 9b) of *D. aspetos*, which extends along the entire length of the specimen adjacent to the lateral longitudinal ciliary bands and the ventro-lateral longitudinal muscle, probably innervating these two structures. Two head dorso-posterior nerve (hdpn, Figs. 5e, 6a, h and 9a), extending along the pharynx, eventually supply the post-pharyngeal ganglion. They may innervate the anterior portion of the dorsal longitudinal muscle. A pair of head diagonal nerves (hdn, Figs. 6a, h and 9a) originates dorso-laterally of the neuropil, decussate dorsal to the pharynx at U22, and each extend ventro-laterally to a single perikaryon at U23. Comparison across specimens suggests that the position of these diagonal nerves corresponds to the position of the pharyngeal dorsal diagonal muscle (pddm, Figs. 2p and 3b, d), which it probably innervates. At U27 and U29, two thin nerves originate from the anterior ventro-median nerve, and form together at U28 a sub-pharyngeal commissure (spc,

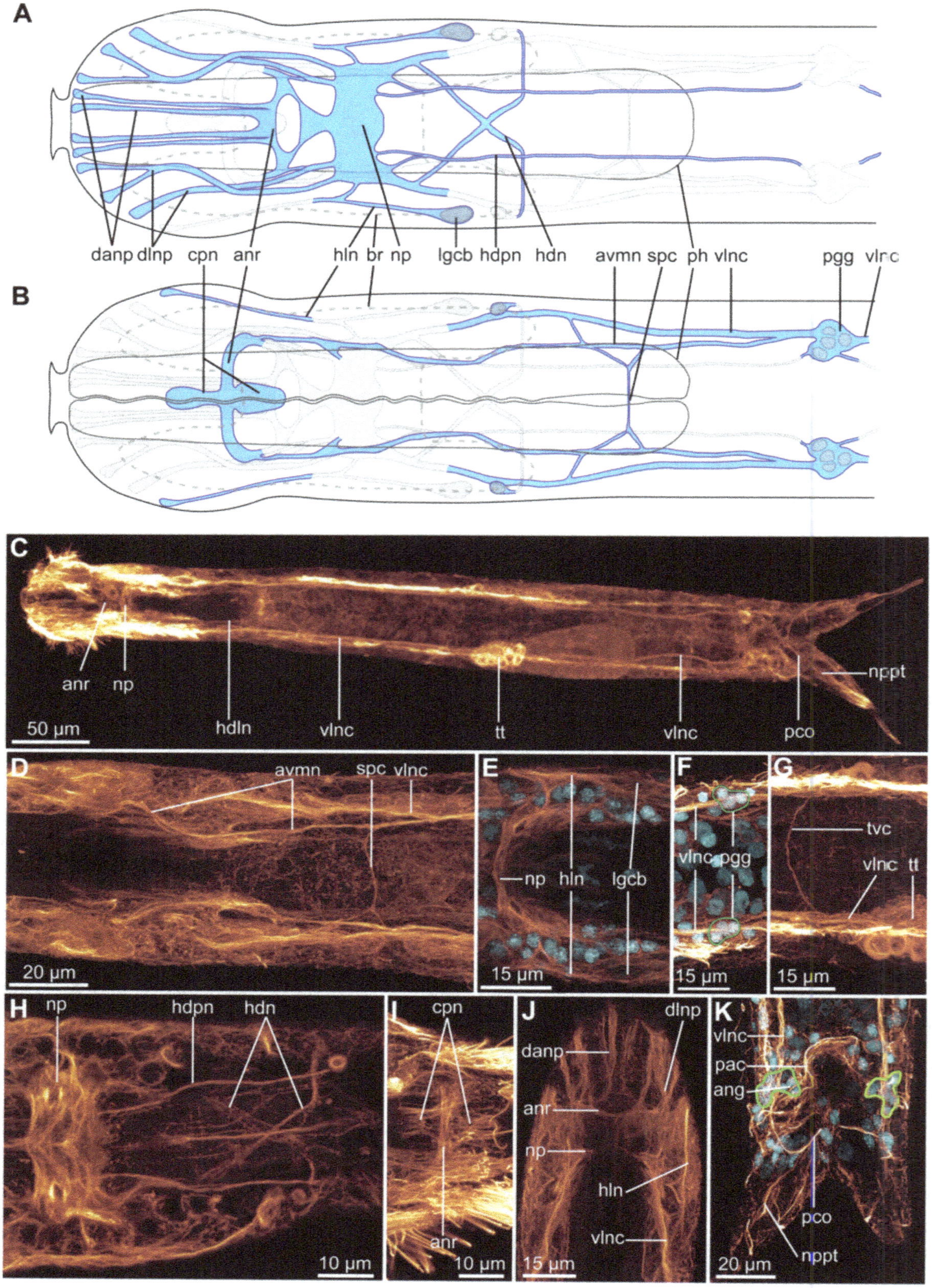

Fig. 6 (See legend on next page.)

(See figure on previous page.)
Fig. 6 Drawing and CLSM of the acetylated α-tubulin-LIR nervous system of *Diuronotus aspetos*. Anterior pointing *left* for **a–i**, and pointing at the *top* for **j** and **k**. **a, b** Schematic drawings of the α-tubulin-LIR of the anterior part of the specimen: nerves in *blue*, nuclei in *grey*, and opposite ventral or dorsal nervous system in *light grey* **a** dorsal **b** ventral. **c** CLSM ventral view of the maximum intensity projection (MIP) of the entire specimen. **d–k** CLSM MIP sub-stacks of various parts of the specimen. Acetylated α-tubulin-LIR in *glow*, and DAPI in *cyan* in all CLSM pictures. **d** Ventro-anterior nervous system. **e** Neuropil side **f** Ventral, post pharyngeal ganglion. **g** Ventral, trunk commissure. **h** Dorso-posterior part of the head **i** ventro-anterior part of the head **j** Dorso-anterior part of the head. **k** Ventro posterior terminal part of the specimen. *ang* anal ganglion, *anr* anterior nerve ring, *avmn* anterior ventro-median nerve, *br* brain, *cpn* ciliated patch nerves, *danp* dorso-median anterior nerve projection, *dlnp* dorso-lateral anterior nerve projections, *hdpn* head dorso-posterior nerve, *hdn* head diagonal nerve, *hln* head lateral nerve, *lgcb* lateral gland cell of the brain, *np* neuropile, *nppt* nerve projection of the primary tube, *pac* pre-anal commissure, *pco* posterior commissure, *pgg* post-pharyngeal ganglion, *ph* pharynx, *spc* sub-pharyngeal commissure, *tt*, testis, *tvc* trunk ventral commissure, *vlnc* ventro-lateral nerve cord

Figs. 6b, d and 9b). At U50, anterior to the testis, a thin trunk ventral commissure (tvc, Fig. 6g) is present. At U84, an anal ganglion (ang, Fig. 6k) of 6–8 cells supplies a pre-anal commissure (pac, Fig. 6k). Posterior to the anus, at U87, the two ventro-lateral nerve cords form the posterior commissure (pco, Fig. 6c, k), from which two nerve projections of the primary tube (nppt, Fig. 6c, k) extend.

Serotonin-like immunoreactivity (serotonin-LIR)

The nervous system shown by serotonin-LIR consists of a dorsal neuropil, the anterior nerve ring, anterior and posterior projections, the two ventro-lateral nerve cords, one posterior commissure as well as several perikarya.

Three main commissures (an anterior (sacn), a median (smcn) and a posterior (spcn)) in the brain neuropil show serotonin-LIR (Fig. 7a, d, e) as well as an isolated patch postero-median to the neuropil (spp, Fig. 7a, d, e). Three longitudinal nerves showing serotonin-LIR are found in the brain: i) the median-most brain nerve (smbn, Fig. 7a, d), ii) the paramedian brain nerve (spbn, Fig. 7a, d), and iii) the lateral brain nerve (slbn, Fig. 7a, d). Four very thin lateral nerves of the posterior commissure of the neuropil (slpn, Fig. 7a, d) form complex connections with the other nerves of the brain as well as to the dorso-median perikaryon (sdmp, Fig. 7a, c, d, e). A postero-lateral nerve node (spln, Fig. 7a, d, e) is present postero-laterally to the neuropil, being formed by the merging of several nerves, and supplies the ventro-lateral nerve cord (slnc, Fig. 7b, c, f). One dorso-lateral perikaryon (sdlp, Fig. 7a, c, d, e) supplies the postero-lateral nerve node, with a short nerve. The median-most brain nerve extends anteriorly from the posterior of the neuropil until U6 (Fig. 7a). The lateral brain nerve is short and extends from the postero-lateral nerve node, being supplied by some of the commissures of the neuropil (Fig. 7a). The paramedian brain nerve originates from the postero-lateral nerve node and supplies the serotonin-like-LI-reactive anterior nerve ring (sanr, Fig. 7a-d), subsequently extending more posteriorly as an anterior nerve projection until U4 (Fig. 7a, d). The anterior nerve ring consists dorsally of two and ventrally of one serotonin-LI-reactive nerves (sanr, Fig. 7a, b). The ventro-lateral nerve cord, consisting of two serotonergic-LI-reactive neurites, extends ventrally to supply a serotonin-LI-reactive para-pharyngeal cluster (sppc, Figs. 7b-e and 9b) with three perikarya, and extends to the posterior end forming the posterior commissure. Additionally, single serotonin-LI-reactive perikarya of the post-pharyngeal ganglion and of the anal ganglion are present respectively at U33 and U93 (spog, and spag, Fig. 7b, c, f) as well as nerve projections of the primary tube (snpt, Fig. 7c, f).

FMRF-amide-like immunoreactivity (FMRF-amide-LIR)

The FMRF-amide-LI-reactive nervous system consists of the brain neuropil, the anterior nerve ring, the anterior ventro-median nerve, the ventro-lateral nerve cord, the sub-pharyngeal commissure and the posterior commissure. Different parts of the nervous system show varying immunoreactivity intensities, as illustrated in Fig. 8.

The neuropil (fnp, Fig. 8a, c, d, f, g) consists of four connectives: two anterior and two posterior, supplied by several posterior and lateral perikarya. Antero-laterally to the neuropil, one pair of perikarya supplies a very short dorso-lateral anterior nerve projections (fpp, Fig. 8a, d, f, g) (corresponding to the base of the acetylated α-tubulin-LI-reactive projections (dlnp; Figs. 6a, j and 9a)). Additionally, three lateral perikarya of the brain (flpb, Fig. 8a, f, g) and a pair of dorso-posterior clusters of the brain (fdpc, Figs. 8a, f, g and 9a) with three perikarya, are present. Comparisons between differently stained specimens, and use of DAPI, enabled us to infer that the postero-median cell of the FMRF-amide-LI-reactive dorso-posterior cluster corresponds to the serotonin-LI-reactive dorso-posterior perikarya (sdmp, Fig.7; fdpc, Figs. 8a and 9a). Anteriorly, the neuropil supplies the anterior ventro-median nerve (fvmn, Figs. 8b, i and 9b), also supplied by two ventro-lateral perikarya of the brain (fvpb, Fig. 8a, e, i) and a ventral perikaryon of the anterior nerve ring (fvpr, Fig. 8b, i). The nerve ring (fanr, Fig. 8a-d, g, h, j) is supplied by one anterior and one posterior unpaired dorso-median perikarya (fpar, Fig. 8a, d, f, h). Ventro-posterior to the neuropil, a pair of tricellular clusters also supply the ventro-median nerve (fvnc, Fig. 8n, i), which extends further posterior until the two FMRF-amide-LI-reactive perikarya of the post-pharyngeal ganglion (fppg,

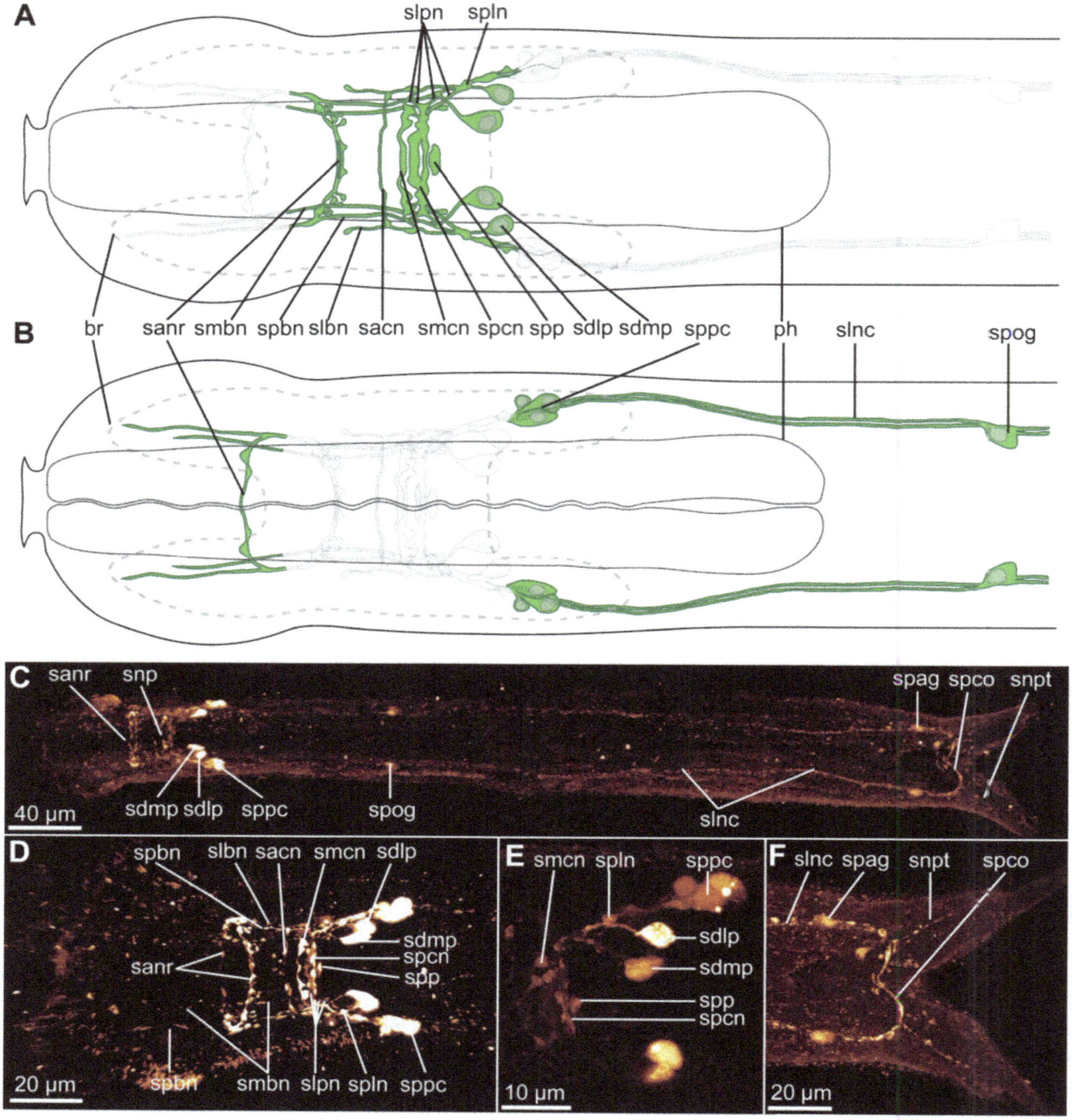

Fig. 7 serotonin-LIR nervous system of *Diuronotus aspetos*. The anterior is pointing *left* for all figures. **a**, **b** Schematic drawings of the serotonin-LIR of the anterior part of the specimen: nerves and perikarya in *green*, nuclei in *grey*, and opposite ventral or dorsal nervous system in *light grey*. **a** Dorsal view, **b** ventral view. **c**–**f** CLSM images with serotonin-LIR in *glow*. **c** CLSM maximum intensity projection (MIP) of the entire specimen. **d** Dorsal MIP of the brain **e** CLSM sub-stack MIP showing details of the brain perikarya **f** CLSM sub-stack MIP of the ventro-posterior terminal part of the specimen. *br* brain, *ph* pharynx, *sacn* serotonin-LI-reactive anterior commissure of the neuropil, *sanr* serotonin-LI-reactive anterior nerve ring, *sdlp* serotonin-LI-reactive dorso-lateral perikaryon, *sdmp* serotonin-LI-reactive dorso-median perikaryon, *slbn* serotonin-LI-reactive lateral brain nerve, *slnc* serotonin-LI-reactive ventro-lateral nerve cord, *slpn* serotonin-LI-reactive lateral nerves of the posterior commissure of the neuropil, *spln* serotonin-LI-reactive postero-lateral nerve node, *smbn* serotonin-LI-reactive median-most brain nerve, *smcn* serotonin-LI-reactive median commissure of the neuropil, *snp* serotonin-LI-reactive neuropil, *snpt* serotonin-LI-reactive nerve projection of the primary tube, *spag* serotonin-LI-reactive perikarya of the anal ganglion, *spbn* serotonin-LI-reactive paramedian brain nerve, *spcn* serotonin-LI-reactive posterior commissure of the neuropil, *spco* serotonin-LI-reactive posterior commissure, *spog* serotonin-LI-reactive perikarya of the post-pharyngeal ganglion, *spp* serotonin-LI-reactive neuropil patch, *sppg* serotonin-LI-reactive para-pharyngeal cluster

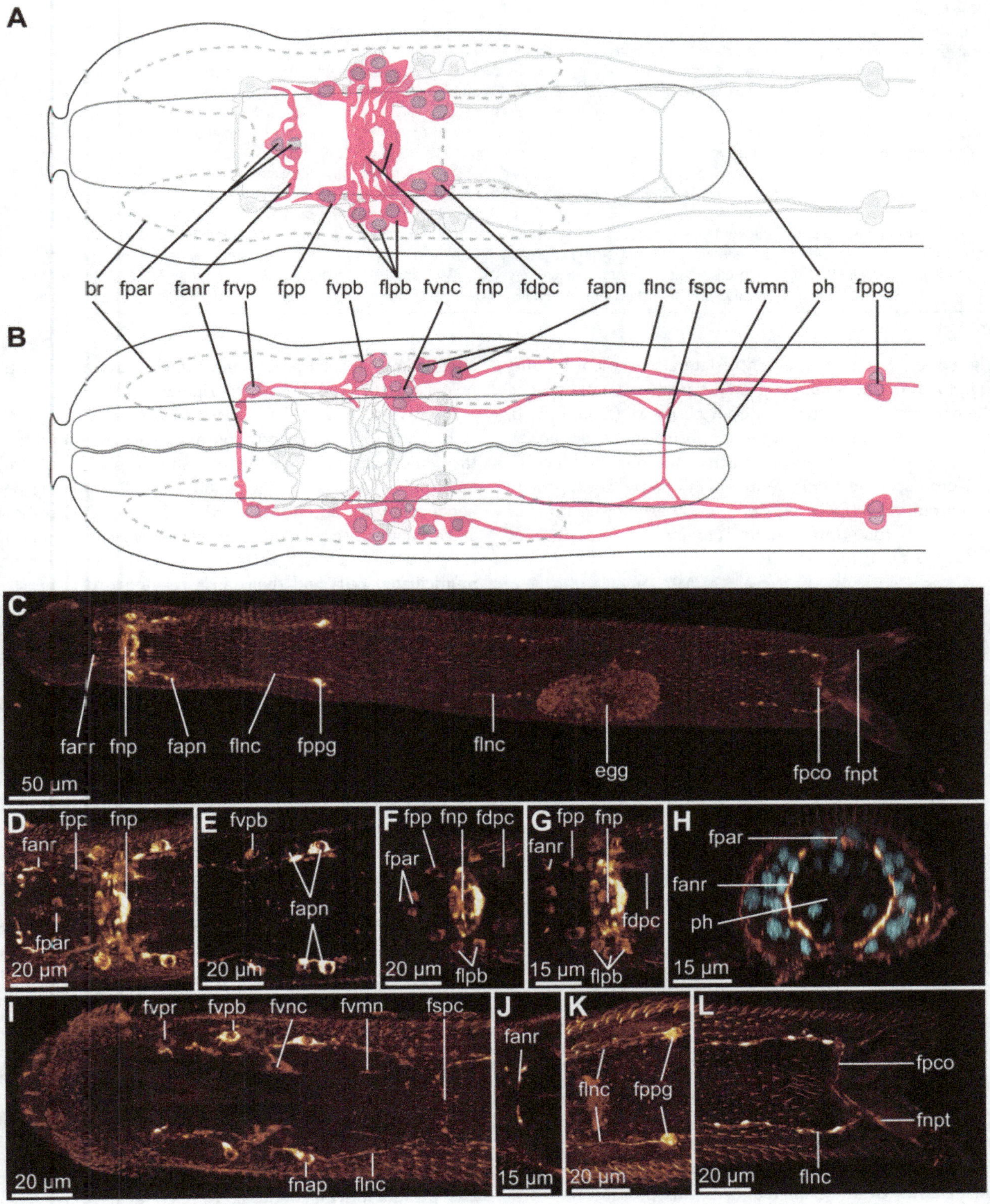

Fig. 8 (See legend on next page.)

(See figure on previous page.)

Fig. 8 FMRF-amide-LIR nervous system of *Diuronotus aspetos*. Anterior is pointing *left* for **a–g** and **i–l** and dorsal pointing at the *top* for **h**. **a, b** Schematic drawings of the FMRF-amide-LIR of the anterior part of the specimen: nerves in *magenta*, nuclei in *grey*, and opposite ventral or dorsal nervous system in *light grey*. **a** Dorsal view, **b** ventral view. **c** CLSM dorsal view of the maximum intensity projection (MIP) of the entire specimen. **d–l** (Except **h**) CLSM sub-stack MIP of various parts of the specimen. FMRF-amide-LIR in *glow*, and DAPI in *cyan* in all CLSM pictures. **d** Dorsal view of the whole neuropil. **e** Ventral part of the brain. **f** and **g** different levels of the dorsal part of the neuropil. **h** CLSM virtual transverse section of the anterior nerve ring. **i** Ventro-anterior part of the head. **j** Ventral commissure of the anterior nerve ring. **k** Ventral post-pharyngeal ganglia. **l** ventro-posterior terminal part of the specimen. Anterior of the specimen on the left for **a-g** and **i-l** and dorsal on top for **h**. *br* brain, *egg* egg, *fanr* FMRF-amide-LI-reactive anterior nerve ring, *fapn* FMRF-amide-LI-reactive anterior perikarya of the ventro-lateral nerve cord, *fdpc* FMRF-amide-LI-reactive dorso-posterior cluster of the brain, *flnc* FMRF-amide-LI-reactive ventro-lateral nerve cord, *flpb* FMRF-amide-LI-reactive lateral perikarya of the brain, *fnp* FMRF-amide-LI-reactive neuropil, *fnpt* FMRF-amide-LI-reactive nerve projection of the primary tube, *fpar* FMRF-amide-LI-reactive dorso-median perikarya of the anterior nerve ring, *fpco* FMRF-amide-LI-reactive posterior commissure, *fpp* FMRF-amide-LI-reactive perikarya of the dorso-lateral anterior nerve projections, *fppg* FMRF-amide-LI-reactive post-pharyngeal ganglion, *fspc* FMRF-amide-LI-reactive sub-pharyngeal commissure, *fvmn* FMRF-amide-LI-reactive anterior ventro-median nerve, *fvnc* FMRF-amide-LI-reactive anterior ventro-median nerve cluster, *fvpb* FMRF-amide-LI-reactive ventro-lateral perikarya of the brain, *fvpr* FMRF-amide-LI-reactive ventral perikarya of the anterior nerve ring, *ph* pharynx

Fig. 8b, c, k). The paired ventro-lateral nerve cords (flnc, Fig. 8b, c, i, k, l) is supplied by the postero-lateral part of the neuropil and by three anterior perikarya (fapn, Fig. 8b, c, e, i) (two anterior and one posterior, separated by 8 μm). The FMRF-amide-LI-reactive sub-pharyngeal commissure (fspc, Fig. 8b, i), ventro-lateral nerve cord, posterior commissure (fpco, Fig. 8c, l), and nerve projections of the primary tube (fnpt, Fig. 8c, l) follow the description of the acetylated α-tubulin-LIR.

Ciliation

The locomotor ciliation consists of a dense ventro-anterior ciliated area and two thin ciliated bands, which are extending to the posterior part of the specimen at U87 (Fig. 10c). This general pattern supports the original description of Todaro et al. [17], although numerous details can be added. CLSM allowed the identification of individual multiciliated cells and determination of their precise pattern.

Dorsally, the muzzle is covered by two transverse rows of multiciliated cells. The anterior row consists of three pairs of relatively small head dorso-anterior ciliated cells (3 μm, hacc, Fig. 10a, e, f), while the posterior row is constituted by a pair of larger head postero-lateral dorsal ciliated cells (hpcc, Fig. CA,E) and a head dorso-median ciliated cell (hmcc, Fig. CA,E) of similar size (6 μm). The pattern of the head lateral ciliated cells (hlcc, Fig. 10a, f) could not be resolved in details due to the dorso-ventral mounting of the specimen. However, at least four cells at the dorso-lateral level are present, and probably the same number at the ventro-lateral level.

The ventral head bears 20 multiciliated cells organized in four paired longitudinal rows and one median row, containing 2,2,3,2,2,2,3,2,2 cells (Fig. 10b). Posterior to the head, ventral to the pharynx, from U9, only two adjacent rows of multiciliated cells are present on each side, containing one large pharyngeal median ciliated cell (pmcc, Fig. 10b, g, 45 μm long) and three pharyngeal lateral ciliated cells (plcc, Fig. 10b, g, h, 20–35 μm long), respectively. Posterior to the pharynx, only one paired lateral row of cells is present, which extends until the posterior trunk, as described originally [17].

The two ventro-median ciliary patches at U7 and U11 (acp, pcp, Fig. 10b, h, i, position of patches measured from the center) are innervated by two short diffuse 5 μm wide longitudinal nerves (cpn, Figs. 6b, i and 7b), joining perpendicularly the anterior nerve ring. Each patch also shows an acetylated α-tubulin-LI-reactive positive ring around the ciliated area. The divergent morphology of these anucleate multiciliated cells and their close relation to the nervous system suggest that they could be sensory structures.

Several sensoria are scattered along the body (ss, Fig. 10d, j), and two pairs of pharyngeal sensory cilia (mk and pk, Figs. 3a-d, i, j, 4c, g and 10i) are located in the pharyngeal region (see the nervous system section for further details).

Two pairs of nephridia are found along the body (Fig. 10d): the anterior pair is situated ventro-laterally and the posterior pair is located dorso-laterally, relatively close to the midline. The anterior pair of protonephridia (apn, Fig. 10d, j) is situated anterior to the testis at U42 and the cilia are 20 μm long. The posterior pair of protonephridia (ppn, Fig. 10d, k) is situated at U74 with 15 μm long cilia. Each nephridium seems to possess two straight coaxial cilia (c, c', Fig. 10j, k), thus resembling the general paucitubulatinian protonephridia with its two adjacent monociliated terminal cells projecting into a non-ciliated canal cell and ending with a nephridiopore epidermal cell [2, 41]. The canal cell and the nephridiopore cell have not been stained, which is why information on the orientation and opening of protonephridia in *D. aspetos* is lacking.

Discussion

Phylogeny

The present phylogenetic analysis confirms that *Diuronotus aspetos* belongs to the monophyletic family Muselliferidae as proposed previously based on morphology [22, 23, 26]. We furthermore find Xenotrichulidae sister group to Muselliferidae (100 % support), opposed to its

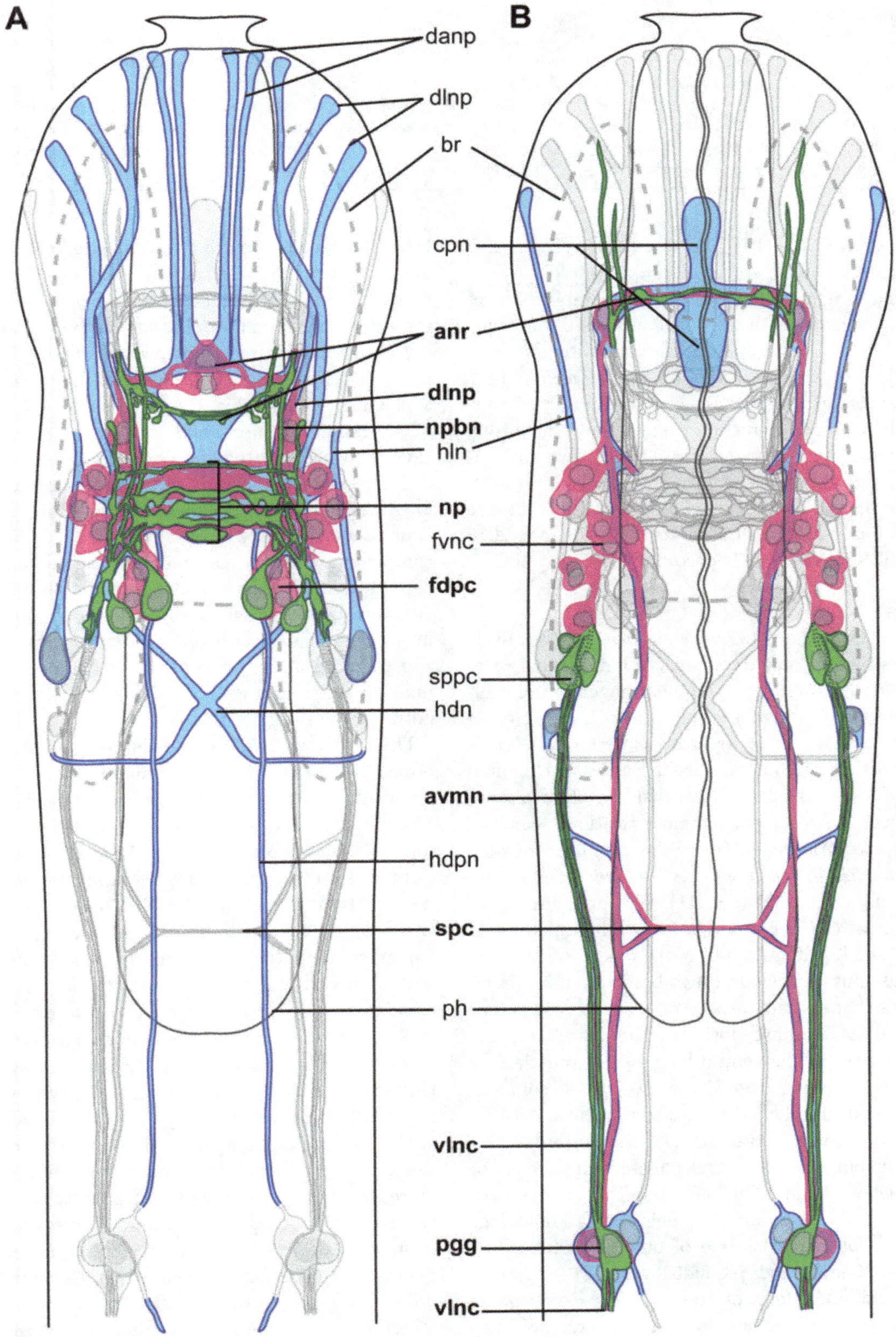

Fig. 9 (See legend on next page.)

(See figure on previous page.)
Fig. 9 Schematic drawing of acetylated α-tubulin-LIR, FMRF-amide-LIR and serotonin-LIR nervous system of *Diuronotus aspetos*, showing the correspondences between the different nervous systems. Anterior pointing at the *top*. Acetylated α-tubulin-LIR nervous system in *blue*. FMRF-amide-LIR nervous in *magenta*, and serotonin-LIR nervous system in *green*. Cell nuclei in *grey* and opposite nervous system in *light grey*. Legends in *bold* indicate structures showing-LIR for at least two molecules tested. **a** Dorsal, and **b** ventral. *anr* anterior nerve ring, *avmn* anterior ventro median nerve cord, *br* brain, *cpn* ciliated patch nerves, *danp* dorso-median anterior nerve projection, *dlnp* dorso-lateral anterior nerve projections, *fdpc* FMRF-amide-LI-reactive postero-lateral brain cluster, *fvnc* FMRF-amide-LI-reactive ventro-median nerve cluster, *hdn* head diagonal nerve, *hln* head lateral nerve, *hdpn* head dorso-posterior nerve, *hpdn* head diagonal nerve, *np* neuropile, *pgg* post-pharyngeal ganglion, *ph*, pharynx, *spc* sub-pharyngeal commissure, *sppc* serotonin-LI-reactive para-pharyngeal cluster, *vlnc* ventro-lateral nerve cord

position next to Group B ("Chaetonotidae" + Dasytydae + Neogosseidae) in Kånneby et al. [26] (69 % PP). Moreover, the placement of *D. aspetos* considerably reduces the internal branch length of Muselliferidae, diminishing the possibility of long-branch attraction, with e.g. *Neodasys* ([49] and present study) or *Dactylopodola* [26], and the support of group B is now maximum. We further note two other interesting points: the sister group relationship between Neogosseidae and Dasydytidae is recovered [50], and the sister group relationship of marine *Aspidiophorus* to the remaining members of the Group B is found again, similarly to Kånneby et al. 2012 [49], but not Kånneby and Todaro [50].

Musculature

The overall musculature of *Diuronotus aspetos* is relatively simple, consisting of only three pairs of longitudinal muscles in the trunk as well as a unique arrangement of multiple dorso-ventral muscles. The number of pairs of longitudinal muscles in *D. aspetos* is inferior to what is found in most Paucitubulatina, having at least five, often six, pairs of longitudinal muscles, that are often distributed as three pairs of splanchnic and three pairs of somatic muscles (*Musellifer, Draculiciteria, Heteroxenotrichula, Xenotrichula, Chaetonotus, Aspidiophorus*, and *Polymerurus*) [22, 31, 32] (Fig. 11). The previously proposed hypothetical ancestral state of musculature in Paucitubulatina [2] shows a split of the dorsal longitudinal muscle (musculus dorsalis) into two branches, not present in *D. aspetos* that instead has a branch of the ventral longitudinal muscle running dorsally. The more complex branching pattern of the ventral longitudinal muscle may be an adaptation to the large size of *D. aspetos*, compensating for the low number of longitudinal muscle.

The helicoidal musculature encircles the dorsal longitudinal muscles but not the ventral longitudinal muscles or the ventro-lateral longitudinal muscles. The relative position of the dorsal longitudinal muscle indicates a homology to the dorsal splanchnic muscle of other Paucitubulatina (Fig. 11) (see Kieneke and Schmidt-Rhaesa [2] for further discussion and limitations of this notion). However, its more dorsal position indicates that it supports the body wall rather than the digestive tract, perhaps furthermore compensating for the missing dorso-dermal muscle branch in *D. aspetos*. The ventro-lateral longitudinal muscle of *D. aspetos* can be homologized with the somatic ventro-lateral muscle (or musculus lateralis) of other Gastrotricha, and the ventral longitudinal muscle resembles those found in the paucitubulatinian *Musellifer delamerei*, *Xenotrichula intermedia* Remane [57] and *Heteroxenotrichula squamosa* Wilke, 1954 [22, 58] (Fig. 11).

The unique semi-circular muscles of *D. aspetos* may aid the oviposition together with the dorso-ventral muscles, hereby functionally replacing the dorso-dermal longitudinal muscle split enveloping the egg in other Paucitubulatina [22]. They likely act as the posterior complete circular muscles found in the region of the sexual organs of *Neodasys* cf. *uchidai* Remane, 1961 [59, 60]. Functionally similar circular muscles are also found in the meiofauna gnathiferan *Gnathostomula armata* Riedl [61] and *Gnathostomula peregrina* Kirsteuer [62] (Gnathostomulida), here arranged in dense pattern around the posterior male organs [63, 64].

The evolution of the dorso-ventral muscles of Chaetonotida as deriving from the circular musculature has been one of the central debates in previous studies [22, 31]. In Macrodasyida and Multitubulatina, the circular musculature consists of splanchnic and somatic elements, the former encircling the intestine and the latter, which derives from splanchnic elements, encircles the ventro-lateral longitudinal muscle on both sides [2, 22, 59]. In Paucitubulatina, the trunk circular muscles are either absent or have a dorso-ventral orientation (i.e. incomplete circular muscles). Such dorso-ventral muscles are found in Xenotrichulidae and Muselliferidae (group A, Fig. 1) [2, 22, 31, 65] (Fig. 11). Hence, if the group A is monophyletic, it means that this arrangement is a synapomorphy of this group, and that the group B lost the dorso-ventral, or circular, muscles (regained in the genus *Polymerurus* [32]). Comparatively, *D. aspetos* is the only Paucitubulatina with more than two sets of dorso-ventral muscles in the transverse axis. The median-most dorso-ventral muscles may be homologous with the visceral circular muscles in other gastrotrichs. The lateral sets of dorso-ventral muscles present varying positions relative to the longitudinal muscles, making homologies difficult to assess. Furthermore, no dorso-ventral muscles are present lateral to the ventro-lateral longitudinal muscle, which would be an

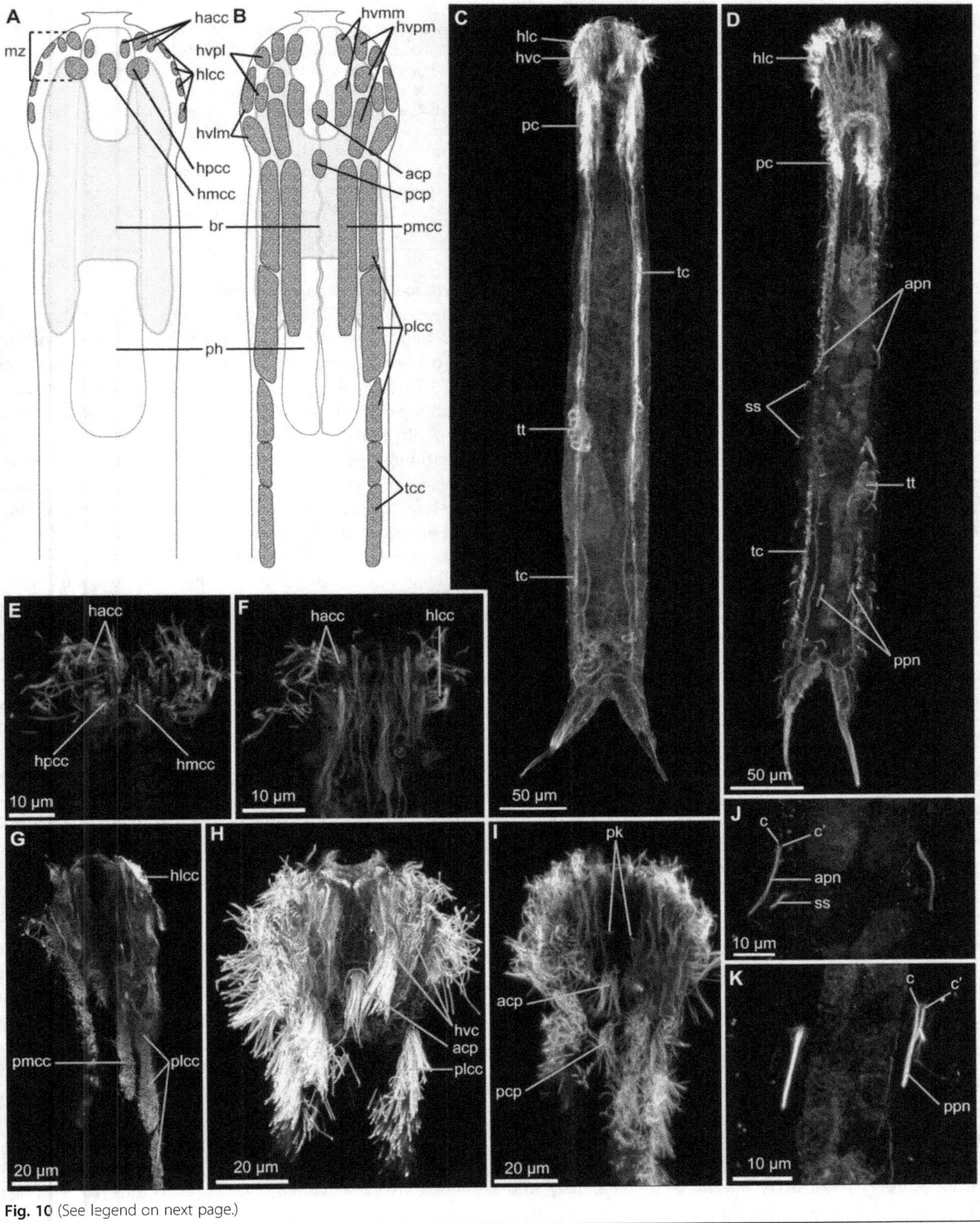

Fig. 10 (See legend on next page.)

(See figure on previous page.)
Fig. 10 Ciliation of *Diuronotus aspetos*. Anterior pointing at the *top* for all figures. **a** and **b** drawings of the locomotory ciliation: **a** dorsal view, **b** ventral view. **c**–**k** CLSM maximum intensity projection (MIP) sub-stacks of the acetylated α-tubulin-LIR. **c** Ventral view of the whole specimen showing the organization of the locomotor ciliation. **d** Dorsal view of the whole specimen showing parts of the locomotor ciliation and the position of the protonephridia. **e** And **f**, dorsal head ciliation. **e** Is more dorsal than **f**. **g**–**i** Ventral head and pharyngeal ciliation: **g** is more dorsal than **h** which is more dorsal than **i**. **j** and **k** details of, respectively, the anterior and the posterior pairs of protonephridia. *acp* anterior ciliated patch, *apn* anterior proto-nephridia, *br* brain, *c, c'* cilia of the proto-nephridia, *hacc* head dorso-anterior ciliated cells, *hlc* head lateral ciliation, *hlcc* head lateral ciliated cells, *hmcc* head dorso-median ciliated cell, *hpcc* head postero-lateral dorsal ciliated cell, *hvc* head ventral ciliation, *hvlm* head ventral lateral-most row of ciliated cells, *hvmm* head ventral median-most row of ciliated cells, *hvpl* head ventral para-lateral row of ciliated cells, *hvpm* head ventral paramedian row of ciliated cells, *mz* muzzle, *pc* pharyngeal ciliation, *pcp* posterior ciliated patch, *ph* pharynx, *pk* posterior pharyngeal kinocilium, *plcc* pharyngeal lateral ciliated cells, *pmcc* pharyngeal median ciliated cell, *ppn* posterior proto-nephridia, *ss* sensoria, *tc* trunk ciliation, *tcc* trunk ciliated cells, *tt* testis

arrangement expected from a derived somatic semi-circular muscle such as found in other Paucitubulatina [20, 22], and see Fig. 11. Consequently, the inner-most dorso-ventral muscles of *D. aspetos* can be homologized with the semi-circular muscles of other Paucitubulatina (Fig. 11), and the arrangement of the other dorso ventral muscles is so far unique in Gastrotricha [2, 22, 32].

The head diagonal muscle of *D. aspetos* may be homologous to the head semi-circular muscle found in *Musellifer delamerei* and *Dactylopodola baltica* (Remane, 1926) [22, 66, 67] showing the same anterior position and shape though a different orientation. It is however probably only homologous to the one found in *Musellifer* since it is the sister group of *Diuronotus*, while *Dactylopodola* is a distantly related Macrodasyida (Fig. 1). The posterior diagonal muscle and the diagonal muscle of the tubes resemble a muscle found in the posterior region of *Heteroxenotrichula squamosa* (Fig. 3a, [22]), but no similar muscle exist in *Musellifer delamerei*, *Xenotrichula intermedia* or *X. punctata* Wilke, 1954. The lack of similar muscle in the closely related taxa within group A indicates that this muscle is an apomorphy of *Diuronotus* (Fig. 11) [22, 58]. The so-called cross-over muscles found in Macrodasyida with a bilobed caudal end has a similar function, being involved in the movement of the posterior tubes, yet with the lack of presence in other Paucitubulatina and different morphology in *D. aspetos* it is most likely of convergent origin [2, 66, 68].

Nervous system

To date, *Xenotrichula intermedia* and *Xenotrichula velox* Remane [69] are the only other Paucitubulatina for which the nervous system has been studied with CLSM [13], therefore the present study adds valuable information. On the other hand, the nervous system of *Neodasys* (Multitubulatina was described in details with CLSM) [12, 39], as well as several Macrodasyida [15, 35, 36, 38, 39, 70, 71]. Furthermore *Cephalodasys maximus* Remane [67] and *Turbanella cornuta* Remane [72] have been described in detail with TEM [37, 73]. This offers a broad, but not comprehensive, literature for comparing the nervous system of *D. aspetos* with other Gastrotricha.

Stomatogastric nervous system

Similar to other Chaetonotida [12, 13, 18], one dorso-median and two ventro-lateral longitudinal nerves constitute the overall pharyngeal nervous system of *Diuronotus aspetos*. However, the present study finds several additional structures previously undescribed for chaetonotids such as: i) five additional symmetric and one asymmetric longitudinal nerves branching off from the main nerves, ii) two previously undescribed commissures (anterior-most buccal nerve ring, middle pharyngeal nerve ring), and iii) a pair of ventro-lateral pharyngeal ganglia (innervating anterior sensory structures).

However, only the pharyngeal nervous system of *Cephalodasys maximus* has been comprehensively described [37] and little is known about the pharyngeal nervous system of Chaetonotida (but see [12, 13]). Nonetheless, ultrastructural studies by Teuchert (1877) [73] and Ruppert [18] provide various details of the pharynx in several gastrotrichs, including some details on *Diuronotus* sp.

In Macrodasyida, the inverted organization of the pharynx generally offers one ventro-median and two dorso-lateral nerves as well as one additional dorso-median nerve [18]. In *Turbanella cornuta*, an additional asymmetric "thick" ventro-lateral nerve is also present in the pharynx [73], which resembles the one short asymmetric dorso-lateral longitudinal nerve found in *D. aspetos* to a certain degree. *Cephalodasys maximus* presents a pair of ventro-lateral asymmetric nerves (one short, one long) in the pharynx, but they originate more posteriorly from the pharyngeal nerve ring [37]. The probable convergent origin of the asymmetric pharyngeal nerves in the morphologically and phylogenetically diverse Macrodasyida *C. maximus* (Cephalodasyidae) and *T. cornuta* (Turbanellidae) contradicts their homology with the one found in *D. aspetos*, but shows that asymmetry in the pharynx of gastrotrichs might have evolved multiple times in Gastrotricha [15, 29].

Macrodasyida and *Neodasys* possess multiple triplets of pharyngeal cilia [18], which according to the basal interrelationship of Gastrotricha suggest that the lack of reports of these structures in Paucitubulatina represents a loss (rather than a primary absence). A single short pair may have either re-appeared in *Diuronotus* (according the

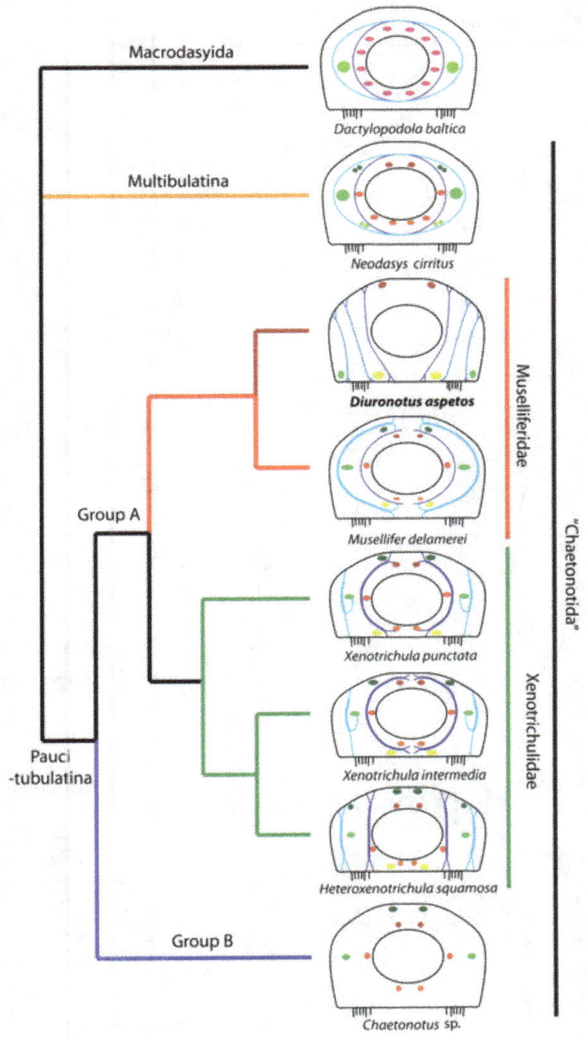

Fig. 11 Evolution of musculature in Paucitubulatina. Each drawing presents the musculature as organized in a transverse section of the trunk with dorsal side pointing at the *top*. Adapted from [22], with addition of missing longitudinal muscles in the original figure, however, mentioned in the article. The musculature of *Neodasys cirritus* is modified from [59], and the phylogeny follows the results of the present study (Fig. 1). Possibly homologous muscles are illustrated with similar colors with the dorso-ventral muscles in *blue*, somatic longitudinal muscles in *green*, and splanchnic longitudinal muscles in *red*. See the discussion for further considerations on the different homologies

topology of our tree (Fig. 1)) or alternatively have been overlooked in previous studies of the sister group Musellifer, or the other members of the group A (Fig. 1), Xenotrichulidae. Ruppert [18] also discusses the presence of discrete glands opening in the mouth of *Chaetonotus* and *Musellifer*, possibly homologous to the anterior pharyngeal glands here described for *D. aspetos*.

Herein is further revealed a presently undescribed pharyngeal canal system within the musculature, occasionally lined by nerves. However, a single transverse TEM micrograph of *Diuronotus* sp. by Ruppert (Figure 14, [18]), purportedly from the level of the ventro-lateral pharyngeal ganglion, reveals a dorso-lateral as well as three ventro-lateral electron-lucent areas, which most likely resemble the canal system. The system may be unique to *Diuronotus* or Muselliferidae, and its function is unknown.

Central nervous system

The overall morphology of the nervous system of *Diuronotus aspetos* is similar to other gastrotrichs [2] consisting of a "dumbbell-shaped" dorsal brain with a dorsal neuropil and a pair of ventro-lateral nerves. However, additional nerves and specific perikarya are found in *D. aspetos*. A summary figure (Fig. 12) shows the evolution of the serotonin-LI-reactive nervous system in Chaetonotida, since this is the most comparable immunostaining across Gastrotricha.

Longitudinal nerves Anterior to the brain neuropil of *Diuronotus aspetos*, four pairs of dorsal nerve projections are found (acetylated α-tubulin-LI-reactive, in addition to several minor neurites left undescribed), most likely related to the anterior sensoria. Similar nerve projections are described in *Neodasys chaetonotoideus* Remane, 1927 [12, 74], *Cephalodasys maximus* [37] and *Thaidasys tongiorgii* Todaro et al. [15] but due to the scarcity of these descriptions, a closer homology cannot yet be stated. Another two pairs of nerves project antero-ventrally from the brain (serotonin-LI-reactive) in *D. aspetos*, one of which may be homologous to the commonly found single pair of serotonin-LI-reactive ventral projections in other gastrotrichs (e.g. *Neodasys chaetonotoideus*, *Dactylopodola* or *Oregodasys cirratus* [12, 36, 38]) (Fig. 12). A similar positioned pair of FMRF-amide-LI-reactive projections is present in *Lepidodasys worsaae* Hochberg and Atherton [70] and *Xenotrichula* [13, 70], and in *Oregodasys cirratus* these are expressing both FMRF-amide-LIR and serotonin-LIR [38], suggesting that the neurotransmitters of these nerves can vary, and that they are a general character of Gastrotricha (cf. nervous system drawing in [2]).

Another striking character found in *D. aspetos* is the paired anterior ventro-median nerve in the anterior trunk. Short, paired anterior ventro-median nerves are also found in *Thaidasys tongiorgii*, *Turbanella* cf. *hyalina* Schultze [75], and extending the entire body length in *Oregodasys cirratus* [15, 38, 39]. However, the exact connection to other nerves and their extension differs from those of *D. aspetos*. Moreover, studies on *Neodasys chaetonotoideus* and on the closely related *Xenotrichula* [12, 13] did not find similar paired anterior ventro-median nerves and we therefore consider the ventro-

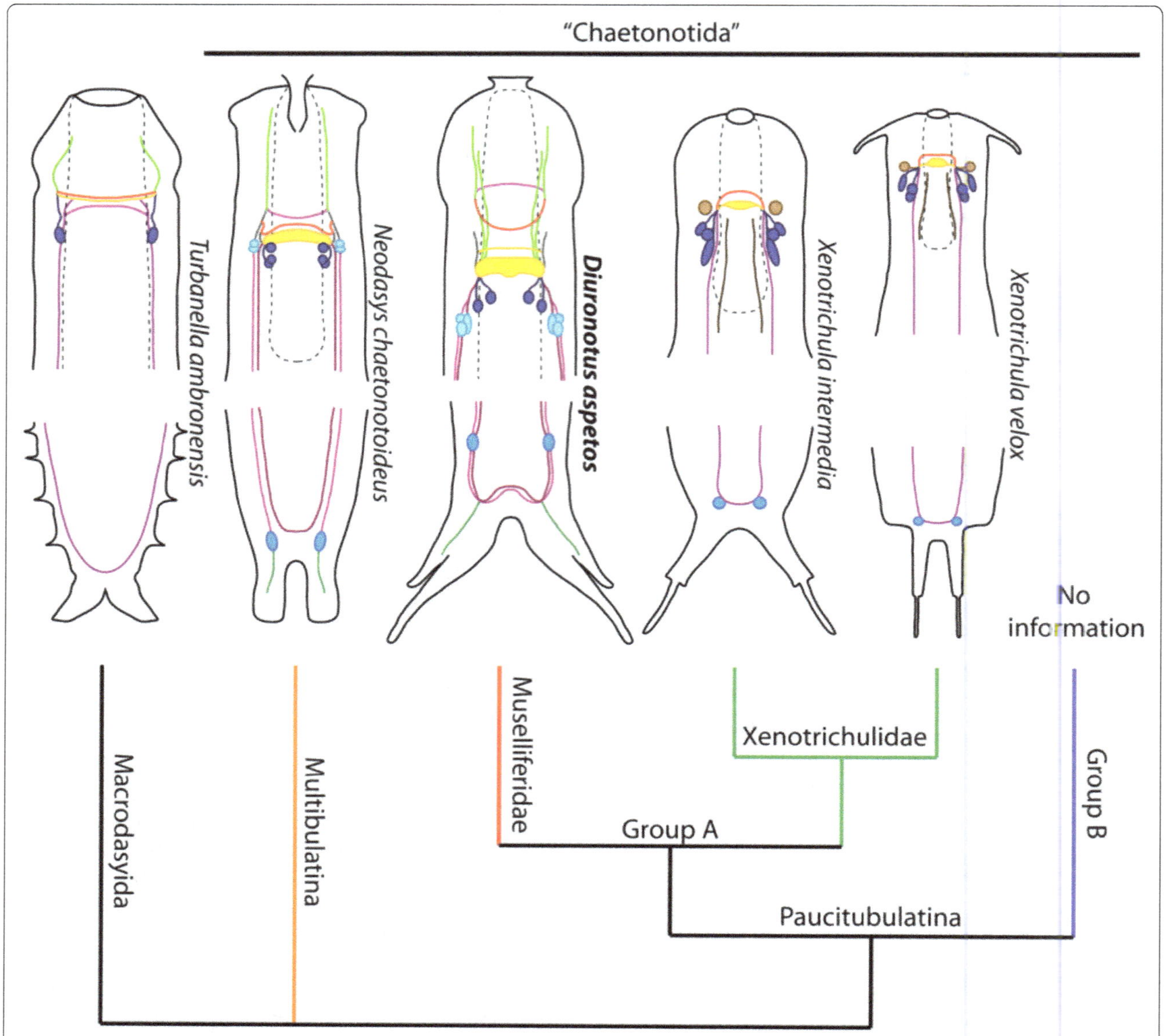

Fig. 12 Evolution of the serotonin-LI-reactive nervous system in Chaetonotida. Schematic presentation of dorsal view, anterior end on *top*, posterior end on *bottom*. *Turbanella ambronensis* adapted from [71], *Neodasys chaetonotoideus* adapted from [12], and *Xenotrichula intermedia* and *X. velox* adapted from [13]. If CLSM pictures are available for representatives of the Group B [77], no formal descriptions exist for this group. The phylogeny follows the results of the present study (Fig. 1). See the discussion for further considerations on the different homologies, and variation between the different immunostainings

median nerves in *D. aspetos* a convergence related to the different ciliation of this species.

The paired short dorso-lateral nerves in *D. aspetos* (hdpn, Figs. 6a, h and 9a) are similar in position and extension to the paired dorsal nerves described in the distantly related Macrodasyida *C. maximus* [37] as well as the dorsal pharyngeal fibers found in the phylogenetically close *Xenotrichula* [13] (and Fig. 1, with Muselliferidae sister group of Xenotrichulidae), of which the latter at least seems to be homologous to the dorsal nerves of *D. aspetos*.

Ganglia and perikarya Several immunoreactive perikarya can be compared to other gastrotrichs, mostly *Neodasys* and *Xenotrichula*. However, immunoreactivity of the perikarya is quite variable, and only a fraction of the brain cells are immunoreactive.

Only five pairs of serotonin-LI-reactive perikarya are found in the brain of *Diuronotus aspetos*, situated postero-laterally to the neuropil. They comprise two dorsal pairs of perikarya, supplying the neuropil, and a ventral pair of para-pharyngeal clusters (spgg, Figs. 7b-e and 9b) with three perikarya each, supplying the ventro-lateral nerve cords. The closely related *Xenotrichula* does not possess a serotonin-LI-reactive equivalent to the ventral clusters, but possesses four dorsal pairs of serotonin-LI-reactive perikarya [13], two of which are likely homologous to the two dorsal pairs found in *D. aspetos* (Fig. 12). *Neodasys chaetonotoideus* possesses three dorso-lateral serotonin-LI-reactive perikarya, which have similar positions and connection to the neuropil than the dorsal serotonin-LI-reactive perikarya of *D. aspetos*. Moreover, *N. chaetonotoideus* possesses a similar paired cluster of para-pharyngeal serotonin-LI-reactive perikarya, associated to the ventro-lateral nerve cords. This suggests that *N. chaetonotoideus* and *D. aspetos* might share some plesiomorphic traits of their serotonin-LI-reactive nervous system, whereas *Xenotrichula* represents a derived condition (Fig. 12). Depending on the position of *Neodasys the* para-pharyngeal cluster could be either a synapomorphy of Chaetonotida or Gastrotricha. In the Macrodasyida investigated to date, the serotonin-LI-reactive brain is generally simpler than in Chaetonotida, comprising only one dorsal commissure and one pair of dorso-lateral perikarya [15, 38, 71] (sometimes two [76]) (Fig. 12), although additional serotonin-LI-reactive perikarya can be found in *Dactylopodola* [36], and *Paradasys subterraneus* Remane [57] (personal observations).

The FMRF-amide-LI-reactive perikarya of the brain of *D. aspetos* are numerous (at least 16 paired and two unpaired of various intensity of the immunoreactivity) and surround the brain neuropil dorsally, ventrally and laterally. Due to the high number and variation of FMRF-amide-LIR of the brain in Gastrotricha, we limit our comparison of *D. aspetos* to the closely related *Xenotrichula* [13]. Homologies of the perikarya depend on whether the anterior dorsal and ventral FMRF-amide-LI-reactive commissures in *Xenotrichula* are homologous to the anterior nerve ring of *D. aspetos*. If so, the two dorso-median FMRF-amide-LI-reactive perikarya found connected to the anterior dorsal commissure in *Xenotrichula*, may be homologous to the two dorso-median found in *D. aspetos*. The two additional described paired lateral and ventral FMRF-amide-LI-reactive perikarya in *Xenotrichula* are difficult to homologize with those of *D. aspetos*. However, one pair of undescribed ventral FMRF-amide-LI-reactive perikarya is found laterally on the ventral commissure of *Xenotrichula* (Figure 4h, [13]) and is possibly homologous to the ventral perikarya of the FMRF-amide-LI-reactive nerve ring in *D. aspetos*. Finally, one of the cells of the FMRF-amide-LI-reactive dorso-posterior cluster of the brain of *D. aspetos* may be homologous to the single pair of perikarya found in *Xenotrichula* in the same position.

In a position similar to the post-pharyngeal ganglion of *D. aspetos*, two pairs of FMRF-amide-LI-reactive (no serotonin-LIR) perikarya are described supplying the ventro-lateral nerve cord in *Xenotrichula* [13]. Between these two pairs, two short transverse FMRF-amide-LI-reactive neurites almost constitute a commissure similar to the one of *D. aspetos*, suggesting that the posterior-most pair of perikarya in *Xenotrichula* is homologous to the ganglia found in *D. aspetos*.

An anal pair of serotonin-LI-reactive perikarya contained in the anal ganglion is found in *D. aspetos* as well as a posterior commissure, similar to what is described for *Xenotrichula* and *Neodasys chaetonotoideus* [12, 13]. Yet, no equivalent is found in any Macrodasyida (Fig. 12), confirming that it is either a synapomorphy of Chaetonotida or Gastrotricha [2], depending of the phylogenetic position of *Neodasys*. Herein observations show that the anal ganglion consists of several cells in *D. aspetos*, contrary to other Chaetonotida [77]. Moreover, we describe an additional pre-anal commissure, originating at the anal ganglion, only revealed by acetylated α-tubulin-LIR and hitherto not found in other gastrotrichs.

Brain commissures *Diuronotus aspetos* does not show a commissure situated directly ventrally to the main brain neuropil contrary to most Gastrotricha documented, including Chaetonotida (e.g. [12, 13, 15, 39]). This character was central in previous discussions on a possible close relationship between Cycloneuralia and Gastrotricha (e.g. [36, 39, 78, 79]), rejected today (e.g. [3, 80]) since, recent interpretations of the brain of Gastrotricha show that it is not truly circular [80]. In *D. aspetos*, the anterior nerve ring is associated to the brain and its ventral portion resembles the ventral brain commissure of other Gastrotricha, although being more anterior (Fig. 12). Furthermore, *Xenotrichula* possesses one ventral FMRF-amide-LI-reactive commissure anterior of the brain [13]. If the FMRF-amide-LI-reactive anterior commissure of the brain and ventral commissure of *Xenotrichula* are continuous, it can be speculated that *Xenotrichula* also possesses an anterior nerve ring, and therefore this arrangement might be a synapomorphy of the group A (Fig. 1).

Ventral ciliation

The main difference from the original description is that the head ventral ciliation forms two medially separated ciliated areas in *Diuronotus aspetos*. Furthermore, a more detailed pattern has been deduced, showing the relevance of CLSM for determining ciliary arrangement [81–83] (but see also Kerbl et al., in prep; Bekkouche and Worsaae, in prep, respectively on Dinophilidae (Annelida) and Micrognathozoa). This also opens the way to a new kind of characters in interstitial animals, which could have a systematics value: the pattern of the multi-ciliated cells. Indeed, preliminary results showing variation in the pattern of the ventral multi-ciliated cells of Thaumastodermatidae support this idea (Bekkouche and Worsaae unpublished). Unfortunately, though the description of the general pattern of the ventral ciliation is common in Chaetonotida, details are rare. In few cases, more details were given, for instance for Neogosseidae (with exact description of the ciliary bands [84]). A few studies have described ciliary patches in the head of some Chaetonotidae (e.g. [85, 86]), but no precise information on the cells themselves has been given, which is why it is unknown whether each patch or band is composed of one or several cells. This limitation of light and electron microscopy can be overcome by employment of CLSM, but due to the lack of similar studies, we cannot yet comment on the evolution of the fine detailed ciliation pattern of Gastrotricha.

Interestingly, some Paucitubulatina show unpaired ciliary patches on the ventral midline of the head, e.g. *Halichaetonotus atlanticus*, Kisielewski [85], *Arenotus strixinoi* Kisielewski [86] or *Kijanebalola devestiva* Todaro et al. [84], but details are insufficient to hypothesize any homology with the ventro-median ciliary patches of *D. aspetos*.

Protonephridial system

Until the present study, all three previously studied Paucitubulatina were known to possess only one pair of protonephridia (*Xenotrichula carolinensis stylensis* Mock [87], *Chaetonotus maximus* Ehrenberg, 1831 [41, 88] and *Polymerurus nodicaudus* (Voigt, 1901) [42, 89]). In this context, *Diuronotus aspetos* is the only Paucitubulatina known to have more than one pair of protonephridia, suggesting, according our phylogeny (Fig. 1), an addition of a pair of protonephridia. However, studies on the protonephridial system of *Musellifer* are needed to confirm if the presence of a single pair of protonephridia has a phylogenetic value or is due to size dependency. Indeed, the number of pairs of protonephridia in other Gastrotricha is variable and seems to be roughly size dependent (e.g. two pairs for the ca. 250 μm long *Dactylopodola baltica,* and 11 pairs for the ca. 1 mm long *Mesodasys laticaudatus* Remane, 1951 [90, 91]).

Conclusion

The present study is the first detailed anatomical description of a member of Muselliferidae, and only the second description of the nervous system within the larger clade Paucitubulitina within Gastrotricha [13]. The key phylogenetic position of *Diuronotus* in Gastrotricha, the newly discovered traits of the nervous, muscular and ciliary system, and the comparison to phylogenetically related lineages reported here lead to new hypotheses on nervous trait homologies and evolution. Our results provide an important step towards understanding the evolution of organ systems within Gastrotricha, and promise new insights for detailed anatomical studies of the major organ systems.

The musculature of *D. aspetos* presents unique traits for Paucitubulatina (reduction of the number of longitudinal muscles, the addition of dorso-ventral muscles) and previously undescribed muscles, showing that the musculature is more varied than expected in Paucitubulatina [22].

Although the nervous system of *D. aspetos* is similar to other gastrotrichs overall, detailed studies revealed numerous unknown minor components of the gastrotrich nervous system (e.g. one pair of anterior ventromedian nerves, one pair of dorso-posterior nerves, two pairs of additional ganglia). This indicates that similar detailed immunohistochemical studies on other species may similarly reveal new elements in the diversity of the gastrotrich nervous system and highlight additional homologies. This is supported by the multiple similarities uncovered from comparison with the immunohistochemistry studies of the closely related *Neodasys* and *Xenotrichula*, suggesting, e.g., homologous brain commissures and perikarya.

The pharynx displays an intriguing canal system and nerves, so far undescribed in Gastrotricha. Additionally, investigation of the ventral ciliation reveals details of the cellular arrangement which refines the previous description [17] and may show to be of broader systematic importance within Gastrotricha.

The many new discoveries in the different organ systems of Gastrotricha unravel an intriguing hidden diversity of morphological traits in Gastrotricha. This immediately stresses the importance of similar studies on additional key lineages of Paucitubulatina such as *Musellifer, Draculiciteria*, and marine *Aspidiophorus*, as well as freshwater "Chaetonotidae," in order to complement this picture and address the surprisingly complex evolution of gastrotrich morphology.

Acknowledgements

The Arctic Station of Qeqertarsuaq, University of Copenhagen provided an excellent working platform with cooling container and we are thankful to the crew of the station as well as R/V Porsild. We also thank Reinhardt Møbjerg Kristensen for help with collecting, and Christopher Laumer for his help on mining the transcriptome of *Diuronotus aspetos*, Michelle Jørgensen for her help with the sequencing, Alexandra Kerbl for her help on proofreading the manuscript, and Laetitia Carrive for her help with the phylogenetic analysis.

Fundings

The fieldwork on Greenland, the lab cost and the salary of the first author was supported by the Villum foundation (Grant no. 102544).

Authors' contributions

NB and KW conceptualized and designed the study, collected the animals, analyzed the data, and wrote the manuscript. NB gathered the immunohistochemical data and made the illustrations. All authors read and approved the final manuscript.

Competing interests
The authors declare that they have no competing interests.

References

1. Edgecombe GD, Giribet G, Dunn CW, Hejnol A, Kristensen RM, Neves RC, Rouse GW, Worsaae K, Sørensen MV. Higher-level metazoan relationships: recent progress and remaining questions. Organ Divers Evol. 2011;11(2):151–72.
2. Kieneke A, Schmidt-Rhaesa A. 1. Gastrotricha. In: Schmidt-Rhaesa A, Editor. Handbook of Zoology, Gastrotricha and Gnathifera. vol. 3. Berlin/Munich/Boston: De Gruyer; 2015.
3. Laumer CE, Bekkouche N, Kerb A, Goetz F, Neves RC, Sorensen MV, Kristensen RM, Hejno A, Dunn CW, Giribet G, et al. Spiralian phylogeny informs the evolution of microscopic lineages. Curr Biol. 2015;25(15):2000–6.
4. Hyman LH. The Invertebrates. Vol. 3, Acanthocephala, aschelminthes, and entoprocta : the pseudocoelomate bilateria, vol. VII. New York: McGraw-Hill Book Company; 1951. p. 572.
5. Bütschli O. Untersuchungen über freilebende Nematoden und die Gattung Chaetonotus. Zeitschrift für wissenschaftliche Zoologie. 1876; 363–413.
6. Rieger RM. Monociliated epidermal cells in Gastrotricha; significance for concepts of early metazoan evolution. Zeitschrift Zool Syst EvolForsch. 1976;14(3):198–226.
7. Zrzavý J, Mihulka S, Kepka P, Bezdek A, Tietz D. Phylogeny of the Metazoa based on morphological and 18S ribosomal DNA evidence. Cladistics. 1998;14(3):249–85.
8. Cavalier-Smith T. A revised six-kingdom system of life. Biol Rev Camb Philos Soc. 1998;73(3):203–66
9. Giribet G. Assembling the lophotrochozoan (=spiralian) tree of life. Philos Trans R Soc B Biol Sci. 2008;363(1496):1513–22.
10. Struck TH, Wey-Fabrizius AR, Golombek A, Hering L, Weigert A, Bleidorn C, Klebow S, Iakovenko N, Hausdorf B, Petersen M, et al. Platyzoan paraphyly based on phylogenomic data supports a noncoelomate ancestry of spiralia. Mol Biol Evol. 2014;31(7):1833–49.
11. Kieneke A, Arbizu PM, Riemann O. Body musculature of *Stylochaeta scirtetica* Brunson, 1950 and *Dasydytes* (*Setodytes*) *tongiorgii* (Balsamo, 1982) (Gastrotiricha : Dasydytidae): A functional approach. Zool Anz. 2008;247(2):147–58.
12. Rothe BH, Schmidt-Rhaesa A, Kieneke A. The nervous system of *Neodasys chaetonotoideus* (Gastrotricha: Neodasys) revealed by combining confocal laserscanning and transmission electron microscopy: evolutionary comparison of neuroanatomy within the Gastrotricha and basal Protostomia. Zoomorphology. 2011;130(1):51–84.
13. Rothe BH, Kieneke A, Schmidt-Rhaesa A. The nervous system of *Xenotrichula intermedia* and *X. velox* (Gastrotricha: Paucitubulatina) by means of immunohistochemistry (IHC) and TEM. Meiofauna Marina. 2011;19:71–88.
14. Todaro MA, Leasi F, Hochberg R. A new species, genus and family of marine Gastrotricha from Jamaica, with a phylogenetic analysis of Macrodasyida based on molecular data. Syst Biodivers. 2014;12(4):473–88.
15. Todaro MA, Dal Zotto M, Leasi F. An Integrated Morphological and Molecular Approach to the Description and Systematisation of a Novel Genus and Species of Macrodasyida (Gastrotricha). PLoS ONE. 2015; 10(7):e0130278.
16. Kolicka M, Dabert M, Dabert J, Kånneby T, Kisielewski J. *Bifidochaetus*, a new Arctic genus of freshwater Chaetonotida (Gastrotricha) from Spitsbergen revealed by an integrative taxonomic approach. Invertebrates Systematics. 2016. (in press).
17. Todaro MA, Balsamo M, Kristensen RM. A new genus of marine chaetonotids (Gastrotricha), with a description of two new species from Greenland and Denmark. J Mar Biol Assoc U K. 2005;85(6):1391–400.
18. Ruppert EE. Comparative ultrastructure of the gastrotrich pharynx and the evolution of myoepithelial foreguts in aschelminthes. Zoomorphology. 1982;99(3):181–220.
19. Ruppert EE. Introduction to the study of meiofauna. Washington, D.C: Smithsonian Institution Press; 1988. p. 302–11.
20. Ruppert EE. Gastrotricha. In: Harrison FW, Ruppert EE, editors. Microscopic anatomy of invertebrates Volume 4: Aschelminthes. New York, Chichester: Wiley-Liss; 1991. p. 41–109.
21. Kieneke A. Record of the 'Arctic' marine gastrotrich *Diuronotus aspetos* (Paucitubulatina) from the southern North Sea. Mar Biodivers. 2015; 45(4):615–6.
22. Leasi F, Todaro MA. The muscular system of *Musellifer delamarei* (Renaud-Mornant, 1968) and other chaetonotidans with implications for the phylogeny and systematization of the Paucitubulatina (Gastrotricha). Biol J Linn Soc. 2008;94(2):379–98.
23. Balsamo M, Guidi L, Ferraguti M, Pierboni L, Kristensen RM. *Diuronotus aspetos* (Gastrotricha): new morphological data and description of the spermatozoon. Helgol Mar Res. 2010;64(1):27–34.
24. D'Hondt JL. Gastrotricha. In: Barnes H, editor. Oceanography and Marine Biology: An Annual Review, vol. 9. London: George Allen and Unwin Ldt; 1971. p. 141–92.
25. Hochberg R, Litvaitis MK. Phylogeny of Gastrotricha: a morphology-based framework of gastrotrich relationships. Biol Bull. 2000;198(2): 299–305.
26. Kånneby T, Atherton S, Hochberg R. Two new species of *Musellifer* (Gastrotricha: Chaetonotida) from Florida and Tobago and the systematic placement of the genus within Paucitubulatina. Mar Biol Res. 2014;10(10):983–95.
27. Todaro MA, Telford MJ, Lockyer AE, Littlewood DTJ. Interrelationships of the Gastrotricha and their place among the Metazoa inferred from 18S rRNA genes. Zool Scr. 2006;35(3):251–9.
28. Paps J, Riutort M. Molecular phylogeny of the phylum Gastrotricha: New data brings together molecules and morphology. Mol Phylogenet Evol. 2012;63(1):208–12.
29. Kieneke A, Riemann O, Ahlrichs WH. Novel implications for the basal internal relationships of Gastrotricha revealed by an analysis of morphological characters. Zool Scr. 2008;37(4):429–60.
30. Hummon WD. *Musellifer sublitoralis* a New Genus and Species of Gastrotricha from San Juan Archipelago Washington. Trans Am Microsc Soc. 1969;88(2):282–6.
31. Hochberg R, Litvaitis MK. The musculature of *Draculiciteria tessalata* (Chaetonotida, Paucitubulatina): implications for the evolution of dorsoventral muscles in Gastrotricha. Hydrobiologia. 2001;452(1-3):155–61.
32. Leasi F, Rothe BH, Schmidt-Rhaesa A, Todaro MA. The musculature of three species of gastrotrichs surveyed with confocal laser scanning microscopy (CLSM). Acta Zool. 2006;87(3):171–80.
33. Kieneke A, Ostmann A. Structure, function and evolution of somatic musculature in Dasydytidae (Paucitubulatina, Gastrotricha). Zoomorphology. 2012;131(2):95–114.
34. Hochberg R, Litvaitis MK. A muscular double helix in gastrotricha. Zool Anz. 2001;240(1):61–8.
35. Hochberg R, Litvaitis MK. Ultrastructural and immunocytochemical observations of the nervous systems of three macrodasyidan gastrotrichs. Acta Zool. 2003;84(3):171–8.
36. Rothe BH, Schmidt-Rhaesa A. Architecture of the nervous system in two *Dactylopodola* species (Gastrotricha, Macrodasyida). Zoomorphology. 2009; 128(3):227–46.
37. Wiedermann A. On the ultrastructure of the nervous system in *Cephalodasys maximus* (Macrodasyida, Gastrotricha). Zur Ultrastruktur des Nervensystems bei *Cephalodasys maximus* (Macrodasyida, Gastrotricha). Microfauna Marina. 1995;10:173–233.
38. Rothe BH, Schmidt-Rhaesa A. *Oregodasys cirratus*, a new species of Gastrotricha (Macrodasyida) from Tenerife (Canary Islands), with a description of the muscular and nervous system. Meiofauna Marina. 2010;18:49–66.
39. Hochberg R. Comparative immunohistochemistry of the cerebral ganglion in Gastrotricha: an analysis of FMRFamide-like immunoreactivity in *Neodasys cirritus* (Chaetonotida), *Xenodasys riedli* and *Turbanella* cf. *hyalina* (Macrodasyida). Zoomorphology. 2007;126(4):245–64.
40. Schöpfer-Sterrer C. *Chordodasys riedli* gen. nov. spec. nov. a macrodasyoid gastrotrich with a chordoid organ. Cah Biol Mar. 1969;10(4):391–404.
41. Kieneke A, Ahlrichs WH, Arbizu PM, Bartolomaeus T. Ultrastructure of protonephridia in *Xenotrichula carolinensis syltensis* and *Chaetonotus maximus* (Gastrotricha : Chaetonotida): comparative evaluation of the gastrotrich excretory organs. Zoomorphology. 2008;127(1):1–20.
42. Kieneke A, Hochberg R. Ultrastructural observations of the protonephridia of *Polymerurus nodicaudus* (Gastrotricha: Paucitubulatina). Acta Zool. 2012; 93(1):115–24.
43. Meyer CP. Molecular systematics of cowries (Gastropoda: Cypraeidae) and diversification patterns in the tropics. Biol J Linn Soc. 2003;79(3):401–59.
44. Giribet G, Carranza S, Baguna J, Riutort M, Ribera C. First molecular evidence for the existence of a Tardigrada + Arthropoda clade. Mol Biol Evol. 1996;13(1):76–84.

45. Brown S, Rouse G, Hutchings P, Colgan D. Assessing the usefulness of histone H3, U2 snRNA and 28S rDNA in analyses of polychaete relationships. Aust J Zool. 1999;47(5):499–516.
46. Vonnemann V, Schrodl M, Klussmann-Kolb A, Wagele H. Reconstruction of the phylogeny of the Opisthobranchia (Mollusca : Gastropoda) by means of 18S and 28S rRNA gene sequences. J Molluscan Stud. 2005;71:113–25.
47. Hall TA. BioEdit: a user-friendly biological sequence alignment editor and analysis program for Windows 95/98/NT. Nucleic Acids Symp Ser. 1999;41:95–8.
48. Geer LY, Marchler-Bauer A, Geer RC, Han L, He J, He S, Liu C, Shi W, Bryant SH. The NCBI BioSystems database. Nucleic Acids Res. 2010;38(Database issue):D492–6.
49. Kånneby T, Todaro MA, Jondelius U. Phylogeny of Chaetonotidae and other Paucitubulatina (Gastrotricha: Chaetonotida) and the colonization of aquatic ecosystems. Zool Scr. 2013;42(1):88–105.
50. Kånneby T, Todaro MA. The phylogenetic position of Neogosseidae (Gastrotricha: Chaetonotida) and the origin of planktonic Gastrotricha. Organ Divers Evol. 2015;15(3):459–69.
51. Todaro MA, Kånneby T, Dal Zotto M, Jondelius U. Phylogeny of thaumastodermatidae (Gastrotricha: Macrodasyida) inferred from nuclear and mitochondrial sequence data. PLoS ONE. 2011;6(3):e17892. 17891-17813.
52. Vaidya G, Lohman DJ, Meier R. SequenceMatrix: concatenation software for the fast assembly of multi-gene datasets with character set and codon information. Cladistics. 2011;27(2):171–80.
53. Huelsenbeck JP, Ronquist F. MRBAYES: Bayesian inference of phylogeny. Bioinformatics. 2001;17:754–5.
54. Rambaut A, Suchard MA, Xie D, Drummond AJ. 2014. Tracer v1.6, Available from http://beast.bio.ed.ac.uk/Tracer. Accessed 23 Aug 2016
55. Hummon WD, Balsamo M, Todaro MA. Italian Marine Gastrotricha 1. 6 New and One Redescribed Species of Chaetonotida. Bollettino Di Zoologia. 1992; 59(4):499–516.
56. Renaud-Mornant J. Présence du genre *Polymerurus* en milieu marin, description de deux espèces nouvelles (Gastrotricha, Chaetonotoidae). Pubblicazioni della Stazione Zoologica di Napoli. 1968;36:141–51.
57. Remane A. Die Gastrotrichen des Küstengrundwassers von Schilksee. Schriften des Naturwissenschaftlichen Vereins für Schleswig-Holstein. 1934; 20(2):473–8.
58. Wilke U. Mediterrane Gastrotrichen. Zoologische Jahrbücher. 1954;82(6): 497–654.
59. Hochberg R. Musculature of the primitive gastrotrich *Neodasys* (Chaetonotida): functional adaptations to the interstitial environment and phylogenetic significance. Marine Biology (Berlin). 2005;146(2):315–23.
60. Remane A. *Neodasys uchdai* nov. spec., eine zwiete *Neodasys* Art (Gastrotrich Chaetoidea). Kiel Meeresforsch. 1961;17:85–8.
61. Riedl R. On the Genus *Gnathostomula* (Gnathostomulida). Int Rev Hydrobiol. 1971;56(3):385–496.
62. Kristeuer E. On some species of Gnathostomulida from Bimini, Bahamas, vol. 2356. 1969. p. 21.
63. Müller MCM, Sterrer W. Musculature and nervous system of *Gnathostomula peregrina* (Gnathostomulida) shown by phalloidin labeling, immunohistochemistry, and cLSM, and their phylogenetic significance. Zoomorphology (Berlin). 2004;123(3):169–77.
64. Tyler S, Hooge MD. Musculature of *Gnathostomula armata* Riedl 1971 and its ecological significance. Mar Ecol. 2001;22(1-2):71–83.
65. Leasi F, Todaro MA. Meiofaunal cryptic species revealed by confocal microscopy: the case of *Xenotrichula intermedia* (Gastrotricha). Mar Biol. 2009;156(6):1335–46.
66. Hochberg R, Litvaitis MK. The muscular system of *Dactylopodola baltica* and other macrodasyidan gastrotrichs in a functional and phylogenetic perspective. Zool Scr. 2001;30(4):325–36.
67. Remane A. Morphologie und verwandtschaftsbeziehungen der aberranten gastrotrichen I. Z Morphol Okol Tiere. 1926;5(4):625–754.
68. Hochberg R, Litvaitis MK. Functional morphology of muscles in *Tetranchyroderma papii* (Gastrotricha). Zoomorphology. 2001;121(1):37–43.
69. Remane A. *Xenotrichula velox* nov. gen. nov. spec., ein chaetonotoides Gastrotrich mit männlichen Geschlechtsorganen. Zool Anz. 1927;71:289–94.
70. Hochberg R, Atherton S. A new species of *Lepidodasys* (Gastrotricha, Macrodasyida) from Panama with a description of its peptidergic nervous system using CLSM, anti-FMRFamide and anti-SCPB. Zool Anz. 2011;250(2):111–22.
71. Rothe BH, Schmidt-Rhaesa A. Variation in the nervous system in three species of the genus *Turbanella* (Gastrotricha, Macrodasyida) Meiofauna Marina. 2008;16:175–84.
72. Remane A. Neue aberrante Gastrotrichen II: *Turbanella cornuta* n. sp. und *T. hyalina* M. Schultze, 1853. Zool Anz. 1925;64:309–14.
73. Teuchert G. The ultrastructure of the marine gastrotrich *Turbanella cornuta* Remane (Macrodasyoidea) and its functional and phylogenetical importance. Zoomorphologie. 1977;88(3):189–246,illust.
74. Remane A. Beiträge zur Systematik der Süsswassergastrotrichen. Zoologische Jahrbücher (Abteilung für Systematik, Ökologie, und Geographie der Tiere). 1927;53:269–320.
75. Schultze M. Über *Chaetonotus* und *Ichthydium* (Ehrb.) und eine neue verwandte Gattung *Turbanella*. Müller's Archiv für Anatomie und Physiologie. 1853;6:241–54.
76. Joffe BI, Wikgren M. Immunocytochemical Distribution of 5-Ht (Serotonin) in the Nervous-System of the Gastrotrich *Turbanella cornuta*. Acta Zool. 1995;76(1):7–9.
77. Schmidt-Rhaesa A, Rothe BH. Gastrotricha. In: Schmidt-Rhaesa A, Harzsch S, Purschke G, editors. Structure and Evolution of Invertebrate Nervous Systems. New York: Oxford University Press; 2016. p. 141–7.
78. Brusca RC, Brusca GJ. Invertebrates, vol. i-xx. 2nd ed. Sunderland: Sinauer Associates, Inc; 2003. p. 1--936.
79. Schmidt-Rhaesa A. The nervous system of *Nectonema munidae* and *Gordius aquaticus*, with implications for the ground pattern of the Nematomorpha. Zoomorphology (Berlin). 1996;116(3):133–42.
80. Schmidt-Rhaesa A. The evolution of organ systems, vol. i-x. Oxford & New York: Oxford University Press; 2007. p. 1–385.
81. Worsaae K, Rouse GW. Is Diurodrilus an annelid? J Morphol. 2008;269(12): 1426–55.
82. Worsaae K, Sterrer W, Iliffe TM. *Longipalpa saltatrix*, a new genus and species of the meiofaunal family Nerillidae (Annelida: Polychaeta) from an anchihaline cave in Bermuda. Proc Biol Soc Wash. 2004;117(3):346–62.
83. Villora-Moreno S. Ecology and distribution of the Diurodrilidae (Polychaeta), with redescription of *Diurodrilus benazzii*. Cah Biol Mar. 1996;37(1):99–108.
84. Todaro AM, Perissinotto R, Bownes SJ. Neogosseidae (Gastrotricha, Chaetonotida) from the iSimangaliso Wetland Park, KwaZulu-Natal, South Africa. ZooKeys. 2013;315:77–94.
85. Kisielewksi J. New records of marine Gastrotricha from the French coasts of Manche and Atlantic. 2. Chaetonotida, with descriptions of four new species. Cah Biol Mar. 1988;29(2):187–213.
86. Kisielewski J. Two new interesting genera of Gastrotricha (Macrodasyida and Chaetonotida) from the Brazilian freshwater psammon. Hydrobiologia. 1987;153(1):23–30.
87. Mock H. Chaetonotoidea (Gastrotricha) der Nordseeinsel Sylt, Akademie Der Wissenschaften Und Der Literatur Mainz Mathematisch-Naturwissenschaftlichen Klasse Mikrofauna Des Meeresbodens. 1979. p. 1–107.
88. Ehrenberg H. Uber die Entwickelung und Lebensdauer der Infusionsthiere; nebst ferneren Beitragen zu einer Vergleichung ihrer organischen Systeme, Abhandlungen der Königlichen Akademie der Wissenschaften zu Berlin. 1831. p. 1–154.
89. Voigt M. Uber einige bisher unbekannte Siisswasserorganismen. Zool Anz. 1901;24:191–5.
90. Remane A. *Mesodasys*, ein neues Genus der Gastrotricha Macrodasyoidea aus der Kieler Bucht. Kiel Meeresforsch. 1951;8:102–5.
91. Neuhaus B. Ultrastructure of the protonephridia in *Dactylopodola baltica* and *Mesodasys laticaudatus* (Macrodasyda): implications for the ground pattern of the Gastrotricha. Microfauna Marina. 1987;3:419–38.

Measurement of refractive indices of tunicates' tunics: light reflection of the transparent integuments in an ascidian *Rhopalaea* sp. and a salp *Thetys vagina*

Hiroshi Kakiuchida[1], Daisuke Sakai[2], Jun Nishikawa[3] and Euichi Hirose[4*]

Abstract

Background: Tunic is a cellulosic, integumentary matrix found in tunicates (Subphylum Tunicata or Urochordata). The tunics of some ascidian species and pelagic tunicates, such as salps, are nearly transparent, which is useful in predator avoidance. Transparent materials can be detected visually using light reflected from their surfaces, with the different refractive indices between two media, i.e., tunic and seawater, being the measure of reflectance. A larger difference in refractive indices thus provides a larger measure of reflectance.

Results: We measured the refractive indices of the transparent tunic of *Thetys vagina* (salp: Thaliacea) and *Rhopalaea* sp. (ascidian: Ascidiacea) using an Abbe refractometer and an ellipsometer to estimate the light reflection at the tunic surface and evaluate the anti-reflection effect of the nipple array structure on the tunic surface of *T. vagina*. At D-line light ($\lambda = 589$ nm), the refractive indices of the tunics were 0.002–0.004 greater than seawater in the measurements by Abbe refractometer, and 0.02–0.03 greater than seawater in the measurements by ellipsometer. The refractive indices of tunics were slightly higher than that of seawater. According to the simulation of light reflection based on rigorous coupled wave analysis (RCWA), light at a large angle of incidence will be completely reflected from a surface when its refractive indices are smaller than seawater. Therefore, the refractive index of integument is important for enabling transparent organisms to remain invisible in the water column.

Conclusion: In order to minimize reflectance, the refractive index should be similar to, but never smaller than, that of the surrounding seawater. The simulation also indicated that the presence or absence of a nipple array does not cause significant difference in reflectance on the surface. The nipple array on the tunic of the diurnal salp may have another function, such as bubble repellence, other than anti-reflection.

Keywords: Tunic, Transparency, Refractive index, Refractometry, Ellipsometry, Rigorous coupled wave analysis (RCWA)

Background

The subphylum Tunicata (= Urochordata) is the sister group of Vertebrata and includes three classes: Ascidiacea, which comprises only sessile species, and Thaliacea and Appendicularia, which are always pelagic. Salpida is an order of the class Thaliacea, and salps are soft-bodied animals apparently lacking effective defense organs. Instead, they are often functionally invisible in the water column due to their transparency, helping them to avoid visual predators. However, their body contours can sometimes be recognized by light reflecting from the body surface. This may indicate that reducing reflection is important for the survival of salps, especially those occurring in epipelagic layers. Some salps accomplish this by dwelling in relatively deeper (i.e., darker) regions during the day and migrating to shallower regions only at night (known as diel vertical migration) [1, 2], but some species remain at a shallow depth throughout the day [3]. Some of these diurnal species, including *Thalia rhomboides* and *Thetys vagina*, have an array of nipple-like protuberances ≤ 50 nm in height on the cuticle surface of their outer

* Correspondence: euichi@sci.u-ryukyu.ac.jp
[4]Faculty of Science, University of the Ryukyus, Nishihara, Okinawa 903-0213, Japan
Full list of author information is available at the end of the article

integument, or tunic [4, 5]. Similar structures are well known in other invertebrate species, such as the corneal nipple array on the surface of the compound eyes of moths, which reduces light reflection [6, 7]. Simulation of light reflection using rigorous coupled wave analysis (RCWA) has indicated that the reflection on the salp nipple array is slightly lower than that on the flat surface at high incidence angles ($\theta > 80°$), suggesting an anti-reflection function of the surface nanostructures [5].

Light reflection occurs at the boundary between two optical media with different refractive indices, but the refractive index of the tunic structures in tunicates is not known. Since the water content of salps is 95% or higher [8], the refractive index should not be considerably different from that of seawater. In the simulation described above, the refractive index of the tunic was assumed to be 0.1 larger than that of seawater, but a tested value rather than an assumption would be essential for producing an accurate simulation of salp light reflectance.

Among the metazoans, the ability of cellulose synthesis is exclusively found in the members of the subphylum Tunicata (or Urochordata) in which epidermal cells secrete cellulose microfibrils to form the tunic in tunicates, including salps [4], except for appendicularians, which form a cellulosic feeding-apparatus called a "house" [9]. Consequently, ascidians and thaliaceans have very unique integument, i.e., tunic, that contains cellulosic matrix and free cells outside the epidermis [10–12]. Although hardness, cellulose content, and water content vary among tunicate species [13–15], salp tunic is usually gelatinous and transparent. The refractive indices of these cellulosic matrices have not been reported to date.

In the present study, we measured the refractive indices of the transparent tunic of a salp, *Thetys vagina*, and an ascidian, *Rhopalaea* sp., using an Abbe refractometer and an ellipsometer. For *T. vagina*, only frozen specimens were examined; for *Rhopalaea* sp., both fresh and frozen materials were compared. The tunics of both species appeared completely transparent, and they are free from debris and epibionts that alter the optical properties in natural condition. We also measured the absorption spectra of the transparent tunic specimens to consider the tunic function as protection against ultraviolet (UV) radiation in shallow waters. Using this information, we proposed a surface structure model of *T. vagina* tunic based on ultrastructural observation. Simulation of light reflectance on the tunic surface with or without nipple array structures was carried out based on the estimated refractive indices to elucidate whether the array provides effective reduction of light reflection.

Methods

Animals

A solitary form of the salp *Thetys vagina* (body length = 16 cm) was collected at Sunakohama port, Otsuchi Bay, Iwate, Japan using a scoop net. A second specimen of the same species was collected using SCUBA at 35 m depth off Cape Zanpa, Okinawajima Island, Okinawa, Japan (Fig. 1a). These specimens were transferred to the laboratory and stored at −20 °C without any cryoprotectants. The specimens were then thawed at room temperature, and 2 × 2 cm sheets of tunic of 3 mm depth were cut from the middle part of the bodies with a razor. The attached mantle tissue below the tunic was removed using forceps. The resulting tunic sample was transparent and often had a warty surface. A flat area was selected for measurement.

Fifteen individuals of the ascidian *Rhopalaea* sp. (ca. 10 cm in length) were collected using SCUBA in Kin Bay (Okinawajima Island, Okinawa, Japan) at approximately 3 m depth (Fig. 1b). Further specimens for electron microscopy were collected off Sunabe (Okinawajima Island, Okinawa, Japan) at approximately 15 m depth. These were reared in seawater or artificial seawater (Marine Art SF-1: Osakayakken Co., Ltd., Minou, Osaka, Japan) for 1–2 days prior to use. Sheets of tunic 2 × 2 cm and 1 mm thick were cut from fresh specimens and the mantle tissue below the tunic was removed with forceps, yielding clear tunic material with a flat surface. Four of these samples were frozen at −20 °C and thawed at room temperature before the measurement.

All procedures conducted in this study comply with the local and institutional ethical guidelines.

Fig. 1 a *Thetys vagina* collected near Cape Zanpa (photo: James-Davis Reimer). **b** *Rhopalaea* sp. found in Kin Bay (photo: Ryosuke Uozumi)

Absorption spectra of tunics

The tunics of *T. vagina* (from a previously frozen specimen) and *Rhopalaea* sp. (from a fresh specimen) were each attached to the wall of a quartz cell, and the 280–800 nm absorption spectra were recorded at 2-nm intervals with a U-4100 spectrophotometer (Hitachi, Tokyo, Japan). As reference for a tunic containing UV-absorbing substances, we used the absorption spectra of *Diplosoma virens*, a colonial ascidian containing symbiotic cyanobacteria measured in a previous study by our group [16].

Refractometry with an Abbe refractometer

Refractive indices of the tunic specimens were measured using an Abbe refractometer (NAR-1 T SOLID: Atago Co., Ltd., Tokyo, Japan) at 24.5–25.0 °C with LED light approximating to the wavelength of D-line (ca. 589 nm). This wavelength originated from the emission spectrum of sodium and has been conventionally used for the measurement of refractive indices as a standard. According to the manufacturer's specifications, the measurement accuracy of this refractometer is ±0.0002 (nD). The outer surface of the tunic, i.e., cuticular side, faced the prism of the refractometer. Refractive indices (nD) of seawater and artificial seawater were also measured at 25 °C, and found to be 1.3392–1.3394. The measured values of these tunic specimens and seawater were compared with Steel-Dwass test using Satcel4 (OMS Publishing Inc., Tokorozawa, Japan).

Ellipsometry

We examined the tunics of *T. vagina* (previously frozen specimens) and *Rhopalaea* sp. (fresh specimens) by ellipsometric measurement. The wavelength dispersion of refractive index was determined in the range between 380 and 1700 nm, using a spectroscopic ellipsometer (M2000, J. A. Woolam Co. Inc., Lincoln, NE, USA). Focusing probes were installed in this ellipsometric system to find a flat surface of the specimens placed on a flat plate. The measurements were carried out at different angles of incidence from 50 to 80° at 5° intervals. The thickness of these samples was large enough to avoid stray light from backside reflection through the transparent body of the specimens. The ellipsometric parameters Ψ and Δ were measured and were analyzed assuming Cauchy expansion [17].

Electron microscopy

Tunic sheets of the thawed *T. vagina* tunic specimen and the fresh *Rhopalaea* sp. tunic specimen were fixed with 2.5% glutaraldehyde (16320-P; Electron Microscopy Sciences, Hatfield, PA, USA) in a 0.45 M sucrose and 0.1 M sodium cacodylate (S008; TAAB Laboratories, Berks, England) buffer (pH 7.5) at room temperature and stored at 4 ° C. The specimens were briefly rinsed with the same buffer, post-fixed with 1% osmium tetroxide (124505; Merck, Darmstadt, Germany) in a 0.1 M cacodylate buffer (pH 7.5) at 4 °C for 1.5 h, and dehydrated through an ethanol series. Specimens were then embedded in an epoxy resin (T024; TAAB Laboratories) for transmission electron microscopy (TEM). Sections of 0.1 μm thick were stained with uranyl acetate and lead citrate and examined under TEM (JEM-1011; JEOL, Tokyo, Japan) at 80 kV. The heights and intervals of the cuticular protuberances were measured from electron micrographs.

Simulation of light reflection with and without a nipple array

The light reflection was calculated at the border between the medium (seawater) and matrix (tunic) with or without the nipple array with rigorous coupled wave analysis (RCWA) using DiffractMOD3.2 software (RSoft Design Group, Inc., Ossining, NY, USA). The differences between refractive indices between the tunic and seawater were assumed to be −0.04, −0.02, 0.02, and 0.04, according to estimates of refractive indices measured with the refractometer and the ellipsometer. Polarized light was used for this simulation, i.e., transverse electric wave (TE wave) and transverse magnetic wave (TM wave). Based on the electron microscope observations, the modeling of the nipple array of *T. vagina* was assumed to be hemispheres (60 nm in radius) in which the distance between adjacent nipples' apices was 140 nm. Parameters for the simulation included the wavelength of light (λ: 633 nm) and the angle of incidence of light (θ: 0–90°).

Results

Absorption spectra of the tunics

In *T. vagina* and *Rhopalaea* sp., the absorption spectra of the tunic showed no specific absorption peaks in the range of visible light (400–760 nm), UV-A (325–400 nm), or UV-B (280–315 nm) (Fig. 2). In *D. virens* tunic containing UV-absorbing substances, the absorption spectra had a prominent absorption peak at around 320 nm (Fig. 2).

Measurement by an Abbe refractometer

When tunic samples of *Rhopalaea* sp. were set in the refractometer, upper and lower boundaries of the refractive index were observed. The refractive indices of the upper boundary ranged from 1.3384 to 1.3408 ($n = 15$, mean = 1.3394, SD = 0.0006), similar to those measured in seawater (1.3392–1.3394). The refractive indices of the lower boundaries were approximately 1.3435, yielding a difference 0.004 between of the upper and lower boundary (Table 1). When *T. vagina* samples were measured, only one boundary was observed. This was comparable to those of the lower boundary in the measurement of *Rhopalaea* sp., and was measured at approximately 1.3416, or 0.002, which differed from the refractive index of seawater (1.3393) (Table 1). As shown in Additional file 1, there were no significant differences among the measured values of the tunic specimens (i.e., fresh and frozen specimens of *Rhopalaea* sp. and frozen

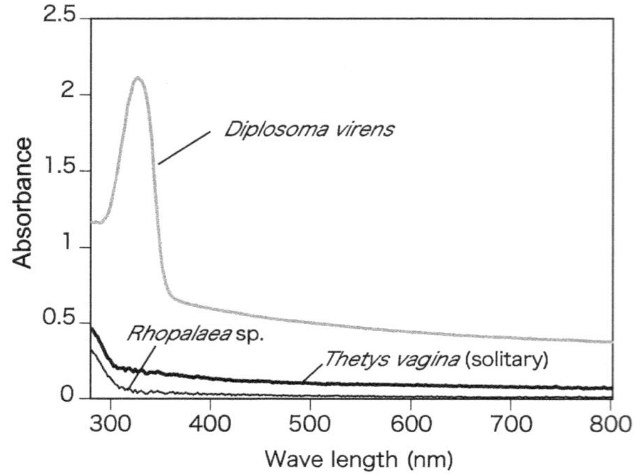

Fig. 2 Absorption spectra of the tunic of *Thetys vagina* (*black thick line*), *Rhopalaea* sp. (*black thin line*), and *Diplosoma virens* (*grey line*). The spectra of *T. vagina* and *Rhopalaea* sp. have no prominent peaks, whereas the spectrum of *D. virens* has an absorption peak due to UV-absorbing substances

specimens of *T. vagina*), whereas these values were significantly larger than the values of the upper boundary, i.e., seawater (Steel-Dwass test: $P < 0.01$ for seawater– fresh *Rhopalaea* sp., $P < 0.05$ for seawater – frozen *Rhopalaea* sp. and seawater – frozen *T. vagina*).

Measurement by an ellipsometer

Ellipsometric parameters Ψ and Δ of *T. vagina* tunic specimens (previously frozen) and *Rhopalaea* sp. (fresh specimens) were obtained at angles of incidence from 50 to 80° at 5° intervals (Additional files 2 and 3). Refractive index (n) was calculated with the constraint of Cauchy expansion, and the values were determined as expressed by the equations below.

$$Rhopalaea\,\text{sp.}: \; n = (1.3447 \pm 0.0012) + (6525.1 \pm 388.0)/\lambda^2$$

$$Thetys\,vagina: \; n = (1.3563 \pm 0.0021) + (4013.4 \pm 588.0)/\lambda^2$$

where the coefficient errors were obtained by the fitting calculations, and are plotted as a function of wavelength (λ) in Fig. 3a. The optical structure of the surface was modeled as a homogeneous bulk for the calculation, and consequently the simplest model was most likely valid because the mean squared error was < 10 in the fitting calculations between the measured and theoretical data in the ellipsometric parameters. At a wavelength of 589 nm (D-line), the refractive indices were estimated as 1.364 in *Rhopalaea* sp. (fresh specimen) and 1.368 in *T. vagina* tunic (previously frozen specimen).

Table 1 Refractive indices of the tunic measured by an Abbe refractometer

Specimens	Number	Refractive indices: Average (± SD)		
		Upper boundary	Lower boundary	Difference between the boundaries
Rhopalaea sp. (fresh)	11	1.3393 (±0.0006)	1.3435 (±0.0017)	0.0042 (±0.0016)
Rhopalaea sp. (frozen–thawed)	4	1.3396 (±0.0002)	1.3434 (±0.0003)	0.0038 (±0.0002)
Thetys vagina (frozen–thawed)	4	Not available	1.3416 (±0.0011)	0.0023 (±0.0011)[a]

[a]Difference of refractive indices between the boundary and the seawater (nD = 1.3393)

Fine structures of tunic surface

In *T. vagina*, the tunic matrix consisting of fuzzy materials was entirely covered with the tunic cuticle, an electron-dense layer approximately 0.2 μm thick (Fig. 4a). There was an array of cone-shaped or hemispherical protuberances on the cuticle surface (Fig. 4a). The height of these protuberances was 0.057 ± 0.011 μm (average ± SD, $n = 18$) and the distance between the apices of adjacent protuberances was 0.138 ± 0.045 μm (average ± SD, $n = 17$). Based on this observation, we hypothesize the surface structure model of the *T. vagina* tunic as hemispheres (60 nm in radius) arranged in 140-nm intervals for the simulation below (Fig. 4b). The tunic of *Rhopalaea* sp. had a very thin cuticular layer (<40 nm thick) that was composed of electron-dense, fibrous materials (Fig. 4c). The cuticular layer was often frayed, suggesting that the layer was not structurally firm. Therefore, the cuticular surface was not always flat, likely due to damage, but there were no regular repetitive structures, such as nipple arrays, on the cuticular surface.

Simulation of light reflection on the tunic surface with and without a nipple array

The ellipsometric measurement indicated that the refractive indices of the tunic are quite similar to seawater, but we assumed that rigorous values could not be determined probably because of the structural unevenness of the tunics. Here, we used four sets of the refractive indices of the tunic, i.e., ± 0.02 and 0.04 of the refractive index of seawater, to simulate the light reflection on the *T. vagina* tunic. The range of these estimates are somewhat larger than the measurement by refractometry (0.002–0.004 at D-line, Table 1) and ellipsometry (0.02–0.03 at D-line, Fig. 3b), in order to account for potential errors in measurement and fluctuations of natural conditions such as temperature and salinity. In the RCWA analysis, the reflectance against incident angle resulted in J-shaped curves (Fig. 5). The reflectance curves were nearly identical, irrespective of the presence or absence of the nipple array in the all sets of the refractive indices, while the curves were quite different among the different refractive indices (Fig. 5a, b). The reflectance curves for TE wave were very similar to those of the same refractive index for TM wave. When the refractive indices were smaller than seawater (i.e., −0.02 and −0.04),

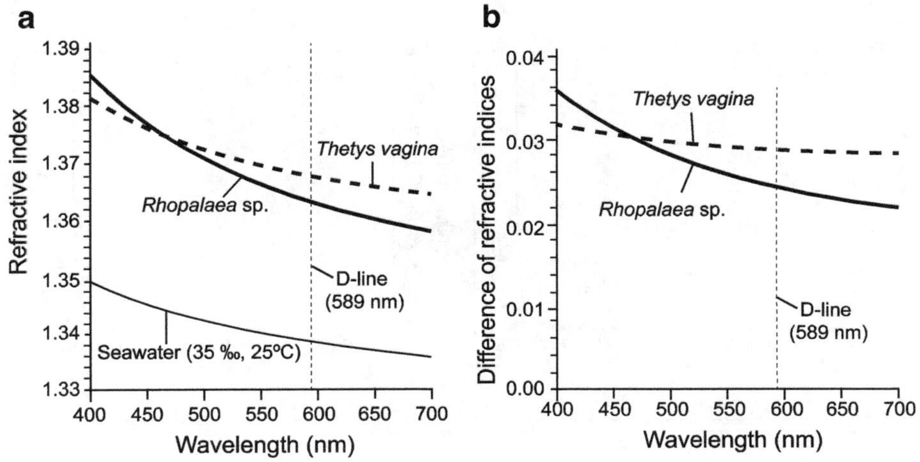

Fig. 3 Refractive indices of tunics and seawater (**a**), and the difference between the two, in the visual light range (400–700 nm) (**b**). Refractive indices of the tunic of *Thetys vagina* (previously frozen specimen) and *Rhopalaea* sp. (fresh specimen) were calculated from ellipsometric parameters Ψ and Δ (Additional files 2 and 3). The *curve* for seawater is based on the equation of Quan and Fry [22] that is valid from 400 to 700 nm. The *vertical dotted line* indicates the wavelength of D-line that was used for the measurement with the Abbe refractometer

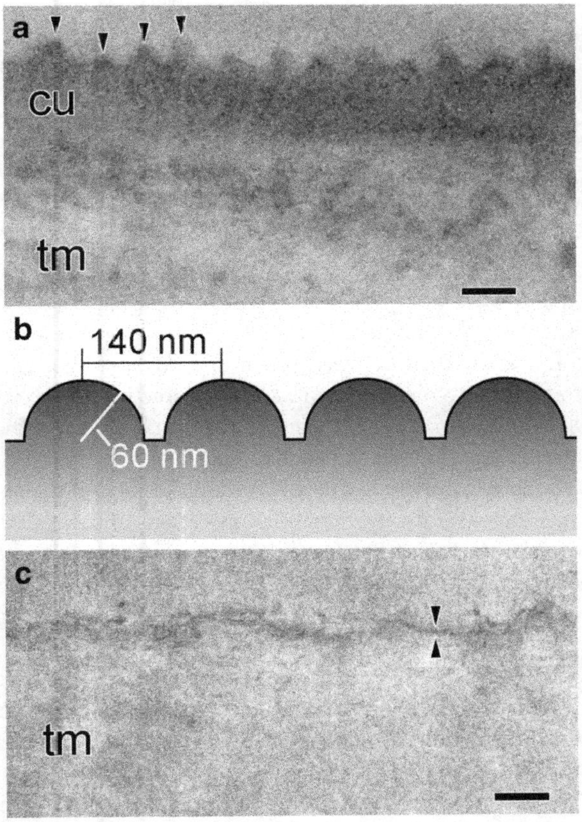

Fig. 4 Fine surface structure of *Thetys vagina* tunic (**a** TEM; **b** model for the simulation of light reflectance) and *Rhopalaea* sp. tunic (**c** TEM). *Arrowheads* in **a** indicate cuticular protuberances, and facing *arrowheads* in **c** indicate the cuticular layer. cu, cuticular layer; tm, tunic matrix. Scale bars in **a** and **c** indicate 0.2 μm

light was completely reflected at angles larger than 76° and 80.1°, respectively. In contrast, when the refractive indices were larger than seawater (i.e., 0.02 and 0.04), complete reflection did not occur at angles below 90°, with 95% reflection occurring at about 89.8°. In the enlargement of the reflection curves (Fig. 5c, d), reflectance on the nipple array was slightly smaller than reflectance on the flat surface. In reflection of TM wave, the reflectance curve had a minimum at around 45° as a result of Brewster's angle.

Discussion

Transparency of the tunics

The apparent transparency of the tunic in *T. vagina* and *Rhopalaea* sp. is consistent with the absorption spectra of these species that have no prominent absorption peaks in the range of 280–800 nm (Fig. 2). The tunic of another diurnal salp, *Thalia rhomboides*, has a similar absorption spectrum lacking prominent peaks [5]. These tunics are therefore transparent to many potential predators, including fish with eyesight sensitive to UV as well as visible light [18]. However, this also indicates that the tunic of these species does not protect inner tissues from UV radiation. As *Thetys vagina* and *Thalia rhomboides* are diurnal salps and *Rhopalaea* sp. is sessile in shallow coral reefs, where there is considerable UV penetration [19], this exposure could be harmful. This may explain the green pigmentation in the alimentary tract of *T. vagina.*

Estimation of refractive indices of the tunics

In the measurement of *Rhopalaea* sp. using an Abbe refractometer, two boundaries were observed, with the refractive indices of the upper boundaries almost equal to those of the seawater or artificial seawater in which

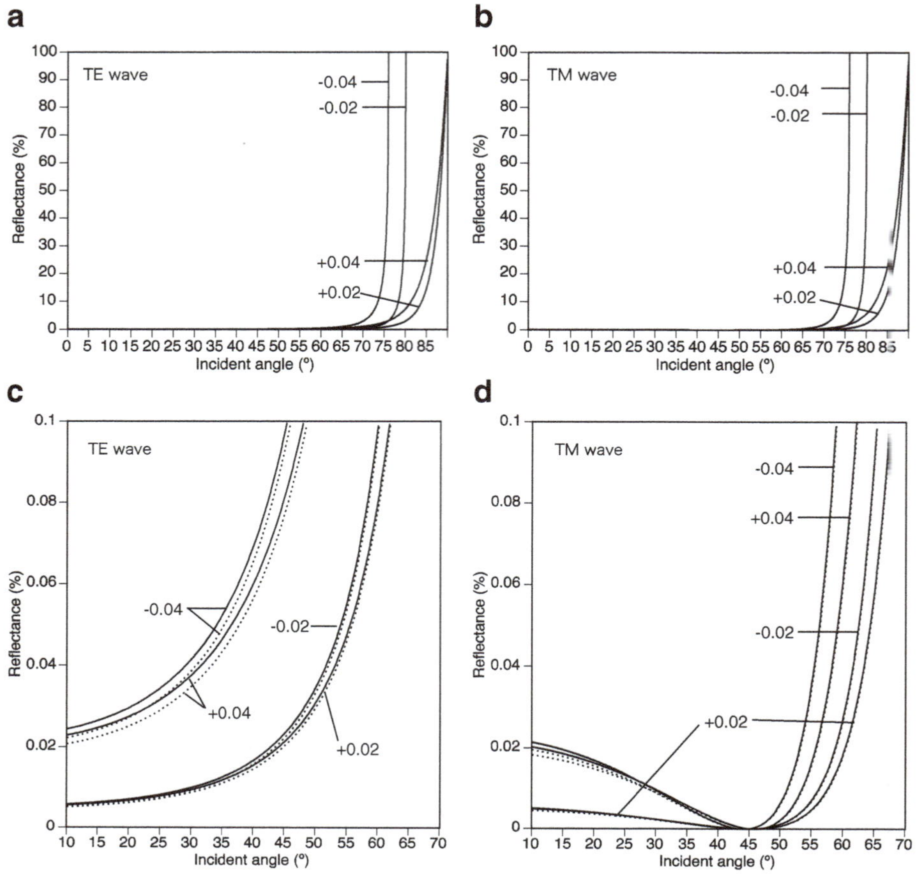

Fig. 5 Reflectance (%) of 633 nm light [TE wave (**a**, **c**) and TM wave (**b**, **d**)] on the flat surface (*solid line*) and on the model of *Thetys vagina* tunic surface (*dotted line*). The difference in refractive index between the tunic and medium (seawater) were assumed to be −0.04, −0.02, 0.02, and 0.04. **c** and **d** show enlargements of **a** and **b**, respectively

Rhopalaea sp. individuals were incubated. Therefore, we assumed that the upper boundary indicates the refractive index of liquid between the tunic and the prism, with this liquid possibly composed of seawater and exudate from the tunic. When the refractive index of the lower boundary is assumed to be that of the tunic surface, the difference of the refractive indices between seawater and tunic is approximately 0.004 (D-line, $\lambda = 589$ nm). Freezing did not cause considerable change of the refractive index of the tunic in *Rhopalaea* sp. In the measurement of *T. vagina* tunic, the refractive index of the boundary was approximately 1.3416. When this value is assumed to represent the refractive index of the tunic, the difference of the refractive indices between seawater and tunic is approximately 0.002 (D-line, $\lambda = 589$ nm).

At the 589 nm wavelength, refractive indices obtained from the ellipsometric analysis were 1.364 in *Rhopalaea* sp. and 1.368 in *T. vagina*. These values are approximately 0.03 larger than the values measured by the Abbe refractometer. We posit that these deviations are the result of evaporation, as samples were exposed to air for approximately one hour during the ellipsometric measurement, while samples were tightly sandwiched between prism and cover during Abbe refractometer measurement. The refractive index of cellulose, the main component of the tunic, is larger than seawater, e.g., 1.4701 at 589 nm light [20], and thus it is reasonable to conclude that the refractive index of the specimen increases during surface evaporation. In some cases, the ellipsometric measurements were unsuccessful when specimens became overly dry. This highlights the technical challenge of examining biological specimens containing water in a cellulose matrix, as water content depends on ambient conditions. The uneven surface of these specimens in comparison to synthetic materials presents additional problems, although flat areas of the specimens were measured as much as possible.

The measured values of the refractive index of tunic samples were 0.002–0.004 larger than seawater in the measurements by Abbe refractometer and 0.02–0.03 larger than seawater in the measurements by ellipsometer at D-line ($\lambda = 589$ nm), which begs the question as to which is the more representative value. As discussed above, the present measurements included some assumptions, and the measured values should be regarded as estimations. Both of the measured values of the two methods were slightly larger than the refractive index of seawater. Considering the possible sources of measurement errors, the true refractive indices of the tunics likely lie between the values measured by the two methods described. That is, Abbe refractometer measurements would underestimate refractive index, owing to the seawater between the tunic surface and the prism, while ellipsometer measurements would overestimate refractive index due to the evaporation of water and the increase of salt concentration in the specimens during the measurement. Accordingly, the refractive indices of the tunics are estimated to be slightly greater (<0.03) than seawater at D-line ($\lambda = 589$ nm).

In the *Rhopalaea* sp. tunic, the refractive indices measured by an Abbe refractometer were almost the same between the fresh specimens and the previously frozen specimens, indicating that the freeze-thaw cycle had no effect. However, we could not verify the effect of the freezing and thawing in the ellipsometric measurement.

Reflectance of the tunic and nipple arrays

As described, the difference in refractive index between tunic and seawater is estimated as 0.03 or less. On the other hand, the refractive index of seawater in situ is variable to some extent, depending on some factors such as salinity and temperature. Therefore, the difference of the refractive index between tunic and seawater might occasionally be larger than the present estimations. In our simulations of light reflection on the tunic, the differences in refractive indices between the tunic and seawater were assumed to be −0.04, −0.02, 0.02, and 0.04. We believe that this range of assumptions is large enough to cover measurement errors and the potential change of refractive index of seawater in situ. As shown in Fig. 5, anti-reflection effect of nipple array, so called moss-eye effect, is larger when the difference of refractive indices is larger. In the case of the moth eye, the difference is 0.5 or larger; the refractive index of an insect's chitinous tissue is > 1.54 [21] and that of air is 1.000 [20] at 589 nm. However, the difference of refractive indices between the tunics and seawater appears so small (<0.03) that the nipple array on the salp tunic rarely causes a significant anti-reflection effect.

The difference in refractive index between tunics (based on ellipsometry) and seawater at 35‰ and 25 °C (based on the equation of Quan and Fry [22]) is larger in shorter wavelength light (Fig. 3a), as is common for the refractive indices of transparent media [23]. Owing to the difference in curve shape, the difference in refractive index between seawater and tunics is larger in shorter wavelength light, particularly between seawater and the *Rhopalaea* tunic (Fig. 3b). Since blue light (470–480 nm) penetrates seawater to greater depths than other visible light wavelengths in marine environments, it has been proposed that many marine animals have visual acuity in this range [24]. The present results indicate that shorter wavelength vision would be more advantageous for predators of salps to see transparent tunic with light reflection in the water column.

Conclusion

Measurements using both Abbe refractometer and ellipsometer showed that the refractive indices of tunics were slightly greater than that of seawater. This is the first report of measurements of refractive indices in tunicate tunics. The present simulation suggests that the refractive index of tunic should not be smaller than the seawater, because total reflection would make the salps and ascidians more conspicuous in the euphotic zone of the water column. We have not previously observed total reflection from the tunic surface in the field or in the laboratory with various angles of illumination. Accordingly, the refractive index of the tunic is hypothesized to be slightly greater than the refractive index of seawater. This is consistent with both refractometer and ellipsometer measurements. Furthermore, the presence or absence of a nipple array does not cause significant difference in reflectance on the surface (Fig. 5).

This raises the question as to the reason for the presence of a nipple array in *T. vagina* tunic. The anti-reflection function of the nipple array is very small when the refractive index of the material (tunic) is close to that of the surrounding medium (seawater). However, even this small difference in reflectance may translate to improved survival. On the other hand, the nipple array on the tunic surface may serve additional and/or alternative functions. Indeed, even if their function is at now undetermined, recent studies demonstrated that a synthetic, hydrophilic nipple array repels bubbles [25] and suppresses cell spreading [26]. These may be important functions for diurnal salps in controlling buoyancy and defending against parasites.

Additional files

Additional file 1: Refractive indices of the tunic specimens and seawater measured by an Abbe refractometer. The refractive indices for seawater are the measured values of the upper boundary in the measurement of *Rhopalaea* specimens. Statistical significance was examined by Steel-Dwass test for nonparametric, multiple comparison.

Additional file 2: Ellipsometric parameter Ψ (A) and Δ (B) of the tunic of *T. vagina* (previously frozen specimens).

Additional file 3: Ellipsometric parameter Ψ (A) and Δ (B) of the tunic of *Rhopalaea* sp. (fresh specimen).

Abbreviations

RCWA: Rigorous coupled wave analysis; TE wave: Transverse electric wave; TM wave: Transverse magnetic wave; UV: Ultraviolet

Acknowledgments
We thank Dr. Hiroshi Miyake (Kitasato University), Dr. James-Davis Reimer and Mr. Ryosuke Uozumi (University of the Ryukyus) for collection of the samples. We are also grateful to Dr. Atsushi Hozumi (Structural Materials Research Institute, National Institute of Advanced Industrial Science and Technology, Japan) for his invaluable advice.

Funding
The present study was supported in part by a University of the Ryukyus Strategic Research Promotion Grant to EH.

Authors' contributions
HK measured the refractive indices. DS carried out the simulation. JN designed the study and prepared the specimens. EH designed the study, prepared the specimens, made microscopic observations, and organized the manuscript draft. All authors read and approved the final manuscript.

Competing interests
The authors declare that they have no competing interests.

Author details
[1]Structural Materials Research Institute, National Institute of Advanced Industrial Science and Technology (AIST), Moriyama, Nagoya, Aichi 463-6560, Japan. [2]Department of Electrical and Electronic Engineering, Kitami Institute of Technology, Koen-cho, Kitami, Hokkaido 090-8507, Japan. [3]Department of Marine Biology, School of Marine Science and Technology, Tokai University, Orido, Shimizu, Shizuoka 424-8610, Japan. [4]Faculty of Science, University of the Ryukyus, Nishihara, Okinawa 903-0213, Japan.

References

1. Madin LP, Kremer P, Hacker S. Distribution and vertical migration of salps (Tunicata, Thaliacea) near Bermuda. J Plankton Res. 1996;18:747–55.
2. Nishikawa J, Tsuda A. Diel vertical migration of the pelagic tunicate, *Salpa thompsoni* in the Southern Ocean during the austral summer. Polar Biol. 2001;24:299–302.
3. Gibbons MJ. Vertical distribution and feeding of *Thalia democratica* on the Agulhas Bank during March 1994. J Mar Biol Assoc UK. 1997;77:493–505.
4. Hirose E, Kimura S, Itoh T, Nishikawa J. Tunic morphology and cellulosic components of pyrosomas, doliolids, and salps (Thaliacea, Urochordata). Biol Bull. 1999;196:113–20.
5. Hirose E, Sakai D, Shibata T, Nishii J, Mayama H, Miyauchi A, Nishikawa J. Does the tunic nipple array serve to camouflage diurnal salps? J Mar Biol Assoc UK. 2015;95:1025–31.
6. Bernhard CG. Structural and functional adaptation in a visual system. Endeavour. 1967;26:79–84.
7. Wilson SJ, Hutley MC. The optical properties of 'moth eye' antireflection surfaces. Opt Acta. 1982;29:993–1009.
8. Curl H. Analysis of carbon in marine plankton organisms. J Mar Res. 1962;20:181–8.
9. Kimura S, Ohshima C, Hirose E, Nishikawa J, Itoh T. Cellulose in the house of the appendicularian *Oikopleura rufescens*. Protoplasma. 2001;216:71–4.
10. Burighel P, Cloney RA. Urochordata: Ascidiacea. In: Harrison FW, Ruppert E, editors. Microscopic anatomy of invertebrates, Vol. 15. Hemichordata, Chaetognatha, and the invertebrate chordates. New York: Wiley-Liss; 1997. p. 221–347.
11. Di Bella MA, Cassara G, Russo D, De Leo G. Cellular components and tunic architecture of the solitary ascidian *Styela canopus* (Stolidobranchiata, Styelidae). Tissue Cell. 1998;30:352–9.
12. Hirose E. Ascidian tunic cells: morphology and functional diversity of free cells outside the epidermis. Invertebr Biol. 2009;128:83–96.
13. Smith M, Dehnel P. The chemical and enyzymatic analyses of the tunic of the ascidian *Halocynthia aurantium* (Pallas). Comp Biochem Physiol. 1970;35:17–30.
14. Smith M, Dehnel P. The composition of tunic from four species of ascidians. Comp Biochem Physiol. 1971;40B:615–22.
15. D'Ancona Lunetta G. Comparative study of the tunics of two ascidians: *Molgula impura* and *Styela partita*. Acta Embryol Morphol Exp. 1983;4:137–49.
16. Hirose E, Ohtsuka K, Ishikura M, Maruyama T. Ultraviolet absorption in ascidian tunic and ascidian-*Prochloron* symbiosis. J Mar Biol Assoc UK. 2004;84:789–94.
17. Jellison Jr GE. Lorentz oscillator model. In: Tompkins HG, Irene EA, editors. Handbook of ellipsometry. New York: Springer; 2005. p. 254–5.
18. Jacobs GH. Ultraviolet vision in vertebrates. Am Zool. 1992;32:544–54.
19. Piazena H, Perez/Rodrigues E, Häder D, Lopez/Figueroa F. Penetration of solar radiation into the water column of the central subtropical Atlantic Ocean-optical properties and possible biological consequences. Deep-Sea Res Pt II. 2002;49:3513–28.
20. Polyanskiy MM. Refractive index database. http://refractiveindex.info. Accessed 19 May 2017.
21. Stavenga DG, Leertouwer HL, Wilts BD. Quantifying the refractive index dispersion of a pigmented biological tissue using Jamin–Lebedeff interference microscopy. Light Sci Appl. 2013;2, e100.
22. Quan X, Fry ES. Empirical equation for the index of refraction of seawater. Appl Opt. 1995;34:3477–80.
23. Born M, Wolf E. Elementary theory of dispersion. In: Born M, Wolf E, editors. Principle of optics, electromagnetic theory of propagation interference and diffraction of light. Oxford: Pergamon; 1980. p. 90–8.
24. Herring P. The biology of the deep ocean. Oxford: Oxford University Press; 2001.
25. Hirose E, Mayama H, Miyauchi A. Does the aquatic invertebrate nipple array prevent bubble adhesion? An experiment using nanopillar sheets. Biol Lett. 2013;9:20130552.
26. Ballarin L, Franchi N, Gasparini F, Caicci F, Miyauchi A, Hirose E. Suppression of cell-spreading and phagocytic activity on nano-pillared surface: in vitro experiment using hemocytes of the colonial ascidian *Botryllus schlosseri*. Inv Surv J. 2015;12:82–8.

10

Comparative morphology and development of extra-ocular muscles in the lamprey and gnathostomes reveal the ancestral state and developmental patterns of the vertebrate head

Daichi G. Suzuki[1*], Yuma Fukumoto[2,3,4], Miho Yoshimura[1], Yuji Yamazaki[5], Jun Kosaka[3,6], Shigeru Kuratani[2] and Hiroshi Wada[1]

Abstract

The ancestral configuration of the vertebrate head has long been an intriguing topic in comparative morphology and evolutionary biology. One peculiar component of the vertebrate head is the presence of extra-ocular muscles (EOMs), the developmental mechanism and evolution of which remain to be determined. The head mesoderm of elasmobranchs undergoes local epithelialization into three head cavities, precursors of the EOMs. In contrast, in avians, these muscles appear to develop mainly from the mesenchymal head mesoderm. Importantly, in the basal vertebrate lamprey, the head mesoderm does not show overt head cavities or signs of segmental boundaries, and the development of the EOMs is not well described. Furthermore, the disposition of the lamprey EOMs differs from those the rest of vertebrates, in which the morphological pattern of EOMs is strongly conserved. To better understand the evolution and developmental origins of the vertebrate EOMs, we explored the development of the head mesoderm and EOMs of the lamprey in detail. We found that the disposition of lamprey EOM primordia differed from that in gnathostomes, even during the earliest period of development. We also found that three components of the paraxial head mesoderm could be distinguished genetically (premandibular mesoderm: *Gsc+/TbxA–*; mandibular mesoderm: *Gsc–/TbxA–*; hyoid mesoderm: *Gsc–/TbxA+*), indicating that the genetic mechanisms of EOMs are conserved in all vertebrates. We conclude that the tripartite developmental origin of the EOMs is likely to have been possessed by the latest common ancestor of the vertebrates. This ancestor's EOM developmental pattern was also suggested to have resembled more that of the lamprey, and the gnathostome EOMs' disposition is likely to have been established by a secondary modification that took place in the common ancestor of crown gnathostomes.

Keywords: Evo-devo, Extra-ocular muscles, Lamprey, Head mesoderm, Head segmentation

* Correspondence: daichi1207@gmail.com
[1]Graduate School of Life and Environmental Sciences, University of Tsukuba, 1-1-1 Tennodai, Tsukuba, Ibaraki 305-8572, Japan
Full list of author information is available at the end of the article

Introduction

The morphological nature and ancestral configuration of the vertebrate head are longstanding topics of interest in comparative morphology and evolutionary biology. A peculiar component of the vertebrate head is the extra-ocular muscles (EOMs), which control the visual field by the moving eyes. These muscles are derived from the paraxial portion of the head mesoderm, and their development has attracted the attention of morphologists in the context of head mesodermal segmentation.

Some researchers believe that these muscles are derived from rostral mesodermal segments of an amphioxus-like ancestor (reviewed in [1–4]). This idea partially stems from the presence of the epithelialized head mesodermal cysts (head cavities) in the shark embryo [5]; the head mesoderm in elasmobranch embryos typically forms three pairs of head cavities, from anterior to posterior, the premandibular (pc), mandibular (mc), and hyoid cavities (hc). These cavities later differentiate into six EOMs innervated by three cranial nerves: the oculomotor (III), trochlear (IV), and abducens nerves (VI) (Fig. 1). Although a complete set of head cavities arises only in cartilaginous fishes, they are occasionally found in osteichthyans including amniotes [6–15]. This indicates that possession of these cavities is a shared, derived characteristic for gnathostomes, although they tend to disappear in many osteichthyan taxa.

In avians, the head cavities have disappeared except for the remnant premandibular cavity [16–18]. In a different line of studies, a series of pseudosegmental blocks, so-called "somitomeres", have been detected by scanning electron microscopy in the early head mesoderm, in which no head cavities are expected to arise (reviewed in [17]). Although the existence of the somitomeres has been questioned (reviewed in [3]), the EOMs do develop from a part of the head mesoderm corresponding to the sites of head cavities in elasmobranchs [11, 18–24].

Classically, the head mesoderm of lampreys, which belong to the most basal group of vertebrates (cyclostomes), was also thought to be segmented along the anteroposterior axis, similar to that in the head cavities in elasmobranch embryos [25–27], and EOMs were thought to differentiate from the three head cavities, innervated by their respective motor nerves [25, 26]. However, scanning electron microscopy-based observations of the Arctic lamprey, *Lethenteron camtschaticum* by Kuratani et al. [28], show no signs of segmental boundaries in the dorsal (paraxial) head mesoderm. Rather, the dorsal head mesoderm was simply secondarily regionalized into preotic and postotic portions by the otic vesicle, and the ventral part segmented passively by the pharyngeal pouches, in accordance with the notion of branchiomery proposed by Sewertzoff [29]). This finding suggests that the ancestral head mesoderm of the vertebrates is likely to have been unsegmented in the paraxial portion, raising the question of how the EOMs developed from this unsegmented head mesoderm.

The anatomical disposition and the innervation patterns of EOMs are highly conserved among gnathostomes, so much so that Neal [26] once noted that "[t]heir 'evolutionary potential' appears to be approximately zero". However,

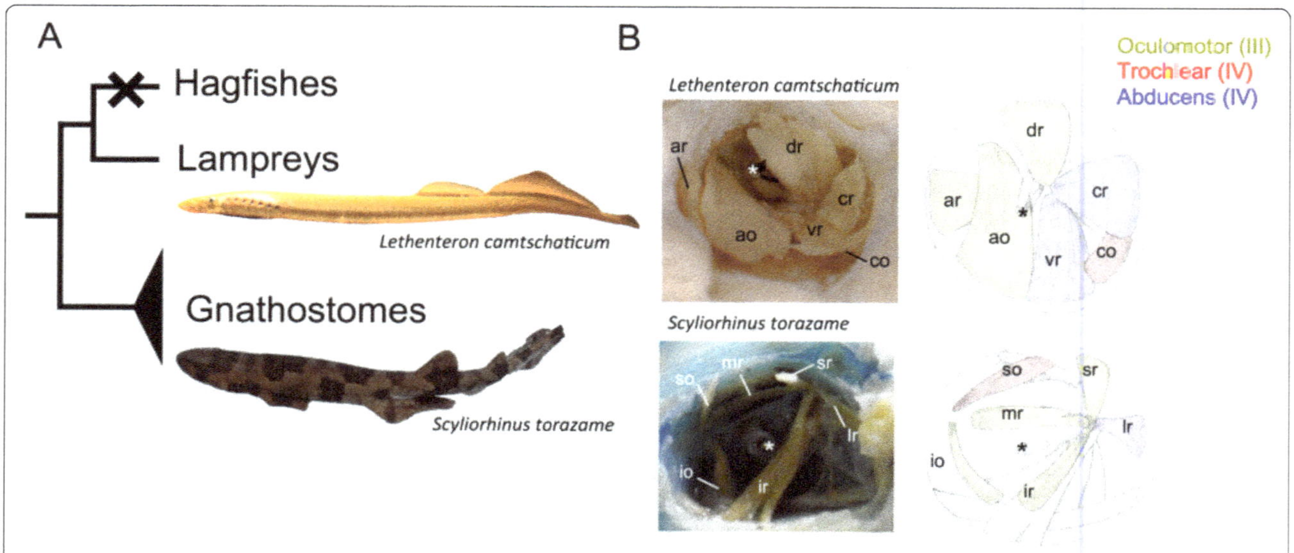

Fig. 1 Phylogenetic relationships among vertebrates and gross anatomy of extra-ocular muscles. **a** A phylogenetic tree of the vertebrates. The hagfishes lack extra-ocular muscles. **b** Gross anatomy of the extra-ocular muscles of the lamprey (*Lethenteron camtschaticum*) and shark (*Scyliorhinus torazame*). Asterisks indicate the optic nerve. The extra-ocular muscles are colored based on their innervation nerve: the oculomotor (III), yellow, trochlear (IV), red; abducens (IV), blue, respectively

the patterns of lamprey EOMs have been known to differ from living gnathostomes (Fig. 1, [30–32]). In the latter, the oculomotor nerve innervates four of the EOMs (medial rectus (mr), superior rectus (sr), inferior rectus (ir), and inferior oblique (io)), whereas the trochlear and abducens innervate only single EOMs (superior oblique (so) and lateral rectus (lr), respectively). In contrast, the lamprey oculomotor nerve innervates only three EOMs (anterior rectus (ar), dorsal rectus (dr), and anterior oblique (ao)), while the abducens innervates two (ventral rectus (vr) and caudal rectus (cr)). Furthermore, the caudal oblique (co) muscles of lampreys, which are innervated by the trochlear nerve, attach to the orbit far more caudally than do those of modern gnathostomes. Because the EOMs have degenerated completely in hagfishes [32], the lamprey is the only key extant animal to speculate ancestral state of the EOMs and its developmental mechanisms.

To understand the evolutionary origin of the vertebrate EOMs and suggest a possible ancestral state of the vertebrate head, we examined the development of the embryonic head mesoderm and EOMs in lampreys. We found that the developmental pattern of EOMs was also conserved in the lamprey, because the muscle originated from three domains along the anteroposterior axis in the dorsal (paraxial) head mesoderm. However, EOM disposition was different between lampreys and gnathostomes, as soon as EOMs were observed as differentiated muscles. These findings indicate that the developmental mechanisms of EOMs from the three subdivisions of the head mesoderm was already established in the common ancestor of vertebrates, and that diversification of the muscle patterns is due to changes during the later phase of development. Based on these findings, we discuss the ancestral state of the vertebrate dorsal head mesoderm and its differentiation.

Materials and methods

Animals

This study was performed in accordance with the Regulations on Animal Experimentation at University of Tsukuba. Approval is not needed for experimentation on fishes under Japanese law, Act on Welfare and Management of Animals.

Adult lampreys (*Lethenteron camtschaticum*, synonym *L. japonicum*) were collected from the Shiribeshi-Toshibetsu River, Hokkaido, Japan. The animals were anesthetized in ethyl,3-aminobenzoate methanesulfonate (MS-222). Mature eggs were squeezed from females and fertilized in vitro by sperm. Embryos were cultured at 16 °C, fixed in 4 % paraformaldehyde in 0.1 M phosphate-buffered saline (PBS) overnight, dehydrated in a graded methanol series, and stored in 100 % methanol at −20 °C. Developmental stages were determined as described by Tahara [33].

As ammocoete larvae were not readily available for *L. camtschaticum*, we used larvae from *Lethenteron* sp. N, related species of *L. camtschaticum* [34, 35]. These larvae were collected from the Kamo River, which flows through the middle of the Shougawa River, Toyama, Japan, in September.

Fertilized eggs of the cloudy catshark (*Scyliorhinus torazame*) were obtained from adults that were bred at 16 ° C in seawater tanks. Shark embryos were staged following Ballard's staging of *Scyliorhinus canicula*, a species closely related to *S. torazame* [36, 37].

Histological analyses

Lethenteron sp. N and *Scyliorhinus torazame* were fixed in Bouin's or Serra's fixative, dehydrated, and embedded in paraffin. Sections were cut at a thickness of 6 μm and stained with hematoxylin and eosin, following a standard protocol.

3D reconstruction

The stained sections of *Lethenteron* sp. N and *S. torazame* were digitized using an Olympus BX60 microscope equipped with an Olympus DP70 camera and the Olympus DP controller software (Olympus, Tokyo, Japan). On the digitized sections, each anatomical component was colored and reconstructed using the Avizo 3D Visualization Framework (Maxnet Co., Ltd, Tokyo, Japan).

Whole-mount immunofluorescence

Whole-mount immunofluorescence with anti-acetylated tubulin (Sigma, T6793) and anti-tropomyosin (Hybridoma bank, CH1) antibodies was performed according to the protocol described by Kuratani et al. [38] with some minor modifications. Briefly, fixed embryos stored in methanol were washed in TBST containing 5 % dimethylsulfoxide (TSTd). The embryos were then blocked with 5 % non-fat dry milk in TSTd (TSTM). They were incubated with the primary antibody (diluted 1:1,000 in TSTM) for 2–4 days at room temperature (RT). After washing with TSTd, samples were incubated with a secondary antibody (Invitrogen, Alexa fluor 555, A21424) diluted 1:200 in TSTM. After a final wash in TSTd, embryos were dehydrated and clarified in a 1:2 mixture of benzyl alcohol and benzyl benzoate (BABB) and then examined using a confocal laser microscope (LSM 510, Zeiss). The data were colored and projected by using a computational graphics editor (Photoshop CS6).

Whole-mount and section in situ hybridization

Gsc was amplified from *L. camtschaticum* by RT-PCR from stage 25 specimens using the primers designed for *Petromyzon* genes (*PmGsc*, HQ248103) [39]. For the other genes, probes were synthesized by using the plasmids in accordance with previously described protocols (*PitxA*: Uchida et al. [40]; *MrfA*, *MA2*: Kusakabe et al. [41]; *Tbx1/10*: Tiecke et al. [42]). Whole-mount in situ hybridization was performed according to the protocol of Ogasawara et al. [43] with minor modifications. For section in situ hybridization, larval lampreys (*L.* sp. N) were fixed for three days in 4 % paraformaldehyde in 0.1 M phosphate-buffered saline (PBS), dehydrated, and embedded in paraffin. Sections were cut at a thickness of 8 μm. After washing out the paraffin, in situ hybridization for cryosectioned materials was performed following the protocol for whole-mount in situ hybridization, except that Tween 20 was not used at any step and proteinase treatment was omitted before hybridization.

Cell labeling

St. 21 embryos were injected with 1 mM DiI, DiD, and DiO solutions (Vybrant Multicolor Cell-labeling kit, Molecular Probes). The embryos were excised from the egg membranes and placed in wells made in solidified agar on a plastic dish. Injections were performed with a fine glass pipette. The embryos were incubated for 10 days until st. 27 was approximately reached and were fixed in 4 % paraformaldehyde in PBS. Observation was made with a fluorescence microscope or confocal microscope (LSM510, Zeiss, Goettingen, Germany).

Results

Development of the EOMs and their innervation

To clarify the disposition and innervation pattern of EOMs in lamprey larvae, we first performed 3D reconstruction of a ca. 100 mm ammocoete larva (Fig. 2a–d). At this stage, six EOMs were already differentiated as distinct muscle primordia attached to the surface of the eye (Fig. 2a). They consisted of four rectus muscles (ar, dr, cr, and vr) and two oblique muscles (ao and co). This muscle organization was the same as that known in the adult lamprey (Fig. 1b, [30–32]). To confirm the muscle identities, we also analyzed the innervation patterns of these EOMs. As reported for adult specimens (Fig. 1b, [30–32]), the oculomotor nerve (III) innervated the 'ar', 'dr', and 'ao' muscles (Fig. 2b), the trochlear nerve (IV) the 'co' muscle (Fig. 2c), and the abducens nerve (VI) the 'vr' and 'cr' muscles (Fig. 2d). Notably, the pathways of the trochlear and abducens nerves partially overlapped those of the trigeminal nerve (V), and the trochlear nerve ramifies into sub-bundles and become fasciculated again near its terminus (Fig. 2c). We confirmed this observation by immunofluorescence analysis using an anti-acetylated tubulin antibody in early larvae, as described below. Furthermore, the attachment site of the 'ao' muscle to the cartilaginous orbital wall was

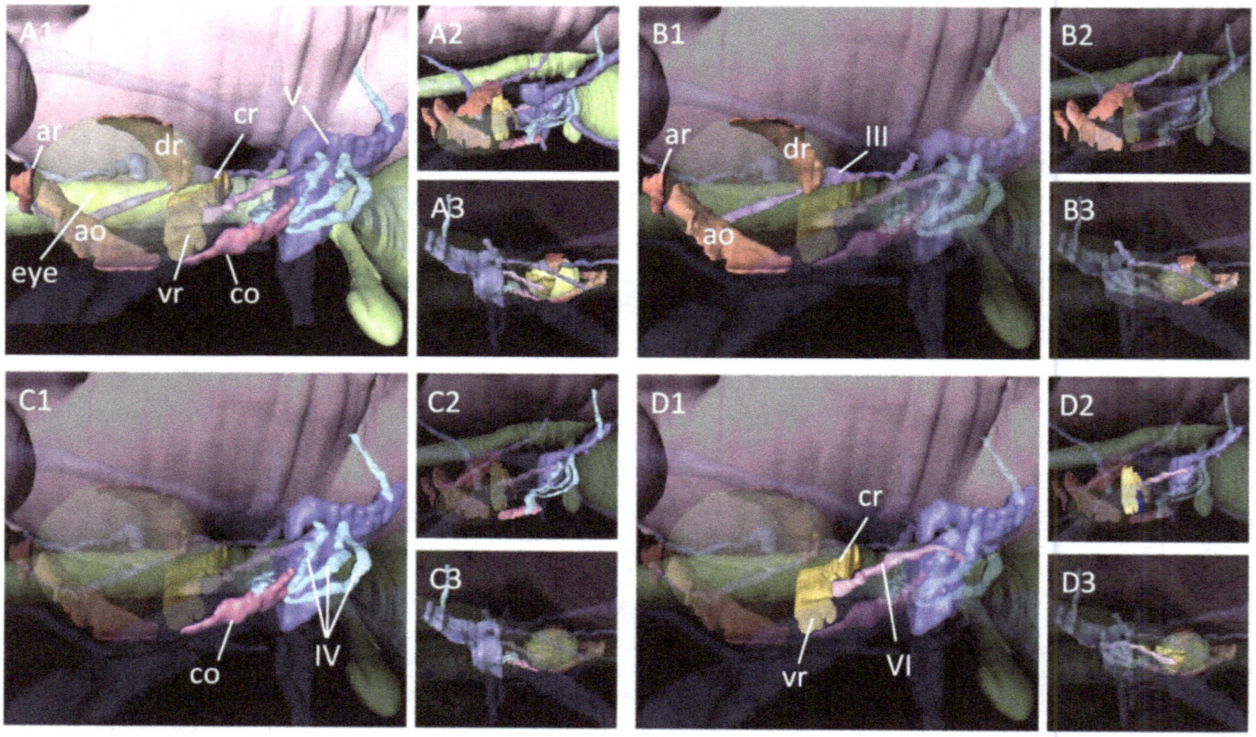

Fig. 2 3D reconstruction of a lamprey ammocoetes larva. **a** Overview. **b** The oculomotor nerve and its innervating extra-ocular muscles. **c** The trochlear nerve and its innervating caudal oblique muscle. **d** The abducens nerve and its innervating extra-ocular muscles. **a1**–**d1**: Lateral; **a2**–**d2**: Dorsal; **a3**–**d3**: Medial view

relatively more ventral (Fig. 2b) than that in the adult, in which the 'ao' muscle crossed over the 'ar' muscle (Fig. 1).

For the comparison, we constructed 3D images of a pre-hatching stage (72 mm long) embryonic shark (*Scyliorhinus torazame*) to represent gnathostomes (Fig. 3a–d). Consistent with the adult anatomy (Fig. 1b) as well as previous descriptions [30–32], the oculomotor nerve (III) innervated the 'mr', 'sr', 'ir', and 'io' muscles (Fig. 3b), the trochlear nerve (IV) the 'so' muscles (Fig. 3c), and the abducens nerve (VI) the 'lr' muscles (Fig. 3d). The ciliary ganglion was observed in the orbit (inset in Fig. 3b1), but no similar ganglion was found in the lamprey (Fig. 2).

To determine whether the disposition of the lamprey EOMs changes during development, we performed a histological analysis by hematoxylin and eosin (HE) staining in stage (st.) 30 prolarvae (Fig. 4), 35 mm (about half a year old, Fig. 5a),100 mm larvae (Fig. 5b), metamorphic (Fig. 5c) and adult lampreys (Fig. 5d). In st.30 prolarvae, we found no muscle fibers suggestive of EOM differentiation, but only mesenchymal cell masses surrounding the eyeball (Fig. 4).

In the 35 mm larvae, EOMs were found as fibrous, distinguishable six-cell clusters (Fig. 5a), identities of which were obvious from their disposition. Rectus muscles were located at the anterior, dorsal, ventral, and caudal parts in the orbit; thus, we named them the anterior, dorsal, ventral, and caudal rectus muscles, respectively. As for two oblique muscles, one primordium originated slightly ventral to 'ar', and was directed caudally; it was therefore identified as the 'ao' muscle. The other, identified as the 'co', originated from the dorsocaudal region in the orbit, was directed ventrally.

In the 100 mm larvae, the EOMs become more clearly compartmentalized and discriminable (Fig. 5b), suggesting the growth of the EOMs of this animal during the larval period, ranging 4–5 years. The topological disposition of the EOMs was the same as that in the 35 mm larvae. In the metamorphic stage, the external part of the EOMs became thinner and wider (Fig. 5c1), suggesting rigid attachment to the eyeball to exert its functional movement. The relatively immature state of larval EOMs may be due to the larval life style of this animal, in which the eyes do not possess image-forming vision [44–47]. Through all of the stages examined, the positions of the EOMs did not show radical changes, and it seemed likely that the EOM morphological pattern is established during the pre-metamorphic stages. However, the little change in the relationship between the 'ar'

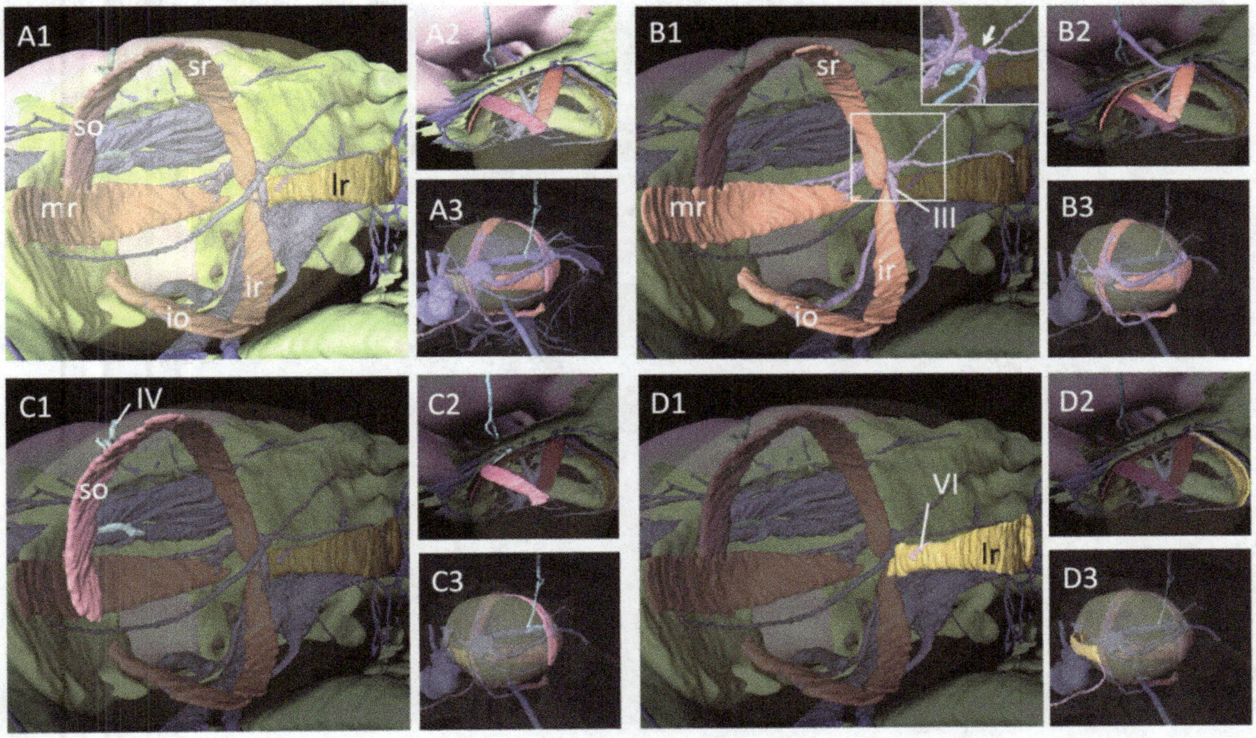

Fig. 3 3D reconstruction on *Scyliorhinus torazame*. **a** Overview. **b** The oculomotor nerve and its innervating extra-ocular muscles. **a** detailed structure of oculomotor nerve (III) is shown in the inset, in which a ciliary ganglion-like structure is indicated by an arrow and highlighted in blue color. **c** The trochlear nerve and its innervating superior oblique muscle. **d** The abducens nerve and its innervating lateral rectus muscle. **a1**–**d1**: Lateral; **a3**–**d2**: Dorsal; **a3**–**d3**: Medial view

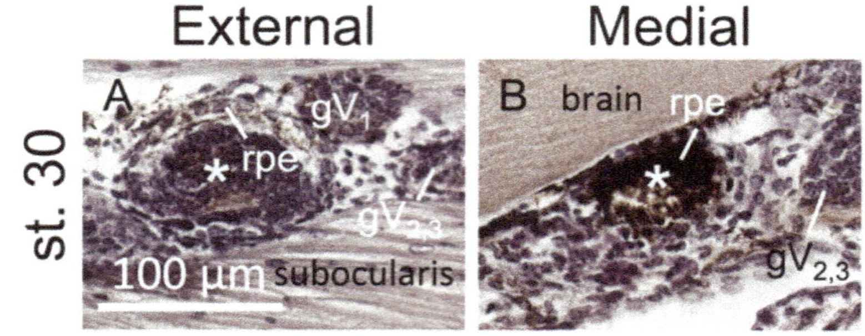

Fig. 4 Histological analysis by the hematoxylin and eosin (HE) staining on the extraocular muscles in st. 30 prolarva. Asterisks indicates the eye. **a** External section. **b** Medial section

and 'ao' muscles was notable. During the larval period, these muscles at first run in parallel to each other (Fig 2a4, Bb), and cross each other in the adults (Figs. 1b and 5d4). This change is likely to occur during metamorphosis.

Developmental mechanism of EOMs and patterning of head mesoderm

To trace further the developmental origin of the lamprey EOMs, immunofluorescence analysis was performed using an anti-tropomyosin antibody in younger lamprey larvae (Fig. 6a–c). We did not detect any EOMs in st. 28 or st. 30 prolarvae, but did detect other muscles, including somatic/branchial muscles; supraocularis, subocularis, elevator labialis ventralis (elv), velocranialis, and constrictor buccalis (Fig. 6a, b, see also [43]).

Next, we traced the developmental origin of EOMs by analyzing more upstream regulatory genes for EOMs. In gnathostomes, the genetic cascade involved in the development of EOMs has already been reported; genes encoding muscle-related factors (MRFs) act as determination and differentiation genes, *Pitx2* acts upstream of MRFs in cranial muscle progenitor cells, and *Pitx2*-null embryos lack EOMs [49]. This cascade is also conserved in sharks, in which *Pitx2* and *Myf5* (a member of the MRF family) are expressed in developing head mesoderm/cavities [50].

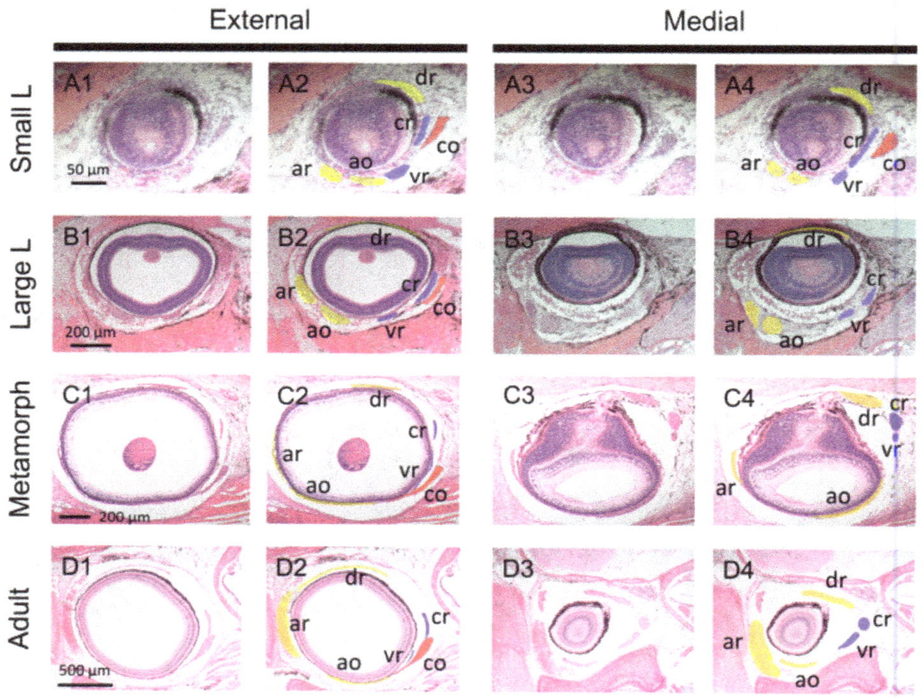

Fig. 5 Histological analysis by hematoxylin-eosin (HE) staining of the extraocular muscles in larval, metamorphic and adult lampreys. **a** Small larva (3.5 cm, about half a year old). **b** Large larva (10 cm). **c** Metamorphic lamprey. **d** Adult lamprey **a1**–**d1**: External sections; **a2**–**d2**: External sections, colored; **a3**–**d3**: Medial sections; **a4**–**d4**: Medial sections, colored

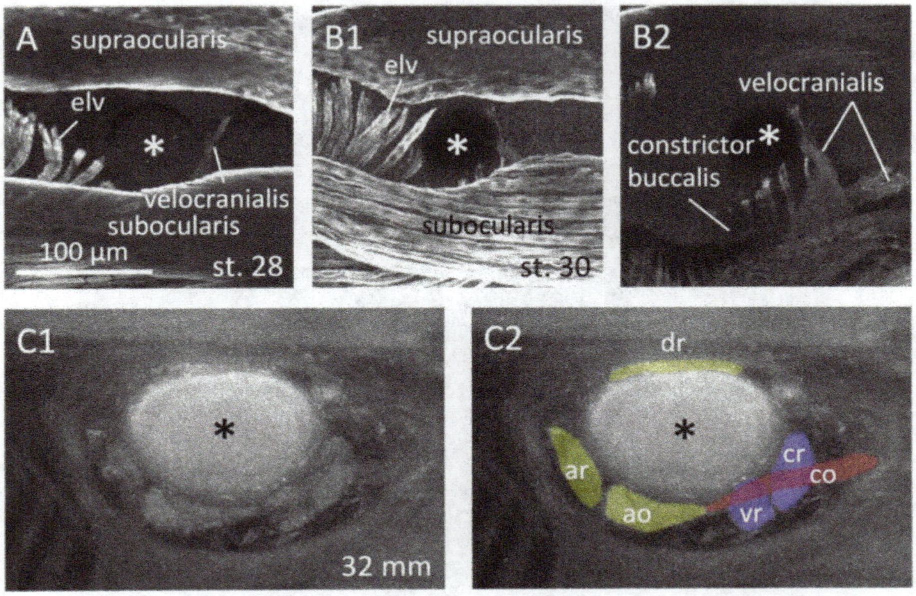

Fig. 6 Whole-mount immunofluorescence with an anti-tropomyosin antibody. Asterisks indicates the eyes. **a** st. 28 prolarvae. **b** st. 30 prolarvae (**b1**: Overview; **b2**: Medial). **c** 32 mm larva (**c1**: Raw; **c2**: Colored)

In a st. 26 prolarva, although *MrfA* (a member of the MRF family) and *MA2* (a muscle differentiation marker) were not expressed [41], we detected *Pitx2* transcripts in the head mesoderm (Fig. 7a). In contrast, in the 90 mm ammocoete larvae, *MrfA* and *MA2* were expressed in EOM prmordia, while the *Pitx2* expression ceased (Fig. 8).

Furthermore, we found that there was distinct genetic regionalization in the dorsal head mesoderm. In gnathostomes, *Gsc* is expressed in the prechordal plate [51], from which the premandibular mesoderm (pm) is thought to arise (lampreys: [28]; sharks: [36]). *Gsc* plays a dominant role as an organizer in head formation, including head muscle differentiation [52]. We found that *Gsc* was expressed in the anteriormost head mesoderm in the st. 26 lamprey prolarvae (Fig. 7b), and expression corresponded to that in the premandibular mesoderm. Simultaneously, *TbxA* transcripts were detected in the paraxial head mesoderm located anterior to the otic vesicle (Fig. 7c). In sharks, an equivalent expression has been observed in the hyoid cavity [50]. In the mouse, *Tbx1* (homolog of the lamprey *TbxA*) regulates craniofacial myogenesis [53]. Thus, *TbxA* expression in the lamprey is expected to represent that in the paraxial portion of the hyoid mesoderm (hm). These results suggest that the dorsal head mesoderm of lamprey, characterized by *PitxA* expression along the anteroposterior axis, is further specified through expression of *Gsc* and *TbxA*, i.e., pm: *Gsc*+, *TbxA*-; mm: *Gsc*-, *TbxA*-; hm: *Gsc*-, *TbxA*+ (Fig. 7d).

Developmental lineage of the head mesoderm: origin of the differentiated EOMs

On examination of expression of *Pitx*, *Gsc*, and *Tbx*, three distinct domains were identifiable in the lamprey dorsal head mesoderm, in a pattern similar to those in gnathostomes. Thus, via immunofluorescence analysis using an anti-acetylated tubulin antibody, we examined differentiation of the three lamprey head mesodermal portions into the specific EOM groups innervated by the respective cranial motor nerves as seen in shark head cavities [5]. In the st. 28 prolarva, although the head mesoderm was not differentiated into the EOMs (Fig. 6a), *PitxA* was expressed in the three components of the dorsal head mesoderm (Fig. 9a). At this stage, the EOM-innervating nerves were already extending their fibers, and the distribution pattern corresponded to each portion of the dorsal head mesoderm: the oculomotor nerve (III) reached the premandibular mesoderm, the trochlear nerve (IV) the mandibular mesoderm, and the abducens nerve (VI) the hyoid mesoderm (Fig. 9b, c). These fibers approached the orbit in the 15 mm larvae (Fig. 9d), and their distribution pattern was maintained through the larval period, by which time the EOMs had already been formed (35 mm; Fig. 9e, see also Fig. 6c). These results indicate that the three components of the dorsal head mesoderm are assigned morphologically to respective nerves in a modern gnathostome pattern, and that nerve innervation is maintained through

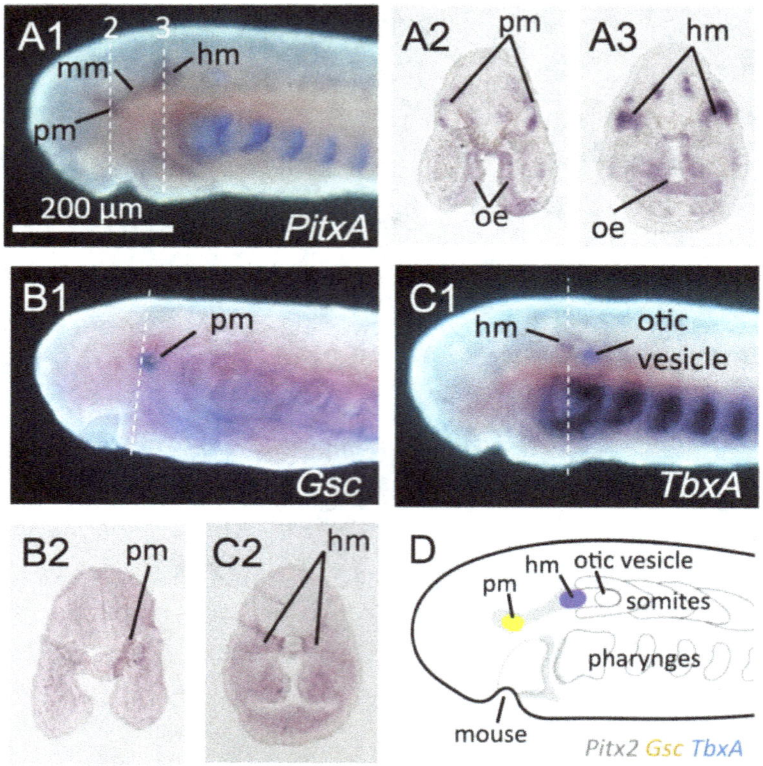

Fig. 7 Whole-mount in situ hybridization in st. 26 lamprey prolarvae. A *PitxA* (**a1**: Lateral view; **a2**, 3: Sections). **b** *Gsc* (**b1**: Lateral view; **b2**: Section). **c** *TbxA* (**c1**: Lateral view; **c2**: Section). **d** Schematic illustration of the *PitxA*, *Gsc*, and *TbxA* expression patterns

differentiation into EOMs, supporting differentiation of the specific paraxial head mesodermal portion into specific subsets of EOMs.

In lamprey development, the premandibular mesoderm is derived from the prechordal plate, and the mandibular and hyoid mesoderm are regionalized rostrocaudally from each other by the growth of the first pharyngeal pouch [28]. Based on our results (Fig. 7), each of these subdivisions appears to correspond to a genetically-specified subdivision, as described above. However, there is another possibility that the mesenchymal cells are mixed and then become re-specified. Thus, we performed cell-labeling experiments to determine whether each head mesodermal portion retained its cohesion from its origin or became mixed. First, only DiO was injected into the prechordal plate region in st. 21 embryos (Fig. 10a) and incubated until st. 27. At st. 27, a DiO signal was observed around the eyeball, although the eyeball itself was also labeled as an artifact (Fig. 10b, c). Subsequently, triple dye injections were performed; DiO was injected into the prechordal plate, DiI into the mandibular mesoderm, and DiD into the hyoid mesoderm in st. 21 embryos (Fig. 10d). The mesodermal portions retained their cohesion and did not mix with each other in almost all of the larvae at st. 27 ($n = 43/48$; no fluorescent signal was detected in the remaining 5 samples), (Fig. 10e). The positions of these mesodermal portions also corresponded to the expression patterns of *Gsc* and *TbxA* as described above (Fig. 7d, e). These results indicate that the above noted dorsal head mesoderm is regionally specified early in development as well, with respect to their developmental fates.

Discussion

Developmental mechanism of the EOMs and the head mesoderm

The aim of the present study was to determine the evolutionary history of vertebrate EOMs. First we compared the disposition and nerve innervation of the EOMs between larval lampreys and shark embryos. The results showed that the disposition and nerve innervation patterns were quite different between these two animals (Figs. 2 and 3). We then traced the developmental process of the EOMs in lampreys using histological analysis. The overall disposition of the lamprey EOMs is established as early at 32 mm larvae (approximately six-months-old, Fig. 6c) in the paraxial (dorsal) head mesoderm, which is regionally and genetically specified during early stages of development (Figs. 4, 5 and 6). These

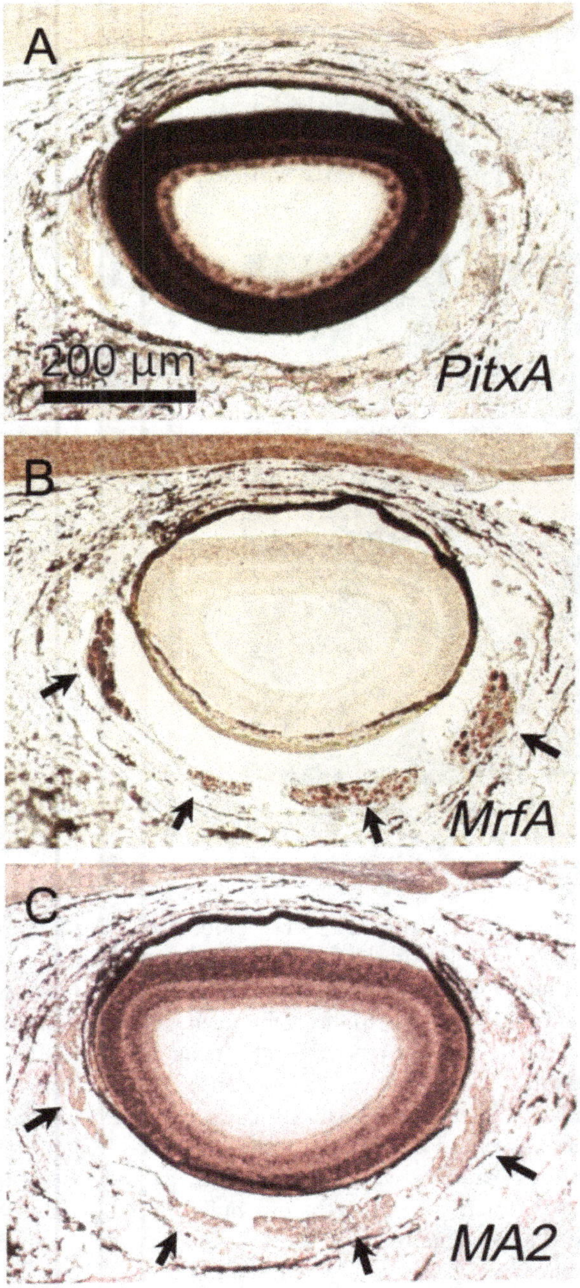

Fig. 8 Sections in situ hybridization in 9 cm ammocoete larvae. **a** *PitxA*.**b** *MrfA*. **c** *MA2*

results indicate that the lampreys and gnathostomes show different distributions of EOMs as soon as they are observed as differentiated muscles (lamprey: Fig. 6c; chick: reviewed in [22]).

In contrast, we found that the genetic cascade involved in development seems to be conserved in the lamprey, and that the dorsal head mesoderm is first marked by the expression of *PitxA* (at st. 26, Fig. 7). Furthermore, the expression patterns of *Gsc* and *TbxA* (Fig. 7) suggest that regionalization as well as specification of the three mesodermal portions underlie the distinct genetic characterizations. *Gsc* expression in the premandibular region is also observed in the zebrafish [54] and mouse [55], and *Tbx* expression has been observed in the dorsal hyoid region of the shark [50], zebrafish [56], *Xenopus* [57], chick [22], and mouse [58]. Thus, these expression patterns appear to be conserved among vertebrates. In addition, these mesodermal portions retain their cohesion and attract their respective innervating nerves (Figs. 9 and 10), similar to those in the shark head cavities [5], even if there is no morphological segmentation in the lamprey dorsal head mesoderm [28].

Since we did not detect any muscle differentiation markers such as the *MA2* gene or the anti-tropomyosin antibody during the developmental stage at which *Pitx*, *Gsc* and *Tbx* expressions were detected, it was unclear whether the *Pitx*-positive head mesoderm truly differentiated into EOMs in the lampreys. We circumvented this problem by examining head mesodermal innervation by motor nerves. We present evidence indicating that, although the muscle markers were not detected at the time *Pitx* expression was detected, motor innervation was observed at this stage. The oculomotor nerve fibers reached the premandibular mesoderm, the trochlear nerve the mandibular mesoderm, and the abducens nerve the hyoid mesoderm (Fig. 9b, c). This innervation pattern supported the hypothesis that lamprey EOMs differentiate from the three components of the dorsal head mesoderm.

Figure 11 shows a comparison of EOM development in lamprey and shark. In both species, the early head mesoderm is primarily uniform with no overt segmental patterns at paraxial levels (Fig. 11a, d, [28, 36]). However, it is specified into three components, each innervated by a single cranial nerve (cranial nerves III, IV, and VI; Fig. 11b, e). In the shark, these components correspond to three pairs of epithelial coeloms called head cavities [36]. Finally, the EOMs are formed in their lineage-specific dispositions (Fig. 11c, f). This comparison indicates that the derivation of EOMs from the three components of dorsal head mesoderm would have already been established by the latest common ancestor of vertebrates (Fig. 12).

Ancestral state of EOMs in vertebrates and subsequent evolutionary modifications in gnathostomes

The comparison of EOM development between lampreys and gnathostomes described above indicated that the evolutionary modification of EOMs would have been introduced into a developmental stage following establishment of the three dorsal head mesodermal portions, but preceding the start of muscle differentiation (corresponding to between st. 28 and the 35 mm early larval stage of lampreys).

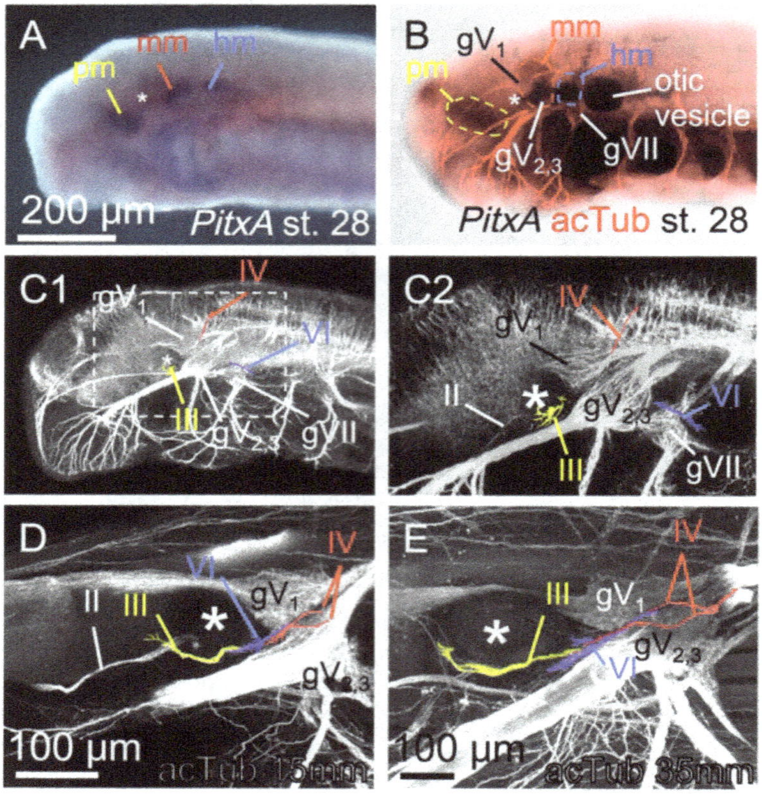

Fig. 9 Nerves innervating head mesodermal portions and EOMs. Asterisks indicate the eyes. **a** *PitxA* expression in st. 28. **b** Double-staining of *PitxA* in situ hybridization (bright field) and an anti-acetylated tubulin antibody immunofluorescence (red). **c**, **d** Single immunofluorescence with the anti-acetylated tubulin antibody. **c** st. 28 (**c1**: Overview; **c2**: magnified dashed box region in **c1**). **d** 15 mm larva. **e** 35 mm larva

However, the type of EOM distribution that represents the ancestral disposition of the vertebrates could not be determined in the present study, as there are only two groups with different EOM patterns (lampreys and extant gnathostomes), and both hagfishes and the outgroup (protochordates) lack EOMs. Nevertheless, previous studies on fossil records [59–62] analyzed the morphology of the orbit including distribution of myodomes (orbital wall depressions indicating muscle insertions) and suggested that the disposition of EOMs in osteostracans and placoderms would have been similar, to some extent, to that of lampreys than the extant gnathostomes. Based on recent phylogenetic analyses [63–65], the pattern of EOM disposition common to chondrichthyans and osteichthyans would have been established as a synapomorphy of jawed vertebrates in some lineages of placoderms (Fig. 12). This modification may be functionally linked with the postorbital connection between the palatoquadrate and neurocranium, which is a synapomorphy of the crown gnathostomes [61]. As the caudal oblique muscle could interfere with the postorbital connection between the palatoquadrate and neurocranium, the position of the caudal oblique muscle may have become anterior within the orbit.

Based on the trochlear nerve innervation, the lamprey caudal oblique is likely to be homologous to the gnathostome superior oblique. However, because the number of oculomotor and abducens nerve-innervated muscles differ between the two taxa, it is difficult to identify one-to-one correspondence of the muscles. Nishi [31] suggested that the lamprey ventral rectus is homologous to the gnathostome lateral rectus, and the lamprey caudal rectus corresponds to the additional EOMs, such as the retractor bulbi in gnathostomes. Nishi also suggested that there are two types of muscle duplication patterns of oculomotor nerve-innervated muscles in gnathostomes based on the branching pattern of the nerve. One is that the lamprey dorsal rectus corresponds to the superior rectus and medial rectus in gnathostomes: these muscles are innervated by the dorsal branch of the oculomotor nerve in sharks and lungfishes. The other is that the lamprey anterior oblique corresponds to the medial rectus and inferior rectus, and the lamprey anterior rectus corresponds to the inferior oblique (all innervated by the ventral branch of the oculomotor nerve) in the other gnathostomes. Based on a neurolabeling analysis, Fritzsch et al. [30] modified these ideas to indicate that the lamprey anterior oblique

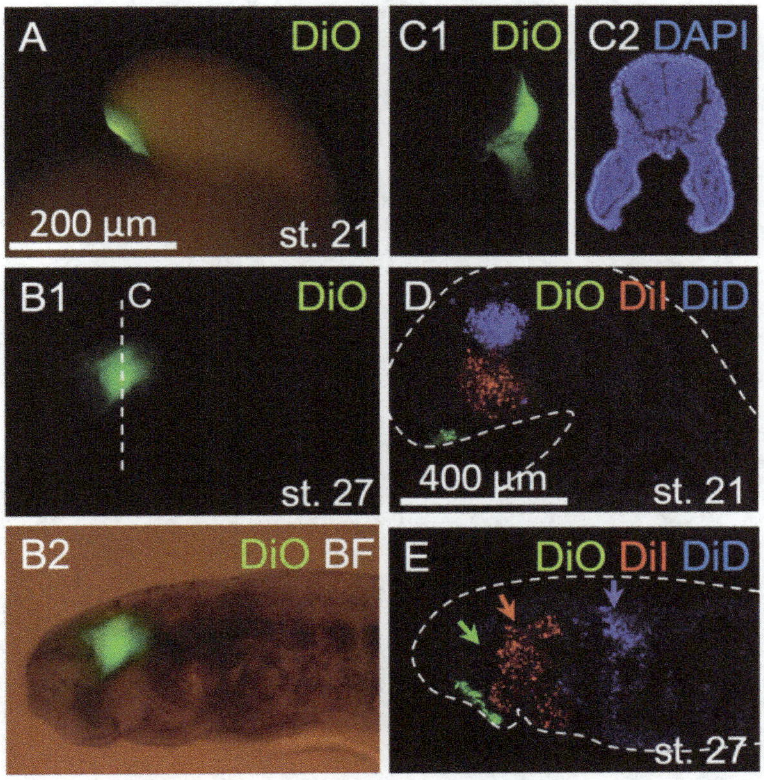

Fig. 10 Dye injections on the head mesoderm. **a** DiO injection into the premandibular mesoderm of the st. 21 embryo. **b** DiO injected sample in st. 27 (**b1**: DiO fluorescence; **b2**: DiO fluorescence and bright field). DiO fluorescence is observed in the periocular region. **c** Section in the dashed line plane in **b1** (**c1**: DiO fluorescence; **c2**: DAPI fluorescence). **d** Three-color dye injections on the three mesodermal portions (premandibular; DiO, mandibular; DiI, hyoid; DiD, respectively). **e** Dye injected sample in st. 27. The three mesodermal portions retained their cohesion

corresponds to the gnathostome inferior oblique, and that the gnathostome medial rectus evolved by duplication of the superior rectus in elasmobranchs and of the inferior rectus in osteichthyans. In the present study, we could not validate the one-to-one correspondence of EOMs, thus we retained the current nomenclature (see also Additional file 1). However, in the lamprey, the dorsal rectus (innervated by the dorsal branch) is differentiated from a cell population distinct from that of the anterior rectus and the anterior oblique (innervated by the ventral branch) (Figs. 5 and 6), providing a clue to determining the correspondence. Also in the chicken [22], there seems to be a close relationship between the medial rectus and inferior rectus, supporting the hypothesis that these two rectus muscles were duplicated in the osteichthyans including amniotes. Further detailed morphological studies on the EOM differentiation in various gnathostomes species are needed to confirm the one-to-one correspondence of EOMs.

Our findings indicate that the latest common ancestors of the vertebrates would have possessed three dorsal head mesodermal portions, although clear morphological segmentation may not have been present. Although the correspondence between the cranial nerves and the mesodermal components is suggestive of a somitomeric type of segmental organization, developmental regionalization and specification do not necessarily indicate the presence of mesodermal segments by themselves. The present data thus support neither presence nor absence of somite-like mesomeres in the ancestral head. Nevertheless, the ancestral dorsal head mesoderm would have also been specified by *Pitx2*, *Gsc* and *Tbx1/10*, and differentiated into EOMs, in a distribution pattern more or less similar to that seen in the modern lampreys.

It remains enigmatic how the head mesoderm and extrinsic eye muscles arose in the vertebrate ancestor. The data from the present study do not resolve this issue. Nevertheless, the present observations suggest that the dorsal head mesoderm is likely to have gone through tripartite regionalization during development by the time of the latest common ancestor of vertebrates. Further comparative developmental studies will be needed to gain insight into the origin of this vertebrate-specific embryonic structure.

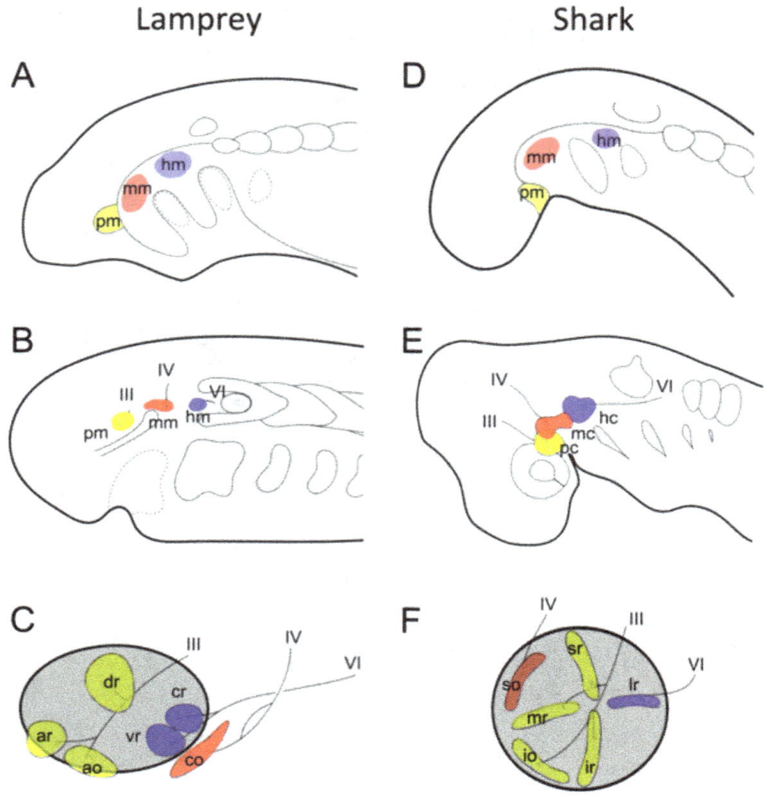

Fig. 11 Schematic illustration of the comparison of the EOMs development. **a**–**c** lamprey. **d**–**f**: shark. **a**, **d** Pharyngeal stages. **b**, **e** Three head mesodermal portions innervated by respective motor nerves. **c**, **f** Differentiated state

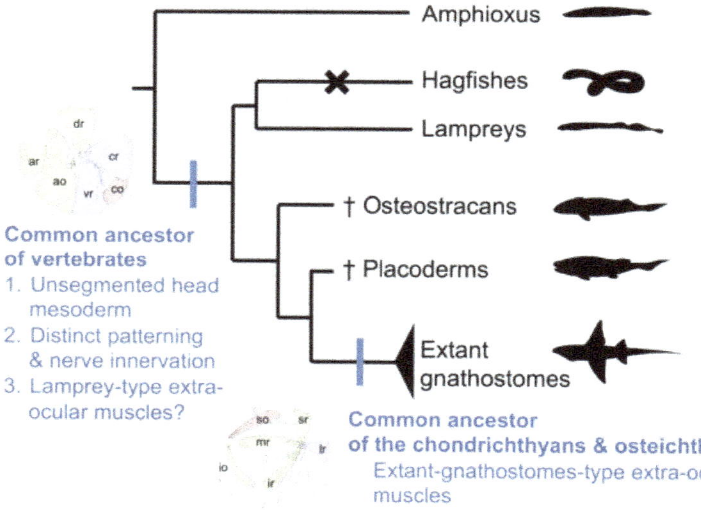

Fig. 12 Hypothetical scenario for the evolution of the EOMs. In the common ancestor of the vertebrates had unsegmented head mesoderm but there were three mesodermal portions with distinct genetic patterning and motor nerve innervation. These had lamprey-type EOMs, which was conserved in Osteostracans and Placoderms (in hagfishes, EOMs are completely degenerated). In the common ancestor of the crown gnathostomes, the disposition of the EOMs changed to the extant-gnathostome-type

Conclusions

We conclude that the EOMs in lamprey developed from three components in the dorsal head mesoderm, which are genetically and regionally specified without segmental boundaries. This developmental mechanism is conserved among vertebrates, indicating that the tripartite origin of EOMs was established in the common ancestor of the vertebrates. Furthermore, our results support the hypothesis that the common ancestor of the vertebrates possessed lamprey-type EOMs, and that the disposition was modified secondarily in the common ancestral lineage of the chondrichthyans and osteichthyans.

Nomenclature

ao anterior oblique
ar anterior rectus
cb constrictor buccalis
co caudal oblique
cr caudal rectus
dr dorsal rectus
elv elevator labialis ventralis
gV_1 ophthalmicus profundus nerve ganglion
$gV_{2,3}$ maxillomandibular nerve ganglion
gVII facial nerve ganglion
hc hyoid cavity
hm hyoid mesoderm
io inferior oblique
ir inferior rectus
lr ateral rectus
mc mandibular cavity
mm mandibular mesoderm
mr medial rectus
oe oral epithelium
pc premandibular cavity
pm premandibular mesoderm
pp phalangeal pouch
rpe retinal pigment epithelium
vr ventral rectus
II optic nerve
III oculomotor nerve
IV trochlear nerve
VI abducens nerve

Competing interests
The authors declare that they have no competing interests.

Authors' contributions
DGS conceived the project, designed and carried out the morphological, immunofluorescence and gene expression studies, and participated in the drafting of the manuscript. YF carried out the 3D reconstructions. MY participated in the gene expression studies. YY participated in the preparation of larval materials, and helped to draft the manuscript. JK, SK and HW participated in its design and coordination, and helped to draft the manuscript. All authors read and approved the final manuscript.

Acknowledgements
We thank the anonymous reviewers for their valuable comments and suggestions that improved the quality of the article. A preliminary part of this study was carried out under the Medical Research Internship program of Okayama University (MRI). Part of this study was also supported by Japan Society for the Promotion of Science (JSPS); Grant numbers: 13 J00621 (to DGS).

Author details
[1]Graduate School of Life and Environmental Sciences, University of Tsukuba, 1-1-1 Tennodai, Tsukuba, Ibaraki 305-8572, Japan. [2]Laboratory for Evolutionary Morphology, RIKEN, Kobe 650-0047, Japan. [3]Department of Cytology and Histology, Okayama University Graduate School of Medicine, Dentistry and Pharmaceutical Sciences, 2-5-1 Shikata-cho, Okayama 700-8558, Japan. [4]Sumitomo Besshi Hospital, 3-1 Oji-cho, Niihama, Ehime 792-8543, Japan. [5]Graduate School of Science and Engineering for Research, University of Toyama, 3190 Gofuku, Toyama 930-8555, Japan. [6]Center for Medical Science, International University of Health and Welfare, 2600-1 Kitakanemaru, Ohtawara, Tochigi 324-8501, Japan.

References

1. Gilland E, Baker R. Conservation of neuroepithelial and mesodermal segments in the embryonic vertebrate head. Acta Anat. 1993;148:110–23.
2. Holland LZ, Holland ND, Gilland E. Amphioxus and the evolution of head segmentation. Integr Comp Biol. 2008;48(5):630–46.
3. Kuratani S. Evolutionary developmental biology and vertebrate head segmentation: a perspective from developmental constraint. Theory Biosci. 2003;122:230–51.
4. Onai T, Irie N, Kuratani S. The evolutionary origin of the vertebrate body plan: the problem of head segmentation. Annu Rev Genomics Hum Genet. 2014;15:443–59.
5. Neal HV, Rand HW. Comparative anatomy. Philadelphia: Blakiston; 1946.
6. Brachet A. Traité d'Embryologie des Verteébrés. Paris: Masson & Cie, Etiteurs; 1935.
7. Fraser EA. The head cavities and development of the eye muscles in *Trichosurus vulpecula*, with notes on some other marsupials. Proc Zool Soc. 1915;22:299–346.
8. Girbert PW. The origin and development of the head cavities in the human embryo. J Morph. 1947;90:149–88.
9. Girbert PW. The premandibular head cavities in the opossum, *Didelphys virginiana*. J Morph. 1954;95:47–75.
10. Girbert PW. The origin and development of the human extrinsic ocular muscles. Cont Embryol. 1957;36:59–78.
11. Jacob M, Jacob HJ, Wachtler F, Christ B. Ontogeny of avian extrinsic ocular muscles. I. A light and electron-microscopic study. Cell Tissue Res. 1984;237:549–57.
12. Kundrát M, Janaáček J, Martin S. Development of transient head cavities during early organogenesis of the nile crocodile (*Crocodylus niloticus*). J Morph. 2009;270:1069–83.
13. Kuratani S, Nobusada Y, Saito H, Shigetani Y. Morphological characteristics of the developing cranial nerves and mesodermal head cavities in sturgeon embryos from early pharyngula to late larval stages. Zool Sci. 2000;17:911–33.
14. Starck D. Die Metamerie des Kopfes der Wirbeltiere. Zool Anz. 1963;170: 393–428.
15. Wedin B. The origin and development of the extrinsic ocular muscles in the alligator. J Morph. 1953;92:303–36.
16. Adelmann HB. The development of the premandibular head cavities and the relations of the anterior end of the notochord in the chick and robin. J Morph Phys. 1926;42:371–439.
17. Jacobson AG. Somitomeres: Mesodermal segments in the head and trunk. In: Hanken J, Hall BK, editors. The vertebrate skull, vol. 1. Chicago: University of Chicago Press; 1993. p. 42–76.
18. Wachtler F, Jacob HJ, Jacob M, Christ B. The extrinsic ocular muscles in birds are derived from the prechordal plate. Naturwissenschaften. 1984;71:379–80.
19. Couly GF, Coltey PM, Le Douarin NM. The triple origin of skull in higher vertebrates: a study in quail-chick chimeras. Dev. 1993;117:409–29.
20. Noden DM. The embryonic origins of avian cephalic and cervical muscles and associated connective tissues. Am J Anat. 1983;168:144–65.

21. Noden DM. Interactions and fates of avian craniofacial mesenchyme. Dev. 1988;103(Suppl):121–40.
22. Noden DM, Francis-West P. The differentiation and morphogenesis of craniofacial muscles. Dev Dyn. 2006;235:1194–218.
23. Noden DM, Trainor PA. Relations and interactions between cranial mesoderm and neural crest populations. J Anat. 2005;207:575–601.
24. Wachtler F, Jacob M. Origin and development of the cranial skeletal muscles. Biblthca Anat. 1986;29:24–46.
25. Koltzoff NK. Entwicklungsgeschichte des Kopfes von *Petromyzon planeri*. Bull Soc Nat Moscou. 1901;15:259–89.
26. Neal HV. The history of the eye muscles. J Morphol. 1918;30:433–53.
27. Damas H. Recherches sur le développement de *Lampetra fluviatilis* L.—Contribution à l'étude de la cephalogénèse des vertébrés. Arch Biol Paris. 1944;55:1–289.
28. Kuratani S, Horigome N, Hirano S. Developmental morphology of the head mesoderm and reevaluation of segmental theories of the vertebrate head: evidence from embryos of an agnathan vertebrate, *Lampetra japonica*. Dev Biol. 1999;210:381–400.
29. Sewertzoff AN. Die Kiemenbogennerven der Fische. Anat Anz. 1911;38:487–95.
30. Fritzsch B, Sonntag R, Dubuc R, Ohta Y, Grillner S. Organization of the six motor nuclei innervating the ocular muscles in lamprey. J Comp Neurol. 1990;294:491–506.
31. Nishi S. Augenmuskulatur. In: Bolk L, Göppert E, Kallius E, Lubosch W, editors. Handbuch der vergreichenden Anatomie der Wirbeltiere, Band 5. Wien: Urban und Schwarzenberg; 1938. p. 453–66.
32. Shimazaki S. Kontribuo al la kompara anatomio de okulmuskoloj če ciklostomoj kaj fîsoj. Acta Anat Nip. 1965;40:354–67.
33. Tahara Y. Normal stages of development in the lamprey, *Lampetra reissneri* (Dybowski). Zool Sci. 1988;5:109–18.
34. Yamazaki Y, Goto A. Genetic structure and differentiation of four *Lethenteron* taxa from the Far East, detected from allozyme analysis. Env Biol Fish. 1998;52:149–61.
35. Yamazaki Y, Yokoyama R, Nishida M, Goto A. Taxonomy and molecular phylogeny of *Lethenteron* lampreys in eastern Eurasia. J Fish Biol. 2006; 68:251–69.
36. Adachi N, Kuratani S. Development of head and trunk mesoderm in the dogfish, *Scyliorhinus torazame*: I. Embryology and morphology of the head cavities and related structures. Evol Dev. 2012;14:234–56.
37. Ballard WW, Mellinger J, Lechenault HA. A series of normal stages for development of *Scyliorhinus canicula*, the lesser spotted dogfish (Chondrichthyes: Scyliorhinidae). J Exp Zool. 1993;267:318–36.
38. Kuratani S, Ueki T, Aizawa S, Hirano S. Peripheral development of cranial nerves in a cyclostome, *Lampetra japonica*: morphological distribution of nerve branches and the vertebrate body plan. J Comp Neurol. 1997;384: 483–500.
39. Cerny R, Cattell M, Sauka-Spengler T, Bronner-Fraser M, Yu F, Medeiros DM. Evidence for the prepattern/cooption model of vertebrate jaw evolution. Proc Natl Acad Sci U S A. 2010;107:17262–7.
40. Uchida K, Murakami Y, Kuraku S, Hirano S, Kuratani S. Development of the adenohypophysis in the lamprey: evolution of epigenetic patterning programs in organogenesis. J Exp Zool B. 2003;300:32–47.
41. Kusakabe R, Kuraku S, Kuratani S. Expression and interaction of muscle-related genes in the lamprey imply the evolutionary scenario for vertebrate skeletal muscle, in association with the acquisition of the neck and fins. Dev Biol. 2011;350:217–27.
42. Tiecke E, Matsuura M, Kokubo N, Kuraku S, Kusakabe R, Kuratani S, Tanaka M. Identification and developmental expression of two *Tbx1/10*-related genes in the agnathan *Lethenteron japonicum*. Dev Genes Evol. 2007;217:691–7.
43. Ogasawara M, Shigetani Y, Hirano S, Satoh N, Kuratani S. *Pax1/Pax9*-related genes in an agnathan vertebrate, *Lampetra japonica*: expression pattern of *LjPax9* implies sequential evolutionary events toward the gnathostome body plan. Dev Biol. 2000;223:399–410.
44. Kleerekoper H. The sense organ. In: Hardisty MW, Potter IC, editors. The biology of lampreys, vol. 2. London: Academic; 1972. p. 373–404.
45. Villar-Cerviño V, Abalo XM, Villar-Cheda B, Meléndez-Ferro M, Pérez-Costas E, Holstein GR, et al. Presence of glutamate, glycine, and γ-aminobutyric acid in the retina of the larval sea lamprey: comparative immunohistochemical study of classical neurotransmitters in larval and postmetamorphic retinas. J Comp Neurol. 2006;499:810–27.
46. Suzuki DG, Murakami Y, Escriva H, Wada H. A comparative examination of neural circuit and brain patterning between the lamprey and amphioxus reveals the evolutionary origin of the vertebrate visual center. J Comp Neurol. 2015;523:251–61.
47. Suzuki DG, Murakami Y, Yamazaki Y, Wada H. Expression patterns of *Eph* genes in the "dual visual development" of the lamprey and their significance in the evolution of vision in vertebrates. Evol Dev. 2015; 17:139–47.
48. Hardisty MW, Rovainen CM. Morphological and functional aspects of the muscular system. In: Hardisty MW, Potter IC, editors. The biology of lampreys, vol. 4A. London: Academic; 1982. p. 137–232.
49. Sambasivan R, Gayraud-Morel B, Dumas G, Cimper C, Paisant S, Kelly RG, Tajbakhsh S. Distinct regulatory cascades govern extraocular and pharyngeal arch muscle progenitor cell fate. Dev Cell. 1999;15:310–21.
50. Adachi N, Takechi M, Hirai T, Kuratani S. Development of the head and trunk mesoderm in the dogfish, *Scyliorhinus torazame*: II. Comparison of gene expression between the head mesoderm and somites with reference to the origin of the vertebrate head. Evol Dev. 2012;14:257–76.
51. De Robertis EM, Fainsod A, Gont LK, Steinbeisser H. The evolution of vertebrate gastrulation. Dev Suppl. 1994;117–24.
52. Sander V, Reversade B, De Robertis EM. The opposing homeobox genes *Goosecoid* and *Vent1/2* self-regulate *Xenopus* patterning. EMBO J. 2007; 26(12):2955–65.
53. Kelly RG, Jerome-Majewska LA, Papaioannou VE. The del22q11.2 candidate gene *Tbx1* regulates branchiomeric myogenesis. Hum Mol Genet. 2004;13:2829–40.
54. Schulte-Merker S, Hammerschmidt M, Beuchle D, Cho KW, De Robertis EM, Nüsslein-Volhard C. Expression of zebrafish *goosecoid* and *no tail* gene products in wild-type and mutant *no tail* embryos. Dev. 1994;120:843–52.
55. Belo JA, Leyns L, Yamada G, De Robertis EM. The prechordal midline of the chondrocranium is defective in *Goosecoid*-1 mouse mutants. Mech Dev. 1998;72:15–25.
56. Begemann G, Gibert Y, Meyer A, Ingham PW. Cloning of zebrafish T-box genes *tbx15* and *tbx18* and their expression during embryonic development. Mech Dev. 2002;114:137–41.
57. Showell C, Christine KS, Mandel EM, Conlon FL. Developmental expression patterns of *Tbx1*, *Tbx2*, *Tbx5*, and *Tbx20* in *Xenopus tropicalis*. Dev Dyn. 2006; 235:1623–30.
58. Garg V, Yamagishi Y, Hu T, Kathiriya IS, Yamagishi H, Srivastava D. *Tbx1*, a DiGeorge syndrome candidate gene, is regulated by sonic hedgehog during pharyngeal arch development. Dev Biol. 2001;235:62–73.
59. Janvier P. Les céphalaspides du Spitsberg. Paris: Centre National de la Recherche Scientifique; 1985.
60. Janvier P. Early vertebrates. Oxford: Clarendon; 1996.
61. Young GC. The relationships of placoderm fishes. Zool J Linn Soc. 1986; 88:1–57.
62. Young GC. Number and arrangement of extraocular muscles in primitive gnathostomes: evidence from extinct placoderm fishes. Biol Lett. 2008;4: 110–4.
63. Brazeau MD. The braincase and jaws of a Devonian 'acathodian' and modern gnathostome origins. Nature. 2009;457:305–8.
64. Davis SP, Finarelli JA, Coates MI. Acanthodes and shark-like conditions in the last common ancestor of modern gnathostomes. Nature. 2012;486:247–50.
65. Zhu M, Yu XB, Ahlberg PE, Choo B, Lu J, Qiao T, Qu QM, Zhao WJ, Jia LT, Blom H, Zhu YA. A Silurian placoderm with osteichthyan-like marginal jaw bones. Nature. 2013;502:188–93.

11

Bivalve-specific gene expansion in the pearl oyster genome: implications of adaptation to a sessile lifestyle

Takeshi Takeuchi[1*], Ryo Koyanagi[2], Fuki Gyoja[1], Miyuki Kanda[2], Kanako Hisata[1], Manabu Fujie[2], Hiroki Goto[2], Shinichi Yamasaki[2], Kiyohito Nagai[3], Yoshiaki Morino[4], Hiroshi Miyamoto[5], Kazuyoshi Endo[6], Hirotoshi Endo[7], Hiromichi Nagasawa[8,9], Shigeharu Kinoshita[10], Shuichi Asakawa[10], Shugo Watabe[10,11], Noriyuki Satoh[1] and Takeshi Kawashima[1,12]

Abstract

Introduction: Bivalve molluscs have flourished in marine environments, and many species constitute important aquatic resources. Recently, whole genome sequences from two bivalves, the pearl oyster, *Pinctada fucata*, and the Pacific oyster, *Crassostrea gigas*, have been decoded, making it possible to compare genomic sequences among molluscs, and to explore general and lineage-specific genetic features and trends in bivalves. In order to improve the quality of sequence data for these purposes, we have updated the entire *P. fucata* genome assembly.

Results: We present a new genome assembly of the pearl oyster, *Pinctada fucata* (version 2.0). To update the assembly, we conducted additional sequencing, obtaining accumulated sequence data amounting to 193× the *P. fucata* genome. Sequence redundancy in contigs that was caused by heterozygosity was removed *in silico*, which significantly improved subsequent scaffolding. Gene model version 2.0 was generated with the aid of manual gene annotations supplied by the *P. fucata* research community. Comparison of mollusc and other bilaterian genomes shows that gene arrangements of Hox, ParaHox, and Wnt clusters in the *P. fucata* genome are similar to those of other molluscs. Like the Pacific oyster, *P. fucata* possesses many genes involved in environmental responses and in immune defense. Phylogenetic analyses of heat shock protein70 and C1q domain-containing protein families indicate that extensive expansion of genes occurred independently in each lineage. Several gene duplication events prior to the split between the pearl oyster and the Pacific oyster are also evident. In addition, a number of tandem duplications of genes that encode shell matrix proteins are also well characterized in the *P. fucata* genome.

Conclusions: Both the *Pinctada* and *Crassostrea* lineages have expanded specific gene families in a lineage-specific manner. Frequent duplication of genes responsible for shell formation in the *P. fucata* genome explains the diversity of mollusc shell structures. These duplications reveal dynamic genome evolution to forge the complex physiology that enables bivalves to employ a sessile lifestyle in the intertidal zone.

Keywords: Pearl oyster, *Pinctada fucata*, Genome, Hox, ParaHox, Heat shock proteins, C1q, Biomineralization

* Correspondence: t.takeuchi@oist.jp
[1]Marine Genomics Unit, Okinawa Institute of Science and Technology Graduate University, Onna, Okinawa 904-0495, Japan
Full list of author information is available at the end of the article

Introduction

Bivalves are the second largest group in the Phylum Mollusca, outnumbered only by gastropods, and represent one of the most common animal groups in both marine and freshwater ecosystems. Notably, some bivalve species are abundant in littoral and shallow water environments, where they experience different types of fluctuating stresses, such as temperature, salinity, and oxygen concentration. Most adult bivalves employ a sessile, suspension-feeding life style, the energetic cost of which is lower than that of browsing [1]. Due to their aquatic habitats and filter feeding, they must defend themselves against microbial invasion by means of innate immune systems. In addition, sedentary bivalves cannot escape predation; therefore, they protect themselves with calcareous shells. These biological features of the bivalve adaptation strategy have resulted from expansion of specific gene families related to environmental response [2], immune defense [2–4], and biomineralization [5]. These findings raise the question of whether these gene expansions occurred in bivalve common ancestor or independently in various lineages.

The Genus *Pinctada* includes the pearl oysters, such as *Pinctada fucata*, *P. margaritifera*, and *P. maxima*, which are distributed in subtropical and tropical portions of the Indo-Pacific Ocean [6]. These species have been commercially farmed for pearl production since Kokichi Mikimoto established the pearl culture industry at the end of 19th century, using *P. fucata* [7]. Recently, transcriptomics [8, 9], proteomics [10–12], and gene knockdown techniques [13–15] have been used to investigate genetic components of shell and pearl biomineralization. Thus, mechanisms of pearl formation in *Pinctada* have been actively investigated for their economic potential, as well as their fascinating biology. Now, pearl oysters are becoming experimental model molluscs for biomineralization research.

In 2012, we decoded the draft genome of *Pinctada fucata* [16], one of the most important species for cultured pearl production in Asia. The genome of *P. fucata* has been thoroughly mined to find genes responsible for biomineralization [5], physiology [17], and reproduction [18]. A broad range of transcription factors [19–21] and signaling molecules [22] has also been investigated. These provide valuable information about lophotrochozoans to better understand evolution of Bilaterian body plans. Soon after publication of the *P. fucata* genome, genomes of the Pacific oyster, *Crassostrea gigas* [2] and the limpet, *Lottia gigantea* [23] were also published. The growing body of molluscan genome data provides an opportunity to characterize general and unique features among molluscs, for which sequence information has, until recently, been scant.

The present study produced a new version of the *P. fucata* genome assembly (version 2.0), which provides longer contigs and scaffolds, and more consecutive gene arrays compared to the previous version. To improve the assembly, additional sequence data were generated, and an advanced assembly strategy addressed the heterozygotic nature of the genome. Along with the establishment of a new genome assembly, we also generated gene model version 2.0. Information on gene annotation done manually by the *P. fucata* research community [24] was used for the gene model prediction.

In this report, we surveyed bivalve-specific genomic changes. We performed molecular phylogenetic analyses for gene families that have been expanded in bivalves, including heat shock protein 70 (HSP70) and C1q domain-containing proteins (C1qDC). In addition, we thoroughly investigated shell matrix protein (SMP) gene clusters, which were partly described in the previous version of the genome assembly [5]. We also verified conserved gene clusters for *Hox*, *ParaHox*, and *Wnt* genes among bilaterians using the new *Pinctada* genome assembly.

Methods

Genome sequencing and assembly

Genomic DNA, which is identical to that obtained in the previous study [16], was prepared for paired-end libraries and sequenced with an Illumina MiSeq and a Genome Analyzer IIx (GAIIx) [25]. Raw reads were quality trimmed using Trimmomatic 0.30 [26]. The whole-genome shotgun (WGS) and paired-end reads sequenced by Takeuchi et al. (2012) [16] and this study were assembled using GS De Novo Assembler, version 2.6 (Newbler, Roche) [27]. After removing redundant sequences from the contig assembly, paired-end and mate-pair sequences were added for scaffolding performed with SSPACE 1.1 [28]. Gaps in scaffolds were filled using GapCloser 1.12 [29]. See Additional file 1: note for more detail.

Transcriptome sequencing and assembly

Transcriptome sequencing used in this study is described in Takeuchi et al. (2012) [16]. Additionally, a cDNA library of early developmental stages and adult tissues, including mantle, was prepared and sequenced with an Illumina GAIIx. All sequences were cleaned and trimmed with Trimmomatic 0.30 [26], and then assembled using Trinity (version r20140413p1) [30].

Gene prediction, annotation, and identification of gene families

The resulting genome assembly (*P. fucata* genome ver. 2.0) and transcriptomic data were used for *de novo* gene model prediction with PASA (version r20130907) [31] and AUGUSTUS 3.0.2 [32] platforms, as described previously [16]. Gene annotation information manually confirmed in Pearl Oyster Annotation Jamborees [24] was added in order to train the gene prediction algorithm and to generate a hint file for AUGUSTUS. Gene models that encoded more than 49 amino acids were retained. Sequences significantly similar to transposable elements were detected with CENSOR 4.2.28 [33] and excluded from the gene model. Gene models of *P. fucata*, *Lottia gigantea* [23], and *Crassostrea gigas* [2] were assigned to the ortholog group of the OrthoMCL Database version 5 [34, 35]. Three molluscan gene models that were not assigned to the OrthoMCL ortholog group were then clustered with local OrthoMCL software in order to identify mollusc-specific gene families. Next, gene models that did not cluster with others ("orphan gene models") were examined with BLASTN against *P. fucata* transcriptomic sequences. Orphan gene models without transcriptomic evidence were excluded from the final gene model set, named gene model, version 2.0.

Gene model version 2.0 was BLASTN-searched against version 1.0 models and the best match model was regarded as synonymous. Hox, ParaHox, Wnt, and SMP genes were annotated by referring to previous studies [5, 21, 22]. SMP genes were also searched with BLASTP against SMPs of *P. margaritifera* and *P. maxima* [10] and were annotated manually. BLASTP searches against the UniProtKB and NCBI non-redundant (nr) databases were also conducted in order to confirm annotations. Results of BLASTP searches (E-value threshold $1e^{-5}$) against protein sequences of the UniProtKB database, were analyzed with PANTHER [36] for GO annotation. GO enrichment of genes conserved between *P. fucata* and *C. gigas*, but not *L. gigantea*, was investigated by calculating a *P* value based on hypergeometric distribution.

Molecular phylogeny of expanded gene families

Conserved domains of proteins predicted in the *Pinctada fucata* genome were searched against the Pfam database (Pfam-A.hmm, release 24.0; http://pfam.xfam.org/) using HMMER 3.0 [37]. Representative animal genomes were also surveyed for comparison and phylogenetic analysis. The following protein sequences from public databases were retrieved for this study; *Acropora digitifera* (http://marinegenomics.oist.jp/), *Lottia gigantea*, *Helobdella robusta*, *Capitella teleta*, *Daphnia pulex*, and *Nematostella vectensis* (http://genome.jgi.doe.gov/), *Crassostrea gigas* (http://gigadb.org), *Hydra magnipapillata*, *Amphimedon queenslandica* (http://www.ncbi.nlm.nih.gov) and *Caenorhabditis elegans*, *Drosophila melanogaster*, *Branchiostoma floridae*, *Strongylocentrotus purpuratus*, and *Homo sapiens* (http://www.genome.jp). Amino acid sequences of *Mytilus galloprovincialis* C1q domain-containing proteins were downloaded from NCBI. Amino acid sequences were aligned with Muscle [38], then maximum likelihood trees were constructed with RAxML 8.1.3 [39]. The best-fit model of amino acid evolution for each tree was selected using ProteinModelSelection.pl script provided in the RAxML package. One hundred bootstrap replicates were generated.

Results and discussion

New genome assembly

To update the assembly, additional sequencing was conducted. Results of genome sequencing are summarized in Additional file 1: Table S1. High-throughput sequencing added more than 1 billion reads, and accumulated sequence data represented ~193× the *P. fucata* genome, which has an estimated genome size of 1.14Gb [16]. The *P. fucata* genome is known to be highly heterozygotic, which complicates assembly [16]. To address this, we removed redundant contigs with low sequence coverage depth before scaffold construction (see Additional file 2: Figure S1 and Additional file 3 for detail). As a result, the final genome assembly (version 2.0) achieved contig and scaffold N50 sizes of 21.3 kb and 167 kb, which are respectively 20× and 11× longer than those of the previous version (version 1.0), (Table 1 and Additional file 1: Table S2). Our results show that contig filtering based on sequence similarity and coverage depth is a very efficient way to reduce redundancy of the assembly and to improve subsequent scaffolding. In consequence of the scaffold elongation, gene clusters and tandem arrangement of genes were clarified, as discussed below.

New gene models

In the new *P. fucata* genome assembly, 30,619 gene models were initially predicted using AUGUSTUS software. Clustering with OrthoMCL shows that 5,096 gene models are not grouped with any other genes in OrthoMCL DB, in the *Lottia*, *Crassostrea*, and *Pinctada* genomes. Among these "orphan" gene models, sequences of 1,266 models were not found in the merged transcriptomic assemblies. We regarded them as faultily predicted gene models and filtered them out of the final gene model. Accordingly, 29,353 models were retained and 23,516 (80.1 %) represent

Table 1 Summary of the *P. fucata* genome assembly version 2.0

Contigs	
Number of contigs	77192
Total length of sequences	760 Mb
Average contig length	9.8 kb
Contig length N50	21.3 kb
Scaffolds	
Number of scaffolds	29306
Total length of sequences	815 Mb
Total gap length	55 Mb
Average scaffold length	27.8 kb
Scaffold length N50	167.0 kb

putative, full-length genes with both start and stop codons (Table 2). The average number of exons per gene was more than double that in the previous version. Among all gene models, 19,533 (66.5 %) were BLASTP hits to the UniProtKB database (e-value cutoff $1e^{-5}$).

The numbers of gene family classified by orthoMCL software are shown in Fig. 1a and b. The numbers of genes included in each gene family are presented in Additional file 2; Figure S2. The new *P. fucata* gene model contains 20,179 genes, assigned to 9,285 gene families using orthoMCL DB Version 5 (Fig. 1a). Of these, 6,660 gene families, which include 15,998 *P. fucata* genes, are shared among the three molluscs (Fig. 1a, Additional file 2: Figure S2a). While orthoMCL DB Version 5 includes as many as 150 organismal genomes, only one lophotrochozoan genome (*Schistosoma mansoni*) is present in the database. In order to identify mollusc-specific gene families, we collected and grouped the *Pinctada*, *Crassostrea*, and *Lottia* genes that were not assigned to an orthoMCL gene family (Fig. 1b). We analyzed the remaining 9,174 gene models (29,353–20,179) that were not assigned in orthoMCL DB, and assigned an additional 5,344 gene models into 2,059 novel gene families (Fig. 2b, Additional file 2: Figure S2d). Finally we categorized the residual 3,830 gene models (9,174–5,344) as orphan gene models since they do not have any sequence similarity to known genes (Fig. 1c). *Pinctada*-specific (779), *Crassostrea*-specific (658), and bivalve-specific (827) gene families were detected, while gene families shared among the three molluscs (328) were less numerous. This feature is more apparent when comparing the number of genes corresponding to each gene family group. In the case of *P. fucata*, only 437 genes belong to shared gene families, while 2,877 genes are unique to *P. fucata* (Additional file 2: Figure S2d). In other words, novel genes that emerged in the common ancestor of bivalves and gastropods are less numerous than genes generated by *Pinctada*-lineage-specific gene expansion. Only 3–8 % of genes in the molluscan genomes are mollusc-specific and shared by molluscs, while more than 20 % are lineage-specific genes (Fig. 1c). Thus, novel genes, acquired and duplicated at the class (Classes Bivalvia and Gastropoda) or lower phylogenetic level, characterize these molluscan genomes.

In order to predict functions of genes conserved between the two bivalves or among all three molluscs, GO annotation was conducted (Fig. 2). Genes that were related to "receptor activity," "response to stimulus," "immune system process," and "extracellular region" were more abundant in bivalves compared to those of genes shared by all three molluscs. This suggests that several gene families related to environmental response and immune system are expanded in bivalves. Similarly, bivalve-specific gene expansion corresponding to "extracellular region" and "extracellular matrix" was evident, and these events were responsible, at least in part, for evolution of bivalve biomineralization-related genes. In addition, genes assigned to "biological adhesion" are significantly enriched in the bivalve lineage, possibly reflecting their sedentary lifestyles.

Table 2 Comparison of gene model versions of *P. fucata*

	Ver. 1.0 Takeuchi et al. [16]	Ver. 2.0 This study
Number of gene models	43760	29353
Number of full length gene models	23257 (53.1 %)	23516 (80.1 %)
Number of exons per transcript	3.16	7.53
Average range per gene model (bp)	6960	12203

Gene expansion of heat shock protein 70 and C1q domain-containing proteins

Recent genomic and transcriptomic surveys have shown that some gene families involved in responses to environmental change or microbial attack, are greatly expanded in bivalves, including *Crassostrea* [2] and *Mytilus* [3, 40, 41]. However, it has been unclear whether gene expansion events are lineage-specific or common to all bivalves.

Heat shock proteins (HSPs) are molecular chaperones that maintain protein folding, and that rescue proteins that have been denatured or damaged by environmental or physiological stresses [42, 43]. The number of HSP70 genes is greatly increased in the oyster genome [2], which may enable them to survive in intertidal zones, where they are exposed to air and to significant temperature changes during tidal cycles. In fact, the HSP70 gene family is also expanded in the pearl oyster genome, while the numbers of genes with other heat shock chaperon domains such as HSP20, HSP90, and DnaJ are comparable to those of gastropods, annelids, and other animals (Fig. 3a, Additional file 1: Table S3).

We reconstructed a molecular phylogenetic tree of HSP70 proteins of five lophotrochozoan (three molluscs and two annelids) and fly genomes (Fig. 3b). The tree clearly shows two distinct groups: one is ancestral and the other is almost completely composed of bivalve genes (Fig. 3b). HSP70 genes of a polychaete, *Capitella teleta*, are also included in the latter group. Although the evolutionary relationship between these bivalve and polychaete genes is indeterminate because of the low bootstrap value, we speculate that the bivalve-dominant gene group was derived from the ancestral gene shared by molluscs and polychaetes. It then expanded in the bivalve lineage while being lost in the gastropod lineage. In the bivalve-dominant gene group, nine pairs of *P. fucata* and *C. gigas* genes are closely associated (Fig. 3b). This topology suggests that they are orthologous pairs and that these nine bivalve-specific HSP70 genes were present prior to the divergence of the

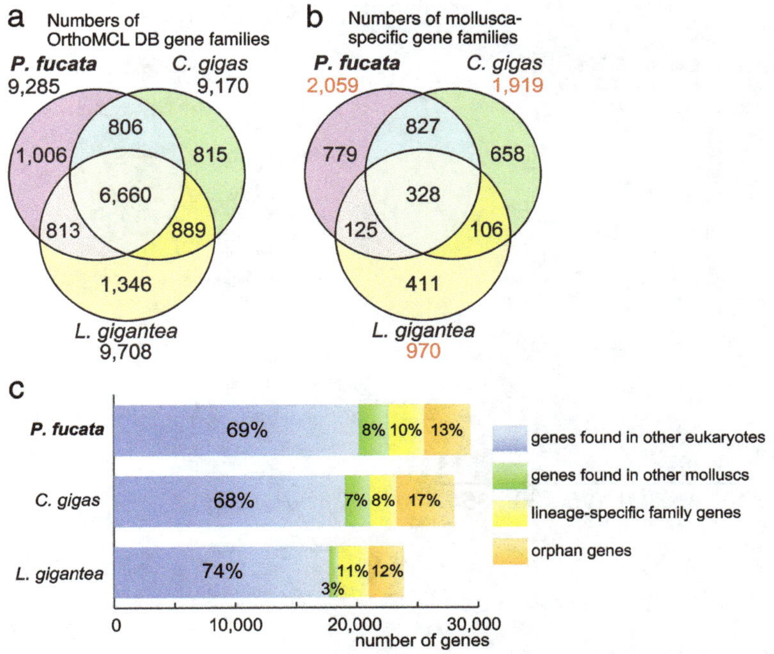

Fig. 1 Lineage-specific genes/gene families are more numerous than conserved gene families among molluscs. **a** The number of common gene families assigned using OrthoMCL DB. **b** The number of gene families not assigned using OrthoMCL DB, but detected among mollusc species. **c** Gene composition of the three mollusc genomes. Bars represent the number of genes that are either members of common gene families (assigned to an OrthoMCL DB gene family; *blue*), shared families (not assigned using OrthoMCL DB, but shared by at least two mollusc species; *green*), lineage-specific families (present only in a single mollusc genome; *yellow*), and orphan genes (with no putative homolog; *orange*)

Pinctada and *Crassostrea* lineages, which date back approximately 400 million years [44]. Furthermore, species-specific gene expansions are also observed (Fig. 3b). At least six gene families in the tree suggest independent gene expansion events in *P. fucata*, while eleven more suggest the same in *C. gigas*. Pearl oysters are mainly found in the subtidal zone, which is a more stable environment than the intertidal zone because it is always inundated. However, the subtidal and intertidal zones are still very stressful environments, and sessile bivalves (the pearl oyster and the Pacific oyster) need to cope with adverse conditions in shallow water, such as salinity and temperature changes caused by weather. Nutrient and oxygen concentrations are also changed by plankton blooms. Therefore, the independently expanded HSP70 gene family in both *P. fucata* and *C. gigas* may underlie adaptation to a sessile lifestyle in such fluctuating and stressful conditions. Moreover, aside from the bivalve-dominant group, eight groups that include HSP70 genes of four or more protostome species, are recognized (Fig. 3b), six of which are supported by high bootstrap values (≥80 %). Therefore, it is likely that the common ancestor of protostomes had approximately eight or nine HSP70 genes, and evolutionary radiation of this gene family has occurred in each lineage.

In the innate immune system, a wide repertoire of proteins is thought to be responsible for detecting pathogens and eukaryotic parasites via direct contact with surface epitopes or pathogen-associated molecular patterns (PAMPs) [45]. For example, microbial recognition capability has been described in many molluscan proteins including C1q domain-containing (C1qDC) proteins [46–48], fibrinogen-related proteins (FREPs) [49, 50], and lectins [51–53]. Bivalve genomes possess a large number of candidate genes that encode functional domains related to recognition of PAMPs (Fig. 4a and Additional file 1: Table S4). In particular, the number of C1q genes is enormously expanded in the *P. fucata* genome, consistent with the *C. gigas* genome [4] and a mussel, *Mytilus galloprovincialis*, transcriptome [3]. A Pfam domain-search detected 296 gene models with C1q domains in the *P. fucata* genome, and 335 in that of *C. gigas*, while only 12 such models were found in the *L. gigantea* genome. Multiple, lineage-specific gene expansions of C1q domain sequences are detected with high bootstrap values in the two bivalve genomes and the transcriptome of *M. galloprovincialis* [3] (Fig. 4b). Moreover, many C1qDC gene-pairs are tandemly arranged in the same scaffolds (Fig. 4b), suggesting that gene duplication events have occurred frequently in bivalve genomes. While the overall tree is supported by a low bootstrap value due to the short sequences (fewer than 120 amino acids) of the C1q domain, three sets of orthologous genes among bivalve species are detected (Fig. 4b), suggesting

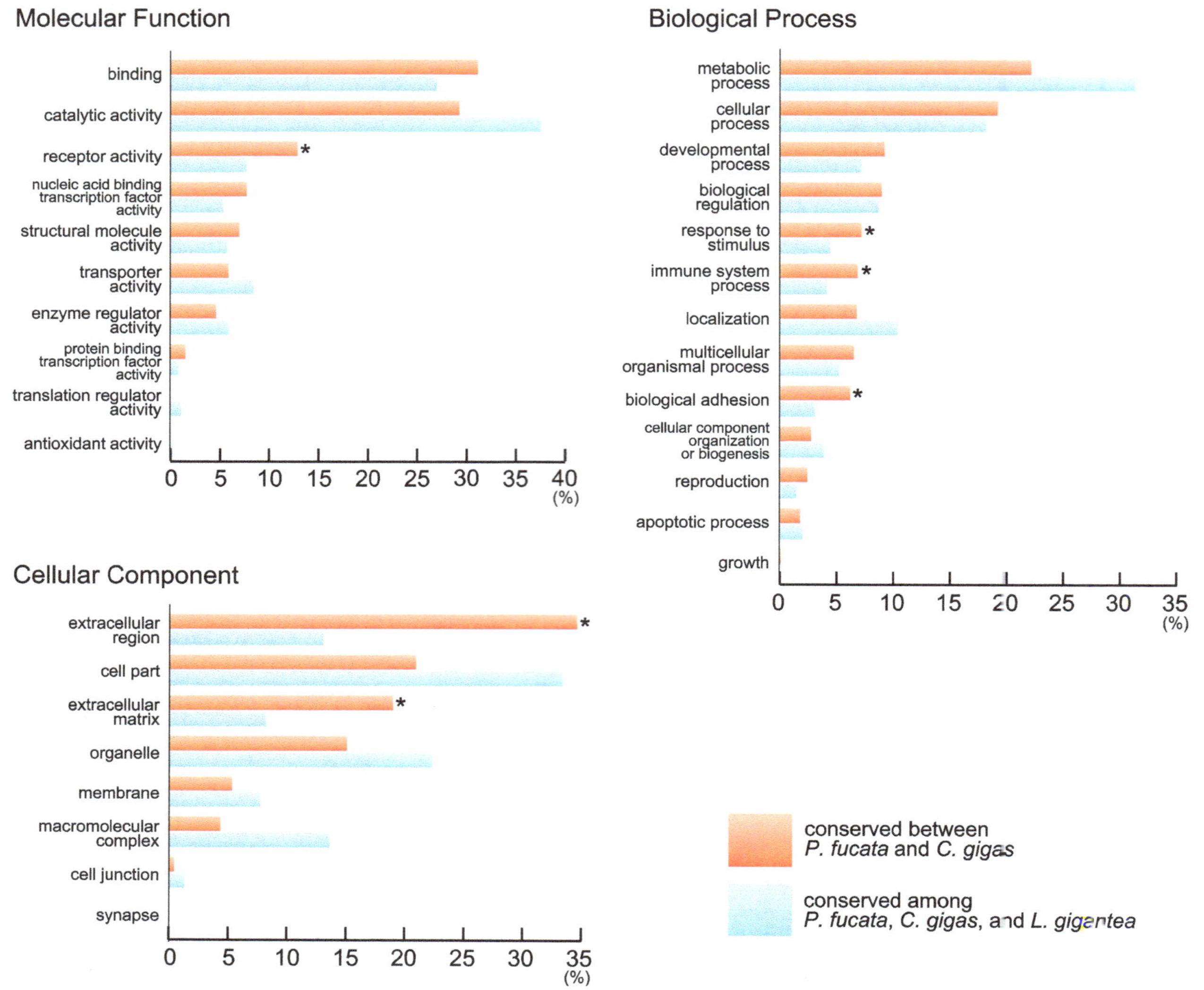

Fig. 2 Enriched GO categories may reflect the sedentary lifestyles of bivalves in the intertidal zone. The percentage of all annotated genes is shown. Asterisks indicate GO categories that are significantly ($p < 0.01$) enriched in bivalve-specific gene models

that these C1qDC genes became duplicated in the common ancestor of the three bivalves after their divergence from the gastropod lineage.

These phylogenetic analyses demonstrate that gene duplication events occurred prior to the divergence of the bivalve species examined here, and that lineage-specific gene expansion is also common in bivalve genomes. This flexible expansion of gene families has generated an immense gene repertoire associated with stress responses and immune defense. By contrast, the limpet, *Lottia*, did not acquire enlarged gene families, despite its similar habitat in the littoral zone. We suggest that because bivalves cannot escape from the adverse conditions, they must respond to the variable environment by having expanded HSP70 gene family. Their suspension-feeding system is protected from wide range of invading microorganisms. Gene expansion allows bivalves to settle in dynamic marine environments, such as the intertidal and subtidal zones.

Tandem duplication of genes responsible for shell formation

Shell formation is one of the unique features of molluscs. The process is highly controlled by the organism by secretion of an organic shell matrix, which generates an organic framework that regulates calcification of the shell [54, 55]. Shell matrix proteins (SMPs) are considered the major components of the organic shell matrix, and SMP evolution is implicated in diversification of mollusc shell characters, including morphology, microstructure, and crystal polymorphism [10, 56].

Using the previous sequence data from *P. fucata*, Miyamoto et al. (2013) generated a comprehensive list of

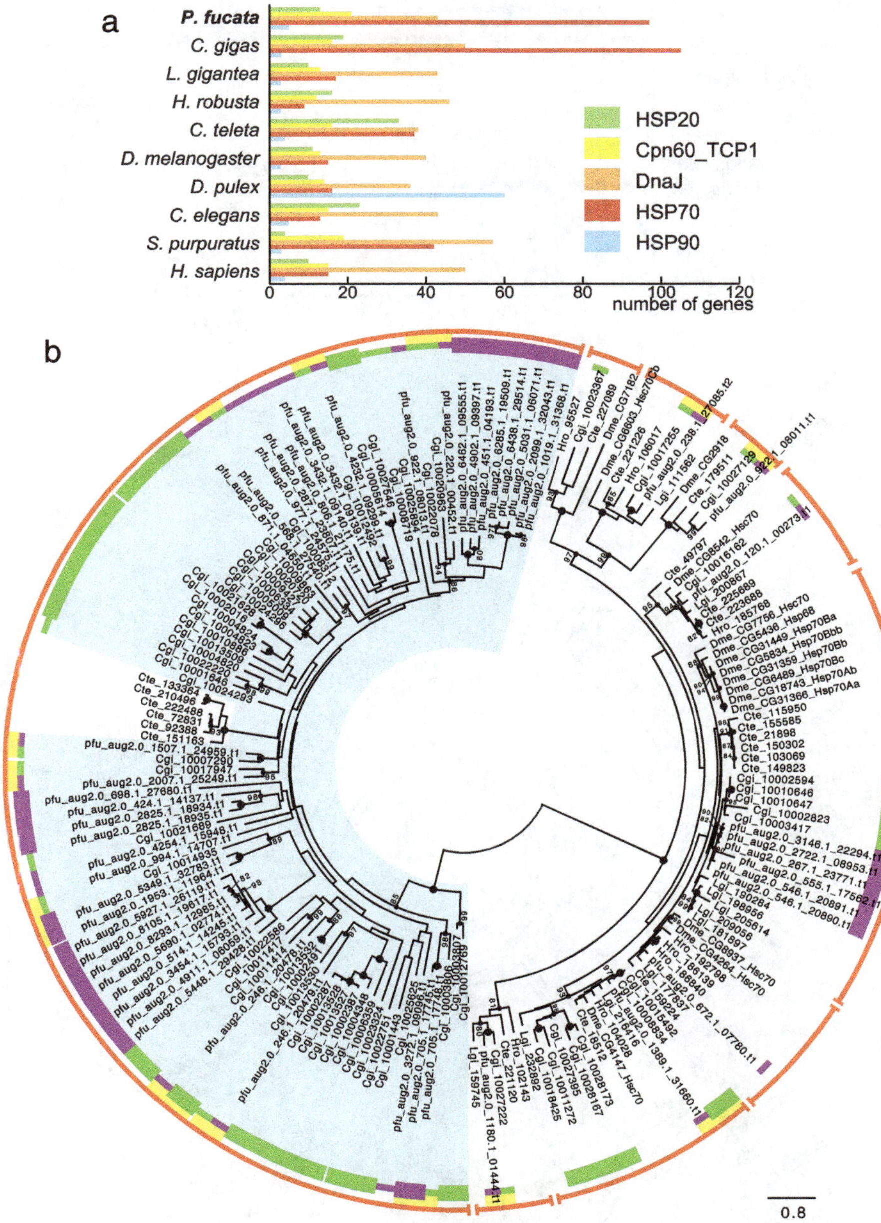

Fig. 3 Expansion of heat shock protein 70 (HSP70) genes has occurred in bivalve genomes. **a** Genes that contain conserved Pfam domains related to heat shock chaperones. See also Additional file 1: Table S4. **b** Unrooted maximum likelihood molecular phylogeny of HSP70 domain sequences of selected animal genomes. The blue area highlights a group of predominantly bivalve genes. Protein sequences of *Pinctada fucata* and *Crassostrea gigas* are marked with purple and green lines, respectively. Thick purple and green lines indicate that the clade comprises only one bivalve species, suggesting lineage-specific gene expansion. Yellow lines designate gene pairs of *P. fucata* and *C. gigas* supported by high bootstrap values (≥80 %), indicating that the genes probably existed in the common ancestor of the two bivalves. Nine outer red arcs, except the largest one on the left, show groups composed of four or more protostome genes. Nodes of the tree supported with high bootstrap values (≥80 %) are marked with black dots with the number, while larger black dots without a number indicate 100 % bootstrap support. The scale bar represents expected substitutions per site in the aligned regions. The first three letters of the protein ID indicate the species name; pfu: *P. fucata*, Cgi: *C. gigas*, Lgi: *Lottia gigantea*, Hro: *Helobdella robusta*, Cte: *Capitella teleta*, and Dme: *Drosophila melanogaster*. HSP70 domain sequences with lengths of 200 amino acids or more were used for analysis with the WAG + GAMMA + F model

shell formation-related genes, and showed that some SMP genes are tandemly arranged in the scaffold [5]. We searched further for SMP gene clusters in the new genome assembly and found an additional 14 duplicated loci for 12 SMP gene families (Fig. 5 and Additional file 1: Table S5). Of these genes, five encoding the Shematrin

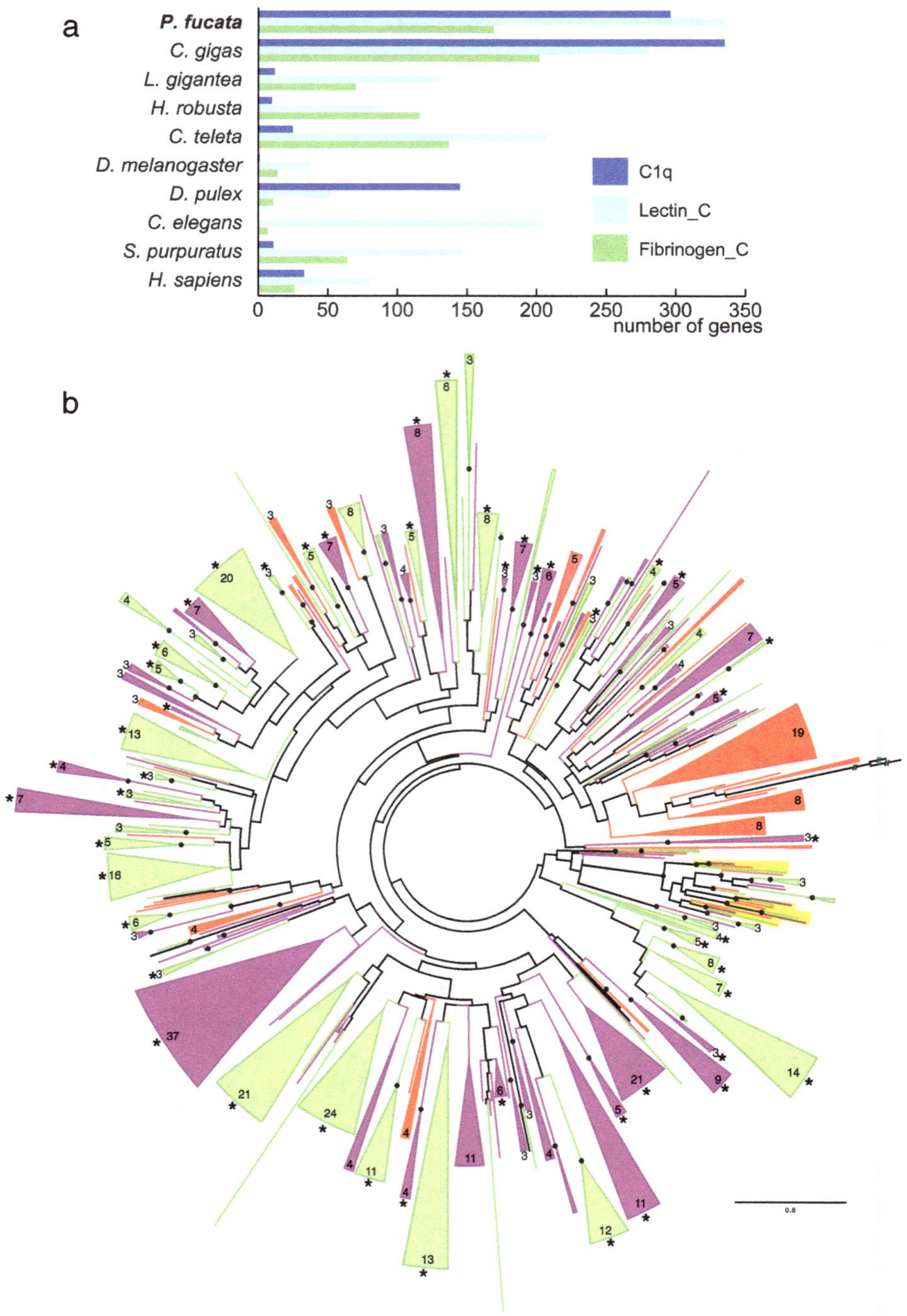

Fig. 4 Tandem duplication and expansion of genes related to innate immune recognition has occurred in all three bivalve lineages. **a** Genes that contain selected Pfam domain candidates related to recognition of non-self antigens. See also Additional file 1: Table S7. **b** Unrooted maximum likelihood molecular phylogeny of C1q domain sequences of three mollusc genomes (*Pinctada fucata*: purple, *Crassostrea gigas*: green, and *Lottia gigantea*: black), and MgC1q proteins of *Mytilus galloprovincialis* (red). Wedges indicate the merged gene group, which is composed of sequences only from bivalve species, and the number of constituent genes shown in the arc. Wedges without numbers indicate groups that include two genes. Merged groups that contain two or more tandemly arranged genes in the genome are marked with asterisks. Branches showing orthologous gene groups of three bivalves are highlighted in yellow. Nodes supported with high bootstrap values (≥80 %) are marked with black dots. Branches with slashes were collapsed in order to fit the tree onto the page. The scale bar represents expected substitutions per site in the aligned regions. C1q domain sequences with lengths of 100 amino acids or more were used for analysis with the WAG + GAMMA + F model

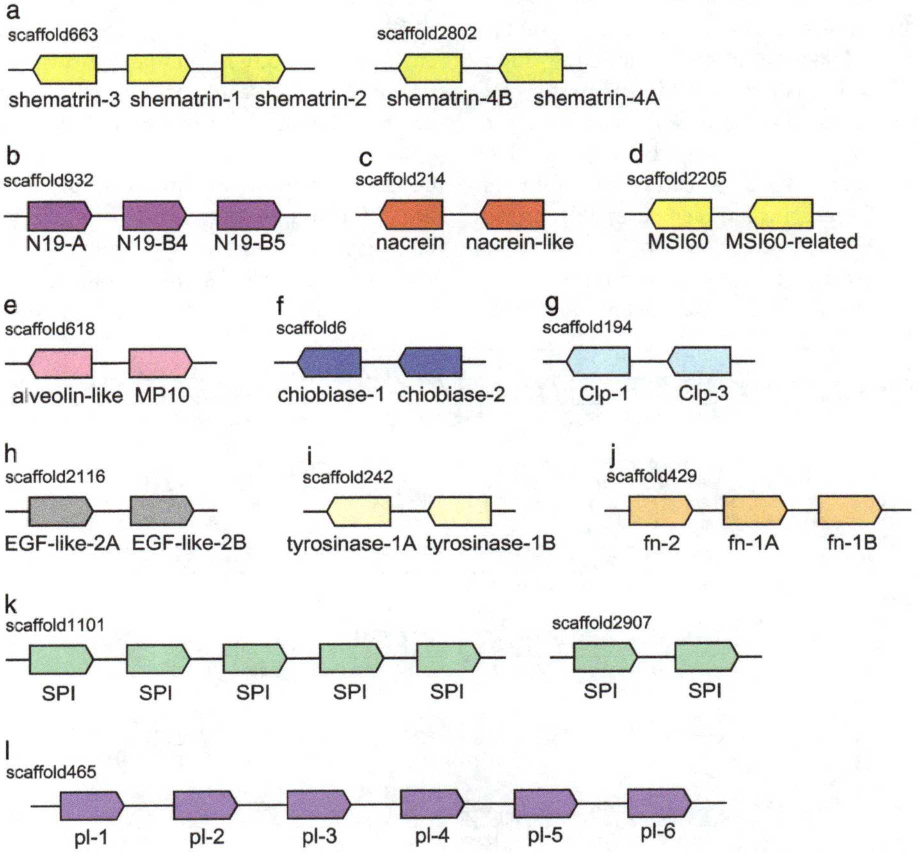

Fig. 5 SMP gene families are clustered in the *P. fucata* genome. Relative position and orientations of genes are indicated. Lengths of scaffolds, genes, and intergenic regions are not to scale. **a** Shematrin. **b** N19. **c** Nacrein and nacrein-like. **d** MSI60 and MSI60-related. **e** Alveolin-like and MP10. **f** Chiobiase. **g** Chitinase-like. **h** EGF-like (**i**) Tyrosinase. **j** Fibronectin domain-containing. **k** Serine protease inhibitor. **l** Peroxidase-like. See also Additional file 1: Table S5 for detail

family, which contains repetitive and glycine-rich proteins [57], are tandemly arranged in two scaffolds (Fig. 5a). Three genes encoding N19 [58] are also clustered in a scaffold (Fig. 5b). Genes encoding Nacrein-like and MSI60-related [5, 59, 60] proteins are detected with BLAST searches, and they are located adjacent to their relatives (Fig. 5c and d). So far, there is no direct evidence that Nacrein-like and MSI60-related proteins are involved in shell formation, and further functional analysis of these proteins is needed.

Orthologous genes that encode SMPs, reported from *P. margaritifera* and *P. maxima* [10], were also investigated. Interestingly, a number of SMP genes are tandemly arranged in the scaffolds (Fig. 5 and Additional file 1: Table S5). Pairs of genes encoding alveolin-like and MP10, chiobiase, chitinase-like protein (Clp), EGF domain-containing protein (EGF-like), and tyrosinase are present (Fig. 5e-i). Likewise, more than two tandem gene clusters of fibronectin domain-containing protein (fn), Kunitz/BPTI serine protease inhibitors (SPI), and peroxidase-like (pl) proteins are identified (Fig. 5j-l). These results show that SMP genes were frequently duplicated in the *Pinctada* genome. Tandemly arranged genes that encode SMPs (EGF domain-containing protein, peroxidase, and uncharacterized proteins) are also evident in the *Lottia* genome [61], indicating that tandem duplication of SMP genes is a common feature of bivalve and gastropods. Although the precise function of these genes in shell formation remains unknown, duplication and rapid molecular evolution [9, 62] of SMP genes may be a key feature for understanding the diversification of mollusc shell structure. It is also possible, however, that molecular functions of duplicated SMPs are redundant, and that increased gene copy number results in larger numbers of transcripts [62], to accelerate shell formation. Indeed, we reported an example in which tandem arrays of SMP genes were expressed in coordinated fashion [63]. In order to resolve the complexity of molecular evolution and functional diversification of SMPs, proteomic analysis of the *P. fucata* shell and a genome-wide gene expression survey of mantle are underway.

Among the SMP gene families discussed above, genes homologous to MSI60, Shematrin, and N19 are absent in the *C. gigas* genome. Likewise, another tandemly duplicated gene family, N16 [5], is not found in the oyster genome. In other words, these SMP gene families are unique to the *P. fucata* lineage. As mentioned before, abundant lineage-specific gene families are a feature of molluscan genomes (Fig. 1b). These gene families emerged and became duplicated in *P. fucata* lineage after the split of the pearl oyster and Pacific oyster lineages. Alternatively, it is possible that these two bivalves share an ancestral SMP gene, and that the gene evolved so rapidly that, at present, SMP genes in two bivalve genomes are significantly different from each other. In either case, rapid molecular evolution and a diverse repertoire of SMPs made possible the great variety of molluscan shell structures.

Conserved clusters of Hox, ParaHox, and Wnt genes in the *P. fucata* genome

Hox, ParaHox, and Wnt gene clusters are the most conserved synteny blocks among bilaterian genomes [64–66]. To ascertain whether the pearl oyster genome preserves

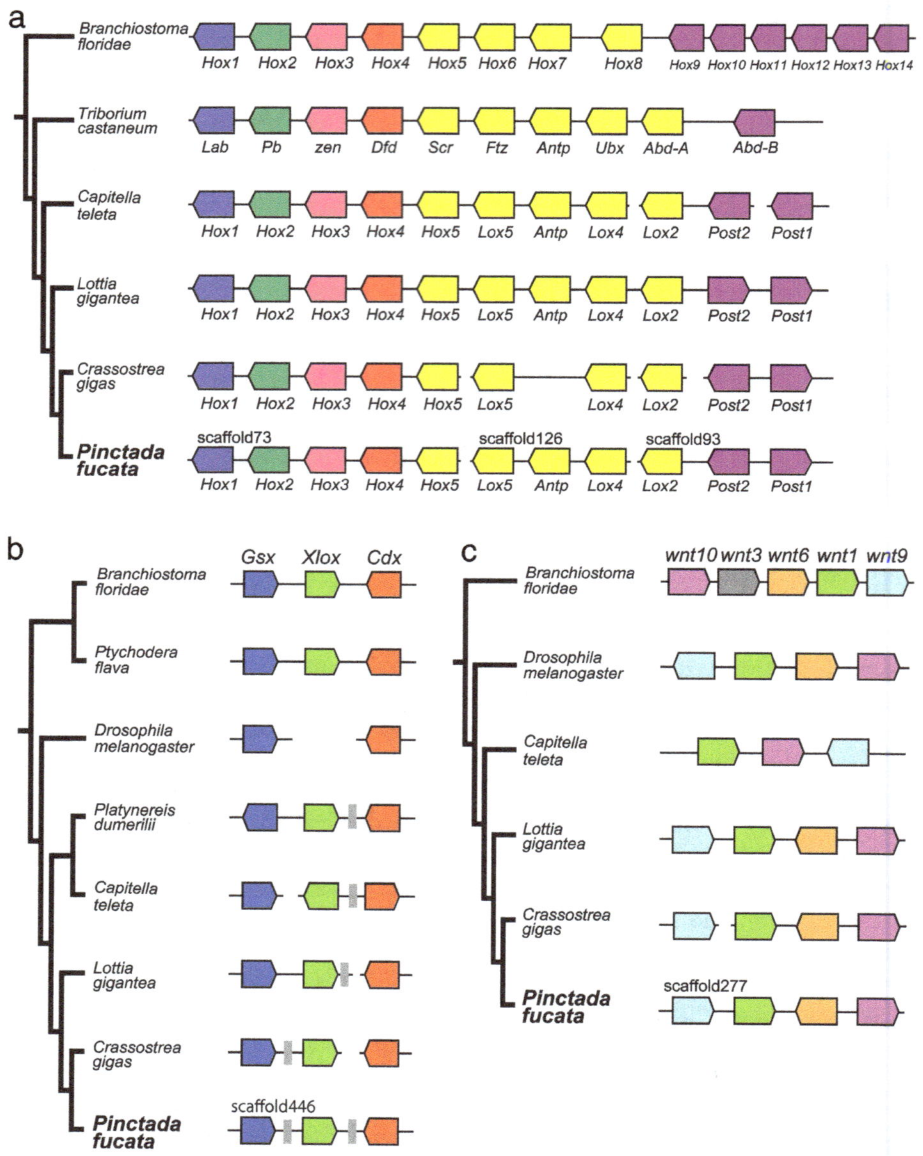

Fig. 6 Hox, ParaHox, and Wnt gene clusters in the *P. fucata* genome resemble those of other protostomes. The relative position and orientation of the genes are indicated. **a** Hox gene cluster. *P. fucata* Hox genes are located on 3 scaffolds. **b** ParaHox gene cluster. *P. fucata* ParaHox genes are aligned on a single scaffold. The gray box indicates a non-ParaHox gene. **c** Wnt gene cluster. *P. fucata* Wnt1, 6, 9, and 10 genes are found on a single scaffold and the gene order is the same as that of *L. gigantea*. Lengths of scaffolds, genes, and intergenic regions are not to scale. See also Additional file 1: Tables S6-S8 for details

these gene clusters, the new genome assembly was thoroughly surveyed. Previously, gain and loss of Hox genes in molluscan classes were reported [67]. All 11 Hox gene transcripts were identified in *Pecten maximus* [68] while *Antp* was lost in the oyster genome [2]. In the *P. fucata* genome assembly, all 11 Hox genes are clustered in three scaffolds (Fig. 6a). The Hox gene, *LoxZ* [21], is located between *Lox5* and *Lox4*. We retrieved a longer sequence of the corresponding gene model from the new genome assembly, and a BLAST search confirmed that the gene actually encodes *Antp* (Additional file 1: Table S6). There are two non-Hox gene models upstream of *Hox5* in scaffold 73, and nine non-Hox gene models are present upstream of *Lox4* in scaffold 126 (Additional file 1: Table S6). As a result, the *P. fucata* Hox cluster is divided into three genomic regions. Two interruptions comprising non-Hox flanking genes between *Hox5* and *Lox5*, and between *Lox4* and *Lox2* are also observed in the *C. gigas* genome (Fig. 6a) [2], indicating that this feature of the Hox cluster occurred in the common ancestor of these two bivalves after their divergence from gastropods.

ParaHox genes, *Gsx*, *Xlox*, and *Cdx* are found in a single scaffold (Fig. 6b), which is the first indication of close linkage of the three ParaHox genes in molluscan genomes. These genes are separated by non-ParaHox gene models, which are also observed in the *Lottia* and *Crassostera* genomes (Fig. 6b and Additional file 1: Table S7). We also found a cluster of Wnt gene family genes (Fig. 6c and Additional file 1: Table S8). The tandem arrangement of *wnt9*, *1*, *6*, and *10* genes is identical to that of the limpet genome [69]; therefore this arrangement is considered the ancestral state in bivalves and gastropods. These results demonstrate that Hox, ParaHox, and Wnt gene clusters in the *P. fucata* genome are comparable to those of other molluscan genomes.

Conclusions

We performed comparative genomic analyses using accessible molluscan genomes with our newly updated genome assembly of the pearl oyster, *Pinctada fucata*. Genes common to the two bivalves include a larger number of genes potentially relevant to extracellular matrix, environmental responses, and immune systems than are seen in a gastropod and other protostomes (Fig. 2). Consistently, protein-domain surveys and molecular phylogenetic analyses reveal extensive gene duplication of stress response genes (HSP70 in Fig. 3, C1qDC in Fig. 4). A survey of gene arrangements confirmed that frequent gene duplication of shell matrix proteins has occurred in bivalves (Fig. 5). All of these results suggest that adaptive changes in extant bivalve genomes have occurred in a species-specific manner. We also confirmed relatively conserved clusters of Hox, ParaHox, and Wnt genes among protostomes (Fig. 6). The revised pearl oyster genome provides insights into the molecular basis of oyster physiology and radiation.

Additional files

Additional file 1: Table S1. Summary of *Pinctada fucata* genome sequence data. **Table S2.** Summary of the *Pinctada fucata* genome assemblies. **Table S3.** Numbers of genes containing functional domains related to heat shock proteins. See also Fig. 3a. **Table S4.** Numbers of genes containing functional domains related to non-self recognition and signaling. See also Fig. 4a. **Table S5.** List of biomineralization-related genes tandemly arranged in the genome. See also Fig. 5. **Table S6.** Hox and neighboring gene models tandemly arranged in the genome. **Table S7.** ParaHox and neighboring gene models tandemly arranged in the genome. **Table S8.** Wnt and adjacent gene models tandemly arranged in the genome.

Additional file 2: Figure S1. Histograms of sequence coverage depth of each contig. (a) Contig coverage distribution in the initial assembly. The peak at lower coverage (near 26.5x) indicates redundant contigs caused by the heterozygotic nature of the *Pinctada fucata* genome. The red line shows the fitted normal distribution, where $\mu = 26.5$ and $\sigma = 3$. (b) The histogram of contig coverage depth after redundant contigs were removed. (c) The histogram of scaffold coverage depth. These data indicate that redundant sequences in the genome assembly were effectively removed by a sequence similarity and coverage-based method. **Figure S2.** The numbers of genes corresponding to grouped gene families presented in Fig. 2a and b in the main text. (a–c) The numbers of genes that were assigned by OrthoMCL DB; (a) *Pinctada fucata*, (b) *Crassostrea gigas*, and (c) *Lottia gigantea*. (d–f) The numbers of genes that were not assigned by OrthoMCL DB, while detected in comparisons among three mollusc genomes; (d) *P. fucata*, (e) *C. gigas*, and (f) *L. gigantea*. Mollusc-specific genes shared by all three molluscs are much less abundant than lineage-specific genes.

Additional file 3: Assembly strategy.

Abbreviations

C1qDC: C1q domain-containing protein; Clp: Chitinase-like protein; fn: fibronectin domain-containing protein; FREP: fibrinogen-related protein; HSP: heat shock protein; PAMP: pathogen-associated molecular pattern; pl: peroxidase-like; SMP: shell matrix protein; SPI: serine protease inhibitor; WGS: whole-genome shotgun.

Competing interests

The authors declare that they have no competing interests.

Authors' contributions

TT and KY prepared samples for genome and transcriptome sequencing. RK, MK, KH, MF, HG, and SY prepared the libraries and sequenced the genome and transcriptomes. TT, TK, RK assembled the genome and transcriptomes and performed bioinformatic analyses. TT, TK, RK, FG, YM, HM, KE, HE, and SK annotated the gene models. TT, KE, HN, SA, SW and NS drafted the manuscript. All authors read and approved the final manuscript.

Acknowledgements

We thank Dr. Steven D. Aird for editing the manuscript. We acknowledge all participants of the Pearl Oyster Genome Jamborees for their enthusiasm and contributions. Super-computing was supported by the IT Section of OIST. This study was supported by Japan Society for the Promotion of Science KAKENHI (no. 23780209) to T. T., and the Japanese Association for Marine Biology (JAMBIO) to K. E. We also gratefully acknowledge support from OIST Graduate University to members of the Marine Genomics Unit.

Author details
[1]Marine Genomics Unit, Okinawa Institute of Science and Technology Graduate University, Onna, Okinawa 904-0495, Japan. [2]DNA Sequencing Section, Okinawa Institute of Science and Technology Graduate University, Onna, Okinawa 904-0495, Japan. [3]Pearl Research Institute, Mikimoto CO. LTD, Shima, Mie 517-0403, Japan. [4]Graduate School of Life and Environmental Science, University of Tsukuba, Ibaraki 305-8572, Japan. [5]Department of Genetic Engineering, Faculty of Biology-Oriented Science and Technology, Kinki University, 930 Nishimitani, Kinokawa, Wakayama 649-6493, Japan. [6]Department of Earth and Planetary Science, Graduate School of Science, The University of Tokyo, Bunkyo-ku, Tokyo 113-0033, Japan. [7]Atmosphere and Ocean Research Institute, The University of Tokyo, Kashiwa, Chiba 277-8564, Japan. [8]Department of Applied Biological Chemistry, Graduate School of Agricultural and Life Sciences, The University of Tokyo, Bunkyo-ku, Tokyo 113-8657, Japan. [9]College of Life Sciences, Zhejiang University, Hangzhou, Zhejiang 310058, People's Republic of China. [10]Department of Aquatic Bioscience, Graduate School of Agricultural and Life Sciences, The University of Tokyo, Bunkyo-ku, Tokyo 113-8657, Japan. [11]Kitasato University School of Marine Bioscience, Sagamihara, Kanagawa 252-0373, Japan. [12]Present Address: Graduate School of Life and Environmental Science, University of Tsukuba, Ibaraki 305-8572, Japan.

References

1. Bayne BL, Newell RC. Physiological energetics of marine molluscs. In: Wilbur KM, Saleuddin ASM, editors. The Mollusca. 4. Cambridge: Academic; 1983. p. 407–515.
2. Zhang G, Fang X, Guo X, Li L, Luo R, Xu F, et al. The oyster genome reveals stress adaptation and complexity of shell formation. Nature. 2012;490(7418): 49–54.
3. Gerdol M, Manfrin C, De Moro G, Figueras A, Novoa B, Venier P, et al. The C1q domain containing proteins of the Mediterranean mussel Mytilus galloprovincialis: A widespread and diverse family of immune-related molecules. Dev Comp Immunol. 2011;35(6):635–43.
4. Gerdol M, Venier P, Pallavicini A. The genome of the Pacific oyster Crassostrea gigas brings new insights on the massive expansion of the C1q gene family in Bivalvia. Dev Comp Immunol. 2015;49(1):59–71.
5. Miyamoto H, Endo H, Hashimoto N, Limura K, Isowa Y, Kinoshita S, et al. The diversity of shell matrix proteins: genome-wide investigation of the pearl oyster, Pinctada fucata. Zoolog Sci. 2013;30(10):801–16.
6. Cunha R, Blanc F, Bonhomme F, Arnaud-Haond S. Evolutionary patterns in pearl oysters of the genus Pinctada (Bivalvia: Pteriidae). Mar Biotechnol. 2011;13(2):181–92.
7. Nagai K. A history of the cultured pearl industry. Zoolog Sci. 2013;30(10):783–93.
8. Kinoshita S, Ning W, Inoue H, Maeyama K, Okamoto K, Nagai K, et al. Deep sequencing of ESTs from nacreous and prismatic layer producing tissues and a screen for novel shell formation-related genes in the pearl oyster. PLoS One. 2011;6(6):e21238.
9. Jackson DJ, McDougall C, Woodcroft B, Moase P, Rose RA, Kube M, et al. Parallel evolution of nacre building gene sets in molluscs. Mol Biol Evol. 2010;27(3):591–608.
10. Marie B, Joubert C, Tayalé A, Zanella-Cléon I, Belliard C, Piquemal D, et al. Different secretory repertoires control the biomineralization processes of prism and nacre deposition of the pearl oyster shell. Proc Natl Acad Sci. 2012;109(51):20986–91.
11. Joubert C, Piquemal D, Marie B, Manchon L, Pierrat F, Zanella-Cleon I, et al. Transcriptome and proteome analysis of Pinctada margaritifera calcifying mantle and shell: focus on biomineralization. BMC Genomics. 2010;11(1):613.
12. Berland S, Marie A, Duplat D, Milet C, Sire JY, Bédouet L. Coupling proteomics and transcriptomics for the identification of novel and variant forms of mollusk shell proteins: a study with P. margaritifera. Chembiochem. 2011;12(6):950–61.
13. Funabara D, Ohmori F, Kinoshita S, Koyama H, Mizutani S, Ota A, et al. Novel genes participating in the formation of prismatic and nacreous layers in the pearl oyster as revealed by their tissue distribution and RNA interference knockdown. PLoS One. 2014;9(1):e84706.
14. Fang D, Xu G, Hu Y, Pan C, Xie L, Zhang R. Identification of genes directly involved in shell formation and their functions in pearl oyster, Pinctada fucata. PLoS One. 2011;6(7):e21860.
15. Suzuki M, Saruwatari K, Kogure T, Yamamoto Y, Nishimura T, Kato T, et al. An acidic matrix protein, Pif, is a key macromolecule for nacre formation. Science. 2009;325(5946):1388–90.
16. Takeuchi T, Kawashima T, Koyanagi R, Gyoja F, Tanaka M, Ikuta T, et al. Draft genome of the pearl oyster Pinctada fucata: a platform for understanding bivalve biology. DNA Res. 2012;19(2):117–30.
17. Funabara D, Watanabe D, Satoh N, Kanoh S. Genome-wide survey of genes encoding muscle proteins in the pearl oyster, Pinctada fucata. Zoolog Sci. 2013;30(10):817–25.
18. Matsumoto T, Masaoka T, Fujiwara A, Nakamura Y, Satoh N, Awaji M. Reproduction-related genes in the pearl oyster genome. Zoolog Sci. 2013; 30(10):826–50.
19. Gyoja F, Satoh N. Evolutionary aspects of variability in bHLH orthologous families: Insights from the pearl oyster, Pinctada fucata. Zoolog Sci. 2013; 30(10):868–76.
20. Koga H, Hashimoto N, Suzuki DG, Ono H, Yoshimura M, Suguro T, et al. A genome-wide survey of genes encoding transcription factors in Japanese pearl oyster Pinctada fucata: II. Tbx, Fox, Ets, HMG, NFκB, bZIF, and C2H2 zinc fingers. Zoolog Sci. 2013;30(10):858–67.
21. Morino Y, Okada K, Niikura M, Honda M, Satoh N, Wada H. A Genome-Wide Survey of Genes Encoding Transcription Factors in the Japanese Pearl Oyster, Pinctada fucata: I. Homeobox Genes. Zoolog Sci. 2013; 30(10):851–7.
22. Setiamarga DHE, Shimizu K, Kuroda J, Inamura K, Sato K, Isowa Y, et al. An in-silico genomic survey to annotate genes coding for early development-relevant signaling molecules in the pearl oyster, Pinctada fucata. Zoolog Sci. 2013;30(10):877–88.
23. Simakov O, Marletaz F, Cho S-J, Edsinger-Gonzales E, Havlak P, Hellsten U, et al. Insights into bilaterian evolution from three spiralian genomes. Nature. 2013;493(7433):526–31.
24. Kawashima T, Takeuchi T, Koyanagi R, Kinoshita S, Endo H, Endo K. Initiating the mollusk genomics annotation community: toward creating the complete curated gene-set of the Japanese Pearl Oyster, Pinctada fucata. Zoolog Sci. 2013;30(10):794–6.
25. Bentley DR. Whole-genome re-sequencing. Curr Opin Genet Dev. 2006;16(6): 545–52.
26. Bolger AM, Lohse M, Usadel B. Trimmomatic: A flexible trimmer for Illumina Sequence Data. Bioinformatics. 2014;30(15):2114–20.
27. Margulies M, Egholm M, Altman WE, Attiya S, Bader JS, Bemben LA, et al. Genome sequencing in microfabricated high-density picolitre reactors. Nature. 2005;437(7057):376–80.
28. Boetzer M, Henkel CV, Jansen HJ, Butler D, Pirovano W. Scaffolding pre-assembled contigs using SSPACE. Bioinformatics. 2011;27(4):578–9.
29. Li R, Fan W, Tian G, Zhu H, He L, Cai J, et al. The sequence and de novo assembly of the giant panda genome. Nature. 2010;463(7279):311–7.
30. Grabherr MG, Haas BJ, Yassour M, Levin JZ, Thompson DA, Amit I, et al. Full-length transcriptome assembly from RNA-Seq data without a reference genome. Nat Biotech. 2011;29(7):644–52.
31. Haas BJ, Delcher AL, Mount SM, Wortman JR, Smith Jr RK, Hannick LI, et al. Improving the Arabidopsis genome annotation using maximal transcript alignment assemblies. Nucleic Acids Res. 2003;31(19):5654–66.
32. Stanke M, Diekhans M, Baertsch R, Haussler D. Using native and syntenically mapped cDNA alignments to improve de novo gene finding. Bioinformatics. 2008;24(5):637–44.
33. Jurka J, Klonowski P, Dagman V, Pelton P. CENSOR–a program for identification and elimination of repetitive elements from DNA sequences. Comput Chem. 1996;20(1):119–21.
34. Chen F, Mackey AJ, Stoeckert CJ, Roos DS. OrthoMCL-DB: querying a comprehensive multi-species collection of ortholog groups. Nucleic Acids Res. 2006;34 suppl 1:D363–8.
35. Li L, Stoeckert CJ, Roos DS. OrthoMCL: Identification of Ortholog Groups for Eukaryotic Genomes. Genome Res. 2003;13(9):2178–89.
36. Mi H, Muruganujan A, Casagrande JT, Thomas PD. Large-scale gene function analysis with the PANTHER classification system. Nat Protocols. 2013;8(8):1551–66.
37. Eddy SR. Profile hidden Markov models. Bioinformatics. 1998;14(9):755–63.
38. Edgar RC. MUSCLE: multiple sequence alignment with high accuracy and high throughput. Nucleic Acids Res. 2004;32(5):1792–7.
39. Stamatakis A. RAxML version 8: a tool for phylogenetic analysis and post-analysis of large phylogenies. Bioinformatics. 2014;30(9):1312–3.

40. Pallavicini A, del Mar CM, Gestal C, Dreos R, Figueras A, Venier P, et al. High sequence variability of myticin transcripts in hemocytes of immune-stimulated mussels suggests ancient host–pathogen interactions. Dev Comp Immunol. 2008;32(3):213–26.
41. Venier P, Varotto L, Rosani U, Millino C, Celegato B, Bernante F, et al. Insights into the innate immunity of the Mediterranean mussel Mytilus galloprovincialis. BMC Genomics. 2011;12(1):69.
42. Sørensen JG, Kristensen TN, Loeschcke V. The evolutionary and ecological role of heat shock proteins. Ecol Lett. 2003;6(11):1025–37.
43. Hartl FU. Molecular chaperones in cellular protein folding. Nature. 1996;381: 571–80.
44. Plazzi F, Passamonti M. Towards a molecular phylogeny of Mollusks: Bivalves' early evolution as revealed by mitochondrial genes. Mol Phylogenet Evol. 2010;57(2):641–57.
45. Medzhitov R, Janeway Jr CA. Decoding the patterns of self and nonself by the innate immune system. Science. 2002;296:298–300.
46. Kong P, Zhang H, Wang L, Zhou Z, Yang J, Zhang Y, et al. AiC1qDC-1, a novel gC1q-domain-containing protein from bay scallop Argopecten irradians with fungi agglutinating activity. Dev Comp Immunol. 2010;34(8): 837–46.
47. Wang L, Wang L, Kong P, Yang J, Zhang H, Wang M, et al. A novel C1qDC protein acting as pattern recognition receptor in scallop Argopecten irradians. Fish Shellfish Immunol. 2012;33(2):427–35.
48. Zhang H, Song L, Li C, Jianmin Z, Wang H, Qiu L, et al. A novel C1q-domain-containing protein from Zhikong scallop Chlamys farreri with lipopolysaccharide binding activity. Fish Shellfish Immunol. 2008;25:281–9.
49. Adema CM, Hertel LA, Miller RD, Loker ES. A family of fibrinogen-related proteins that precipitates parasite-derived molecules is produced by an invertebrate after infection. Proc Natl Acad Sci. 1997;94(16):8691–6.
50. Zhang S-M, Zeng Y, Loker ES. Expression profiling and binding properties of fibrinogen-related proteins (FREPs), plasma proteins from the schistosome snail host Biomphalaria glabrata. Innate Immun. 2008;14(3):175–89.
51. Zheng P, Wang H, Zhao J, Song L, Qiu L, Dong C, et al. A lectin (CfLec-2) aggregating Staphylococcus haemolyticus from scallop Chlamys farreri. Fish Shellfish Immunol. 2008;24(3):286–93.
52. Huang M, Wang L, Yang J, Zhang H, Wang L, Song L. A four-CRD C-type lectin from Chlamys farreri mediating nonself-recognition with broader spectrum and opsonization. Dev Comp Immunol. 2013;39(4):363–9.
53. Takahashi KG, Kuroda T, Muroga K. Purification and antibacterial characterization of a novel isoform of the Manila clam lectin (MCL-4) from the plasma of the Manila clam, Ruditapes philippinarum. Comp Biochem Physiol B Biochem Mol Biol. 2008;150(1):45–52.
54. Mann S. Biomineralization. London: Oxford University Press; 2001.
55. Lowenstam HA. WS. On Biomineralization. New York: Oxford University Press; 1989.
56. Marin F, Luquet G, Marie B, Medakovic D. Molluscan shell proteins: Primary structure, origin, and evolution. Curr Top Dev Biol. 2008;80:209–76.
57. Yano M, Nagai K, Morimoto K, Miyamoto H. Shematrin: a family of glycine-rich structural proteins in the shell of the pearl oyster Pinctada fucata. Comp Biochem Physiol B Biochem Mol Biol. 2006;144(2):254–62.
58. Yano M, Nagai K, Morimoto K, Miyamoto H. A novel nacre protein N19 in the pearl oyster Pinctada fucata. Biochem Biophys Res Commun. 2007; 362(1):158–63.
59. Miyamoto H, Miyashita T, Okushima M, Nakano S, Morita T, Matsushiro A. A carbonic anhydrase from the nacreous layer in oyster pearls. Proc Natl Acad Sci. 1996;93(18):9657–60.
60. Sudo S, Fujikawa T, Nagakura T, Ohkubo T, Sakaguchi K, Tanaka M, et al. Structures of mollusc shell framework proteins. Nature. 1997;387(6633):563–4.
61. Marie B, Jackson DJ, Ramos-Silva P, Zanella-Cléon I, Guichard N, Marin F. The shell-forming proteome of Lottia gigantea reveals both deep conservations and lineage-specific novelties. FEBS J. 2013;280(1):214–32.
62. McDougall C, Aguilera F, Degnan BM. Rapid evolution of pearl oyster shell matrix proteins with repetitive, low-complexity domains. J R Soc Interface. 2013;10(82):20130041.
63. Takeuchi T, Endo K. Biphasic and dually coordinated expression of the genes encoding major shell matrix proteins in the pearl oyster Pinctada fucata. Mar Biotechnol (N Y). 2006;8(1):52–61.
64. Garcia-Fernandez J. The genesis and evolution of homeobox gene clusters. Nat Rev Genet. 2005;6(12):881–92.
65. Holland PWH. Beyond the Hox: how widespread is homeobox gene clustering? J Anat. 2001;199(1-2):13–23.
66. Ikuta T, Chen Y-C, Annunziata R, Ting H-C, Tung C-h, Koyanagi R, et al. Identification of an intact ParaHox cluster with temporal colinearity but altered spatial colinearity in the hemichordate Ptychodera flava. BMC Evol Biol. 2013;13(1):129.
67. Iijima M, Akiba N, Sarashina I, Kuratani S, Endo K. Evolution of Hox genes in molluscs: a comparison among seven morphologically diverse classes. J Molluscan Stud. 2006;72(3):259–66.
68. Canapa A, Biscotti MA, Olmo E, Barucca M. Isolation of Hox and ParaHox genes in the bivalve Pecten maximus. Gene. 2005;348:83–8.
69. Cho S-J, Vallès Y, Giani VC, Seaver EC, Weisblat DA. Evolutionary dynamics of the wnt gene family: a lophotrochozoan perspective. Mol Biol Evol. 2010; 27(7):1645–58.

12

Mechanisms of wound closure following acute arm injury in *Octopus vulgaris*

Tanya J. Shaw[1,3*], Molly Osborne[1], Giovanna Ponte[2], Graziano Fiorito[2] and Paul L.R. Andrews[1]

Abstract

Background: Octopoda utilise their arms for a diverse range of functions, including locomotion, hunting, defence, exploration, reproduction, and grooming. However the natural environment contains numerous threats to the integrity of arms, including predators and prey during capture. Impressively, octopoda are able to close open wounds in an aquatic environment and can fully regenerate arms. The regrowth phase of cephalopod arm regeneration has been grossly described; however, there is little information about the acute loca response that occurs following an amputation injury comparable to that which frequently occurs in the wild.

Methods: Adult *Octopus vulgaris* caught in the Bay of Naples were anaesthetised, the distal 10 % of an arm was surgically amputated, and wounded tissue was harvested from animals sacrificed at 2, 6, and 24 h post-amputation. The extent of wound closure was quantified, and the cell and tissue dynamics were observed histologically, by electron microscopy, as well as using ultrasound.

Results: Macroscopic, ultrasonic and ultrastructural analyses showed extensive and significant contraction of the wound margins from the earliest time-point, evidenced by tissue puckering. By 6 h post amputation, the wound was 64.0 ± 17.2 % closed compared to 0 h wound area. Wound edge epithelial cells were also seen to be migrating over the wound bed, thus contributing to tissue repair. Temporary protection of the exposed tip in the form of a cellular, non-mucus plug was observed, and cell death was apparent within two hours of injury. At earlier time-points this was apparent in the skin and deeper muscle layers, but ultimately extended to the nerve cord by 24 h.

Conclusions: This work has revealed that *O. vulgaris* ecologically relevant amputation wounds are rapidly repaired via numerous mechanisms that are evolutionarily conserved. The findings provide insights into the early processes of repair preparatory to regeneration. The presence of epithelial, chromatophore, vascular, muscle and neural tissue in the arms makes this a particularly interesting system in which to study acute responses to injury and subsequent regeneration.

Keywords: Arm, Cell death, Cephalopod, Healing, Invertebrate, Microscopy, *Octopus vulgaris*, Regeneration, Wound repair

Background

The limbs (arms and tentacles) are one of the most striking external features of the approximately 700 known extant species of cephalopods ranging from the numerous tentacles of nautiloids (e.g. *Nautilus pompilius)* to the eight arms and two tentacles of cuttlefish (e.g. *Sepia officinalis*) and squid (e.g. *Loligo pealii*) and the eight arms of octopoda (e.g. *Octopus vulgaris*, [1]). Due to the sea floor habitat of most species of octopus and the extensive use made of the arms for exploration, predation, and defence, the distal parts of the arms are the part of the body most likely to encounter a noxious challenge, and hence to be physically damaged. Octopuses are also known to remove arms by autophagy if they become trapped or damaged [2], and some species (*Abdopus aculeatus*, *Ameloctopus litoralis*) are able to autotomize their arms if traumatised [1, 3–5]. Surveys of

* Correspondence: tanya.shaw@kcl.ac.uk
[1]Division of Biomedical Sciences, St George's University of London, Cranmer Terrace, London SW17 0RE, UK
[3]King's College London-Centre for Molecular and Cellular Biology of Inflammation, New Hunt's House, Great Maze Pond, London SE1 1UL, UK
Full list of author information is available at the end of the article

octopuses caught in the wild show that arm damage is common. For example, in *O. vulgaris* 50 % of the animals caught in the Bay of Naples had some damage to one or more arms [6] and in the Pacific pygmy octopus, *Octopus digueti*, 26 % of animals had arm damage [7]. The high probability of damage to an appendage in the wild is one of the key evolutionary pressures for regeneration [8]. Although limb regeneration is a well-established phenomenon in members of all four orders of coleoid cephalopod (e.g. *S. officinalis* [9], *Sepioteuthis lessoniana* [10]; squid, *Omnastrephes bartrami* [11], *Octopoteuthis deletron* [12]; *O. vulgaris* [13]; and deep sea octopus, *Vampyroteuthis infernalis* [14]), it is rarely mentioned in reviews of comparative aspects of regeneration, despite the broad acknowledgment that knowledge of the underlying evolutionary processes would better inform our understanding of the fundamental mechanisms of regeneration [8, 15].

The arms in octopus function as "muscular hydrostats" [16] capable of marked changes in length under neural control. In *O. vulgaris* about 60 % of the neurones in the nervous system are located in the arms [17], with a centrally located axonal nerve cord comprising two adjacent axonal tracts containing motor and sensory nerves, and a more ventrally located chain of ganglia clustered in the region of each sucker to which they are connected [18]. In contrast to most other models of external appendage regeneration (e.g. axolotl forelimb, zebrafish fin), the arm in octopus provides a model without any skeletal reference point, and arguably a more complex behavioural repertoire for investigation of the relationship between physical and functional recovery.

To date, there have only been a few macroscopic and histological descriptions of the regenerative process of the octopus arm [13, 19] and healing of the mantle [20] following surgical injury. The redevelopment of appendages in other cephalopods [9, 21, 22] has also been described, and it is anticipated that many features of the tissue dynamics underpinning regeneration in these species will be shared. Despite detailed descriptions of appendage regrowth, there is a scarcity of information about the acute wound response to tissue damage (e.g. the first 24 h after injury), although it is known from other animal models that the acute response "sets the scene" for subsequent regeneration [23, 24]. In 1920, Lange provided a careful and eloquent description of these early tissue repair events in *O. vulgaris*; however, our current knowledge regarding the cellular basis of regeneration, together with improved microscopy and image acquisition capabilities, equip us to revisit this important topic. The inclusion of cephalopods in Directive 2010/63/EU regulating the use of animals for scientific purposes [25] also motivates an improved understanding of octopus wound repair, since this information will be informative in establishing guidelines for best surgical practice and post-operative care when cephalopods are used as research models [26]. The present study characterises wound closure over the first 24 h after removal of the distal 10 % of an arm in *O. vulgaris*. As with some vertebrate wounds, tissue contraction is confirmed as the key mechanism by which rapid closure is achieved; yet active cell migration by the epithelial cells at the wound edge was also initiated within 6 h of injury. A cellular plug overlaid the exposed arm tip, which is predicted to provide protection in a manner functionally analogous to a blood clot in vertebrates, as well as cells and extra-cellular matrix scaffolding for subsequent regenerative events.

Methods

Animals

Octopus vulgaris, Cuvier 1797 of both sexes caught in the Bay of Naples, Italy were used in this study. A total of 19 animals, 14 males and five females, were used in the main study, with a body weight range from 158 to 449 g (Gaussian distribution; mean ± standard error (SEM), 320 ± 20 g). Experimental groups were balanced for sex to the extent possible, and there was no significant difference in body weight between the groups at the three experimental time points (Kruskal-Wallis test, data not shown). As age determination is not possible in live cephalopods [27], it was not possible to account for this in the design. Studies were performed in September and February when the water temperature ranged between 17–21 °C. One male animal (body weight 570 g) was studied in June for in vivo high-resolution ultrasound imaging (details below). Animals were housed individually in tanks (30 × 50 × 100 cm) containing a den, constantly supplied with fresh sea water [28], and were fed with a small live crab (*Carcinus mediterraneus*) three times per week.

At the time the studies were performed, no national legislation was in place in Italy for regulation of research involving cephalopods. The care and welfare assessment of animals was consistent with best practice at the time the study was undertaken [29, 30] and took into account the then impending changes in the European Union legislation bringing cephalopods within the scope of regulations covering research involving living animals (Directive 2010/EU/63, [25, 26, 31]). Experimental design and animal care was discussed with the institution veterinarian who also monitored the study.

Anaesthesia, surgery and humane killing

Immediately prior to surgery and to humane killing, the latency to attack a crab placed in the tank was measured to provide an indication of welfare following anaesthesia and surgery [32]. For surgery, animals were removed from their home tank by fine mesh net between 10.00 h and 12.00 h and anaesthetised in 5 L of 3.5 % magnesium chloride hexahydrate (Sigma-Aldrich) in sea water [33, 34]. Surgical anaesthesia occurred after 15–20 mins and was defined by an absence of a response to a noxious stimulus (mechanical), skin pallor (indicative of reduced central drive to chromatophores), muscle flaccidity, loss of righting reflex and suppression/absence of ventilation [29]. Animals were then placed in a shallow dish containing gassed 3.5 % magnesium chloride, arm L2 (second arm on the left side) was identified, the length measured (beak to tip; L2 mean arm length (±SEM): 405 ± 14 mm; $n = 16$, normal distribution verified). L2 lengths for two animals were not recorded, and in another animal R2 was used due to existing damage to L2; inclusion of this measurement results in a mean length of 404 ± 14 mm, $n = 17$. The distal 10 % was removed from the extended arm by transection using a new alcohol-sterilised Gillette Super Silver blade with the cut positioned to cut as vertically as possible to the longitudinal axis of the arm [13], whilst avoiding transection of a sucker. The cut face of the transected arm was immersed over a length of ~1 cm in local anaesthetic (3 % mepivacaine; Carboplyina, Curaden Healthcare SRL) for 5 min. Previous studies have shown that nerve block can be produced in *O. vulgaris* by topical application of the local anaesthetic agent xylocaine [35] and topical application of a related agent mepivacaine (3 %) blocks transmission in the axial nerve cord (unpublished observations). It should be noted that there are no systemic analgesics with demonstrated efficacy in cephalopods (reviewed in [29, 31]) and that previous studies using surgical transection of arms or tentacles in cephalopods have not provided any post-surgical analgesia or local anaesthesia [19, 22, 36]. Animals were then placed in a container with fresh sea water and monitored until ventilation and the righting reflex had recovered (15–30 min), after which animals were returned to their home tank until they were killed 2 h ($n = 6$), 6 h ($n = 6$) or 24 h ($n = 7$) from the time of transection. Animals were inspected regularly in their home tank after surgery. When the animals were returned to the home tank there was no obvious behaviour change and the latency to attack a crab was not significantly different between the three groups prior to lesion, and following lesion.

Animals were killed 2, 6, or 24 h post-lesion by prolonged (>30 min) immersion in a 3.5 % solution of magnesium chloride hexahydrate in sea water [34]. Breathing as indicated by mantle contraction normally ceased within 15 min [37]. Death was confirmed by mechanical destruction of the brain except in studies where the brain was required for subsequent analysis, in which case the dorsal aorta and systemic heart were transected. This methodology is consistent with current guidelines [31].

Macroscopic and ultrasound assessments

Immediately following transection (0 h) and at the time of sacrifice (2, 6, 24 h) the cut face of the arm was photographed (Canon Eos 600D, Sigma 105 mm 2:2.8 DG Macro lens). The entire transected face of the arm was white in appearance due to exposure of the arm musculature with the central nerve cord visible and the skin partially retracted from the cut edge. Wound healing was assessed macroscopically by quantification of the white area visible at 6 h and 24 h. The area was measured from the high-resolution photographs using Image J software (National Institutes of Health, USA) and expressed as a percentage of the area at 0 h (100 %). Data are expressed as mean values with standard deviation (SD). A Kruskal-Wallis test (GraphPad Prism 6) was used for statistical analyses, and significance was inferred at $p < 0.05$.

For one animal, the transected tip of an arm was imaged under general anaesthesia immediately following surgery and again 24 h later in the same animal using high-resolution ultrasound (Vevo 2100 Visualsonics, The Netherlands). Ultrasound has previously been used to image the structure of the arm in *O. vulgaris* with and without general anaesthesia [37].

Histology

For all electron microscopy (EM), tissues were fixed in 2.5 % gluteraldehyde in seawater, post-fixed in 1 % osmium tetroxide, and washed in 0.2 mol/L sodium cacodylate buffer. For scanning EM, samples were then dehydrated through a series of ascending concentrations of ethanol, critical point dried using an Emitech K850, mounted on stubs, and sputter-coated with gold (Bio-Rad E5000). Samples were viewed with a Cambridge Stereoscan 360 and images acquired with Image Access (V.3). For transmission EM, following post-fixation and washing, samples were stained en-block for 30 min in 3 % alcoholic uranyl acetate, and then dehydrated through an ascending series of ethanol. Incubation in propylene oxide for 30 min was followed by overnight infiltration in 50/50 propylene oxide resin mix (TAAB low viscosity resin). Finally, samples were infiltrated with 100 % resin (2 × 1 h) and embedded/polymerised overnight at 60 °C. Semi-thin sections (2 μm) were stained with alkaline toluidine blue (1 % dye in 1 % borax) to establish regions of interest and tissue orientation. Thin sections were cut at 90 nm (diamond knife), mounted on copper grids (uncoated 200-mesh), pretreated with 1 % aqueous potassium permanganate, stained with uranyl acetate and Sato's lead citrate (for contrast). Viewing was achieved with an Hitachi 7100 TEM with a GATAN multi-scan digital camera. Digital Micrograph software was used for image acquisition.

For analysis by standard microscopy, tissue fixed in 4 % paraformaldehyde in sea water was processed for routine wax embedding and sectioned at 10 μm. Sections were de-waxed and rehydrated through a series of descending concentrations of ethanol and stained using haemotoxylin and eosin (H&E), periodic acid-Schiff (PAS), or for Terminal deoxynucleotidyl transferase dUTP nick end labelling (TUNEL) according to manufacturers' instructions (Roche) to reveal dying cells.

Results

Macroscopic assessment of wound closure

In order to study the acute wound response in *O. vulgaris*, the distal 10 % of an arm (primarily L2) was amputated under general anaesthesia. There was no arm withdrawal in response to the transection, but the proximal arm recoiled from the transected distal tip by a few millimetres, indicating that the arm was under some longitudinal tension. The skin did not appear to retract from the wound site, nor did the central muscle or nerve cord protrude; histologically however (description below), there was evidence of skin wrinkling, again indicating that the tissue was under some tension that was lost upon transection. No overt bleeding was observed, although this is difficult to detect as cephalopod blood is virtually colourless when de-oxygenated and pale blue when oxygenated [38].

To quantify and plot the time-course of wound closure, we exploited the white appearance of the exposed arm musculature/central nerve cord of the transected face of the arm at 6 and 24 h and compared this with the area of the wound at 0 h (immediately after amputation). Qualitatively, at 2 h there was already a notable reduction in wound size (Fig. 1a). At 6 h, significant wound closure was observed (64.0 ± 17.2 % closure; mean ± standard deviation (SD), $n = 6$), but interestingly, only modest additional progress was made in the 6–24 h time-frame (71.5 ± 11.1 % closure, $n = 7$). This may be in part attributable to the inter-animal variability in rate of closure, as the data revealed two distinct populations of "faster" and "slower" healers, at both 6 and 24 h (Fig. 1b).

Wound contraction and cell migration contribute to rapid closure

Gross analysis of the amputation wounds at 24 h revealed a puckered appearance of the skin at the wound margin (Fig. 1a), suggesting that a purse-string-like contractile mechanism may contribute to wound closure. High-resolution ultrasound imaging of a 24 h wound confirmed that the marginal skin was raised and drawing closed over the musculature and nervous tissue constituting the core of the arm (Fig. 2a). Equally, scanning electron microscopy (EM) at the same time-point showed tissue puckering at the wound margin (Fig. 2b). Higher magnification analysis revealed: 1) that each tissue fold consisted of two or more cell widths (Fig. 2c); 2) that the wave-like advancing epithelial margin is at least two cell heights (Fig. 2c); and 3) a belt-like structure that may constitute a contractile cable, could be observed between some of the cells (Fig. 2d). Notable was the incomplete contractile response around the circumference of the wound, with very little gathering observed on the ventral (sucker) side (Fig. 2b).

Again using scanning EM, analysis of the cells at the advancing wound margin at 6 h post-amputation showed cell protrusions indicative of active cell migration. The epithelial cells at the leading edge had abundant and obvious attachments to the wound bed at 24 h (Fig. 2e).

A dense cell-rich plug provides wound coverage

Macroscopically, we did not consistently observe protrusion of the central muscle and nervous tissue beyond the surrounding skin. At 2 h post-amputation, there was

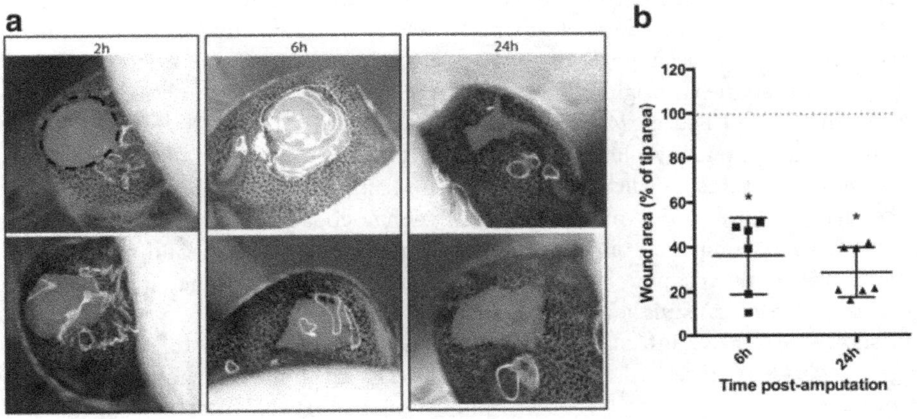

Fig. 1 Dynamics of wound closure. **a** Images of transected stumps at 2, 6, and 24 h were acquired using a dissection microscope, and 2 biological replicates at each time-point are shown. **b** Wound area was calculated using ImageJ software, and is presented as a percentage of the area of the stump at the time of amputation (calculated using width (diameter) of an assumed "circular" arm). Calculations were performed by two independent reviewers, mean ± SD was plotted, and compared using a Kruskal-Wallis test (*$p < 0.05$, significant compared to 0 h wound size)

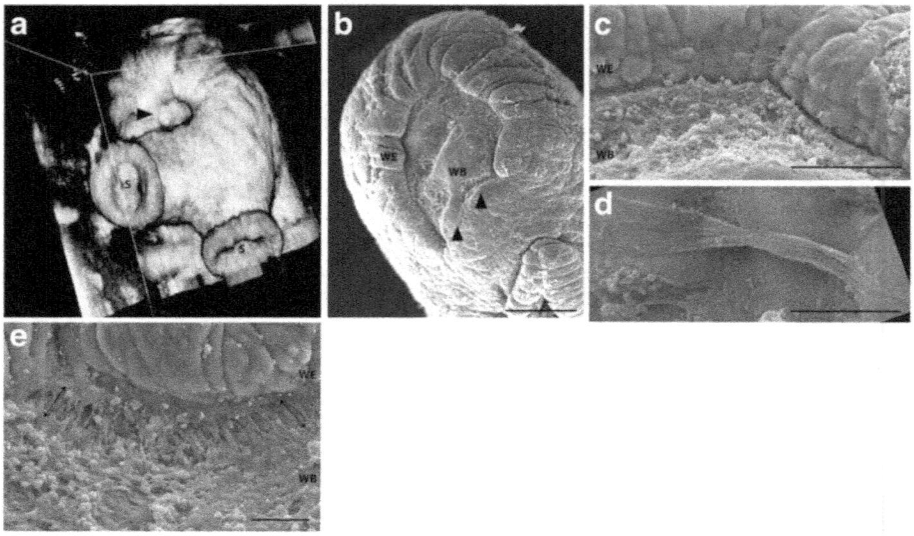

Fig. 2 Wound closure mediated by purse-string contraction. **a**) High-resolution ultrasound imaging of a 24 h wound confirmed that the marginal skin was drawing closed over the musculature and nervous tissue constituting the core of the arm (S, suckers; arrowhead, wound). **b** Equally, scanning EM showed tissue puckering at the wound margin, which was incomplete ventrally (arrowheads). Higher magnification analysis revealed: 1) that each tissue fold consisted of two, three or more cell widths **c**; 2) that the wave-like advancing epithelial margin (wound-edge epithelium, WE) is at least two cell heights **c**; and 3) a belt-like structure that may constitute a contractile cable, could be observed between some of the cells **d**). **e** At 24 h, basal wound edge epithelial cells (WE) displayed continued adhesion to their neighbours, lamella on the leading edge, and cell adhesions to the underlying wound matrix (arrows), suggestive of collective cell migration. WB, wound bed. Scale bars: **b**, 500 μm; **c**, 100 μm; **d**, 10 μm; **e**, 50 μm

a varied degree of extension, with some showing significant exposure, and others none at all (Additional file 1: Figure S1). This appeared to affect rate of closure, with the least protrusion correlating with greatest closure in the 2 h time-frame. For all 2 h specimens, the most distal skin appeared somewhat detached from the underlying muscle layer (Additional file 1: Figure S1), an observation that was no longer obvious at 6 h (Fig. 3). Variably at 2 h, but consistently by 6 h, an abundance of cells coalesced over the exposed tip of the amputated arm, and even accumulated in the plane between the exposed muscle and the overlying skin. A distinct advancing simple epithelial cell layer over a portion of the plug (Fig. 3d) suggests a non-epithelial origin of the plug cells. Scanning EM of the wound bed at 24 h provided some insight into the composition, revealing abundant extracellular matrix and structures smaller than cells that may be extracellular vesicles and/or mucous. Certainly the bulk of the plug was comprised of cells (based on the haematoxylin-positive nuclei, and lack of PAS staining of mucus in this area). Transmission EM of the cells in this region showed extensive interdigitation, indicating the strength of this structure and its ability to fulfil a protective function.

Occurrence of cell death in multiple tissue layers

As early as 2 h after amputation, there was a spread of cell death (TUNEL-positive cells) within the skin at the leading edge of the wound, across the surface of the wound bed, and a few cells apparent within the most distal portion of cut muscle (Fig. 4). At this time-point, the dying cells appeared restricted to the wound edges, and were not yet present in the tissue constituting the centre of the arm. By 6 h, cell death had spread to the centre of the wound bed, and evidence of dying cells in the cleft between the inner muscle tissue and the migrating epidermis persisted (Fig. 4). Cell death spanned the breadth of the wound bed and appeared in the central nerve cord and distal ganglion by 24 h. Although a few cells with fragmented DNA were seen in proximal musculature, it was localised to the most distal portion of the stump.

Discussion

In Lange's first detailed description of regeneration of cephalopods in 1920, three stages were identified: wound healing, degeneration, and renewal [13]. Since then, and with a recent resurgence in this topic, there have been a few and increasingly molecular descriptions of arm regeneration in the octopus [19] and other cephalopods [9, 21, 22]; however, the acute wound response has not been the focus of these investigations. The present study focused on the initial wound healing and degeneration events occurring in the first 24 h after removal of the distal 10 % of an arm in *O. vulgaris.* This model represents an extent of injury that would be commonly

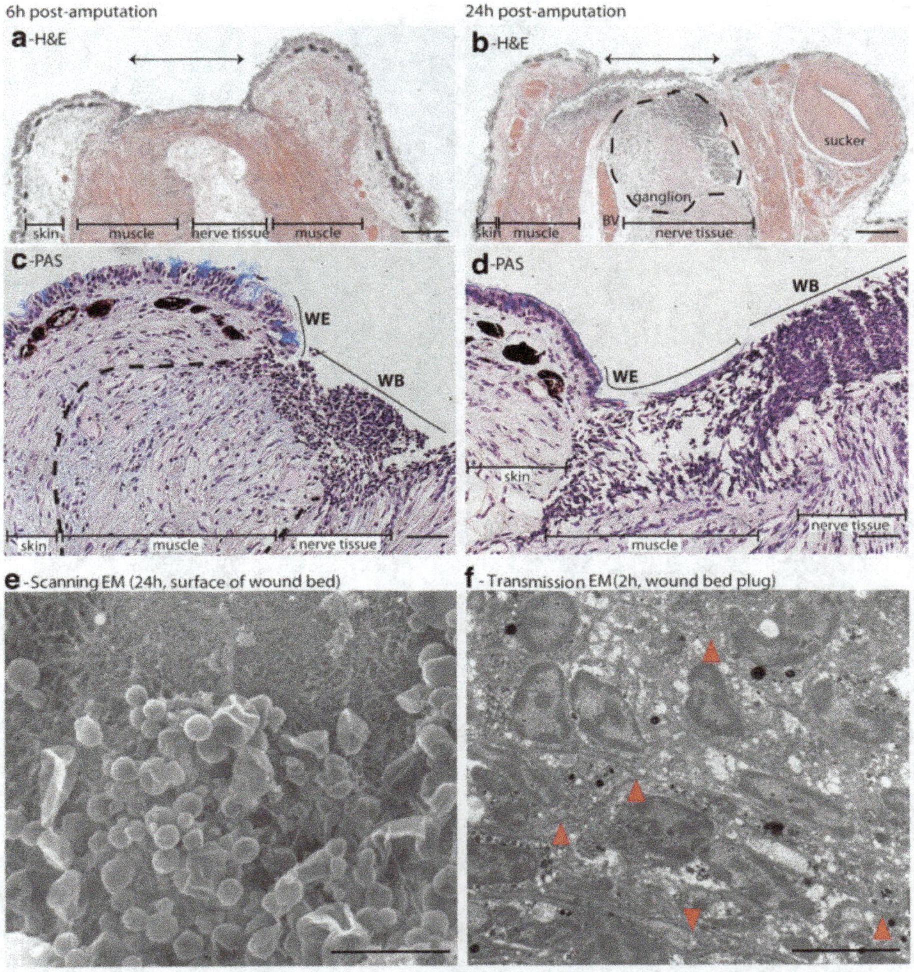

Fig. 3 A cellular plug rapidly covers exposed tissue. (**a**, **b**) H&E and **c**, **d** Periodic Acid-Schiff staining of healing amputation wounds. An increasing cellularity of the wound bed was seen during the time-course studied. PAS confirmed that this is not simply mucous secretions. Arrow overlies central tissue and highlights width of wound. WE, wound-edge epithelium; WB, wound bed. **e** Scanning EM showed diverse sub-cellular-sized structures on the surface of the plug (e.g. extracellular vesicles, mucus). **f** Transmission EM revealed the densely packed and tightly interdigitated cells (arrowheads show finger-like processes of cell membrane) comprising the deeper plug. Scale bars: **a&b**, 200 μm; **c&d**, 50 μm; **e**, 5 μm; **f**, 10 μm

experienced in the wild [6], thus it is considered ecologically relevant [8]. Moreover, the events occurring in this time frame are functionally essential in "setting the scene" for subsequent limb regeneration [23, 24].

Immediately following injury, the arm recoiled slightly from the dissected tip, suggesting some tension in the tissue; however, there was no consistent protrusion of either the central nerve cord or muscle as previously described [9, 13]. This may be attributable to differences in anaesthesia, the magnitude of the injury, or the state of tension of the arm muscles and/or skin, which may be related to the proximity of the nearest sucker and could potentially affect subsequent repair dynamics.

Assessment of the temporal dynamics of gross wound closure over 24 h showed dramatic contraction of the skin tissue over the wound as early as 2 h, and that translated into a remarkable >60 % closure by 6 h. This was only marginally increased in the subsequent 18 h, with approximately 70 % of the tip covered after one day. The extensive closure within one day in octopus aligns well with those previously reported in other cephalopods. For example, 3–5 days [9] or 5–7 days [21] were required for complete re-epithelialisation following arm amputation in cuttlefish (*S. officinalis*), a time-course which may be influenced by species, water temperature, and animal age, state of health, size and source (e.g. wild *vs* captive bred, [9, 21]). Interestingly, our specimens appeared to segregate into two groups, with approximately equal numbers of "faster" and "slower" healers. Although the biological replicates are low in number, this appeared independent of the sex or weight of the animals, although the influence of sucker proximity and tissue tension, nor age of the animal could be assessed [27].

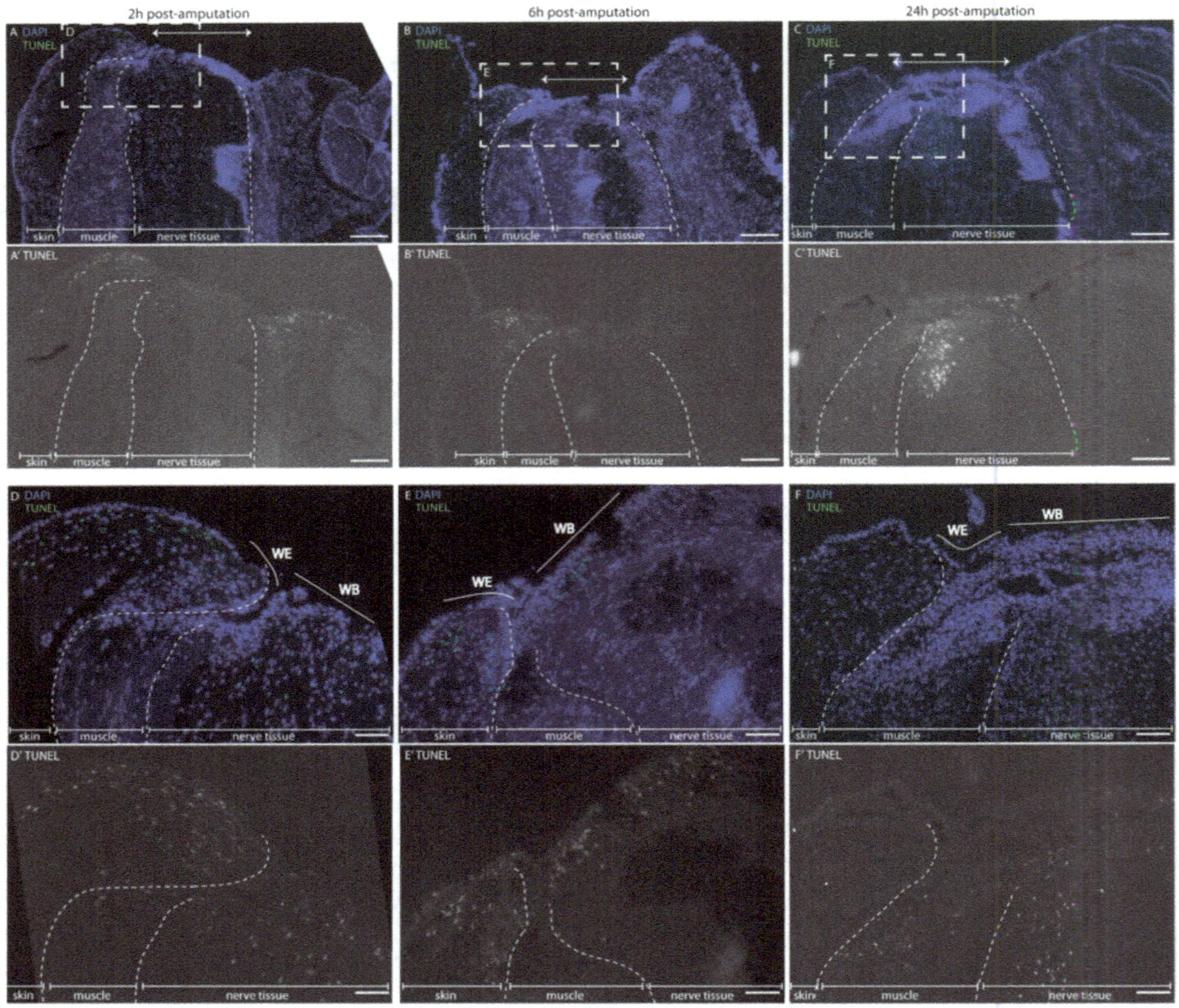

Fig. 4 Cell death in wound-edge skin and stump. Wound margins at 2 h (**a**, **d**), 6 h (**b**, **e**), and 24 h (**c**, **f**) post amputation were stained for TUNEL (terminal deoxynucleotidyl transferase dUTP nick end labelling; green in upper panels, greyscale in lower panels), and counterstained DAPI (blue). Hatched white boxes (**a**, **b**, **c**) are magnified in (**d**, **e**, **f**). Arrow overlies central tissue and highlights width of wound. WE, wound-edge epithelium; WB, wound bed. Scale bars: **a**-**c**: 200 μm; **d**-**f**: 50 μm

Ultrasound and scanning EM assessment of the wound revealed contraction as a significant mechanism underlying the rapid closure. The reaction of the skin musculature is thought to contribute to these tissue movements [39], but an epithelial component was also observed. As first described by Lange [13] and re-iterated by Feral (in the study of cuttlefish arms, [9]) and Polglase et al. (in the study of mantle repair in octopus, [20]), we saw a crest of tissue, a "raised lip", which could protect the exposed tip. Furthermore, scanning EM allowed us to observe a belt-like structure at the wound perimeter that we suggest may be likened to a contractile actin purse-string. Characterisation of the composition of this apparatus in future studies (for example with phalloidin staining for filamentous actin) would provide insight into the mechanism of this phase of closure, and has the potential to highlight an evolutionarily conserved feature of wound repair [40]. The incomplete contractile apparatus around the circumference of the wound (lacking ventrally) was intriguing in that this difference in tension may permit a dorsal shift in the wound, pulling suckers over the end of the arm. This "abnormal positioning of the suckers" [13] is predicted to restore some level of functionality to the arm very quickly, despite the shortened length [21]. We suggest that this early asymmetric wound contraction also underpins the consistently dorsal orientation of regrowth described for amputation injuries caused experimentally [13], or which occurred in the wild [6].

Tissue contraction made a significant contribution to wound closure in the first 6 h, but then its involvement in further healing lessened. As contraction plateaued, cell protrusions suggestive of active cell migration at the epithelial margin were seen. Tissue contraction, cell migration, and cell morphology changes together are thought to be sufficient for closure of a reasonably-sized wound, in the absence of proliferation [9, 13, 19, 21]. This is perhaps analogous to mechanisms of epithelial restitution during repair of mammalian gastrointestinal mucosa [41], and also to the tissue dynamics observed during dorsal closure and wound repair in *Drosophila* embryos [42]. Although cell proliferation has not been analysed in our samples, based on cuttlefish studies, cell division is thought to become important during the regenerative phase starting around day 5 after injury [21].

Histological analysis of the wounds revealed a coalescing of cells over the exposed tip of the amputated arm. These cells were impressively interdigitated, creating a seemingly impermeable temporary protective covering. This may be equivalent to the interdigitation of platelets during the haemostatic response after vertebrate skin wounding [43]. Of course this structure is distinct from the vertebrate clot, in that its destiny is not to be sloughed off, but retained, likely contributing to the blastema [9, 13]. The identity of the cells comprising this plug has historically been assumed to be blood cells [9, 13], and specifically not muscle cells [13]. The prevailing view is that amoebocytes/haemocytes make a significant contribution [39, 44, 45]; however, our histology data (e.g. Fig. 3b) suggest that muscle cells may also play a part. Whether muscle de-differentiation contributes to regeneration appears to be highly species–specific (e.g. occurring in newt, but not axolotl [46]). Coupling information about cell proliferation in the plug with emerging molecular signatures of distinct cell populations and culture strategies should allow us to study both the origins [47] and plasticity of these cells.

The matrix composition of the plug is now widely accepted to be functionally important for regeneration in many species, from mouse ear [48] to axolotl limb [49]. Interestingly, Lange [13] almost 100 years ago may have been one of the first to hypothesize that matrix susceptibility to proteolytic degradation (or resistance, as in fibrosis) could affect regenerative capacity [50]; octopus arm may be an ideal model for the study of wound-associated extracellular matrix features permissive of regeneration.

From the first descriptions of cephalopod response to limb amputation injury, it is known that there is "degeneration" or "disintegration" of muscle and nerve tissue [9, 13, 21]. Since that time, work on other regenerating species has revealed that this is likely to be attributable to apoptotic cell death, which is essential for blastema formation and activating the proliferative response required for regeneration [51]. In this study, the mechanism of cell death was not confirmed, however, cell death based on TUNEL-positivity was seen in skin, muscle and nerve cells, with varying temporal dynamics, but all within six hours.

As recently revealed by a study of axolotl skin [52], a comparison of wound repair in a regenerating tissue (e.g. octopus arm) with a site which only displays wound healing (e.g. octopus mantle [20]) could improve our understanding of conserved cellular and molecular mechanisms, and possibly also the impact of tissue functionality in driving evolution of regeneration [8]. In both the transected arm and the wounded mantle, skin tissue infolding and contraction was a major contributor to early closure. Mantle repair may be distinct from our observations in the arm in that the cellular plug was partially excluded during re-epithelialisation, and cell death was specifically described to be necrotic [20]. These differences are thought to reflect the absence of a blastema in the mantle, and we are intrigued as to whether or not absence of chromatophore regeneration in the mantle [20] contributes to the "healing" as opposed to "regenerative" environment.

Conclusion

Herein, we have described the acute response of *O. vulgaris* to amputation injury. Repair to an arm wound occurred rapidly, with almost complete closure occurring in some individuals in 6–24 h. The observed mechanisms of tissue repair included tissue contraction, epithelial cell migration, and a coalescing of cells at the site of injury, which are notably evolutionarily conserved features of the process. Using this model, which mimics an injury relevant to the life of the animal, we have identified key early processes of repair preparatory to regeneration. The complexity of the tissue architecture of the *O. vulgaris* arm that includes epithelial, chromatophore, vascular, muscle and neural tissue makes this a particularly interesting system in which to study acute responses to injury and subsequent regeneration.

Ethics approval and consent to participate

The following text appears in the Methods section of the manuscript:

At the time the studies were performed, no national legislation was in place in Italy for regulation of research involving cephalopods. The care and welfare assessment of animals was consistent with best practice at the time the study was undertaken [29, 30] *and took into account the then impending changes in the European Union legislation bringing cephalopods within the scope of regulations covering scientific research involving living animals (Directive 2010/EU/63,* [25, 26, 31]*). Experimental design and animal care was discussed with the institution veterinarian who also monitored the study.*

Additional file

Additional file 1: Figure S1. Variable tissue reaction to amputation, without consistent protrusion. (A) Gross observations illustrate the range of response (two biological replicates). (B) Histological assessment at 2 h (H&E; three biological replicates) showed protrusion of the central muscle and nervous tissue in 1 of 3 samples (arrow overlies central tissue and highlights width of wound). Scale bars: A, 2 mm; B, 200 μm.

Abbreviations

EM: electron microscopy; H&E: haematoxylin and eosin; PAS: periodic acid-schiff; SD: standard deviation; SEM: standard error of the mean; TUNEL: terminal deoxynucleotidyl transferase dUTP nick end labelling.

Competing interests

The authors declare that they have no competing interests.

Authors' contributions

TJS, PLRA and GF designed the study. TJS, PLRA, MO and GP performed the experiments and data analysis. TJS and PLRA drafted the manuscript. All authors discussed the paper and gave approval for publication.

Acknowledgements

We thank the Director at the Stazione Zoologica Anton Dohrn for access to facilities and thank P. Amodio and Virginia Acerra for their assistance. We also acknowledge the contribution of Ray Moss and Maria McGlynn of the Imaging Resource Facility (St George's, University of London). We are grateful to Dr. Andrea Affuso, for veterinary supervision. High Resolution Ultrasound Imaging was achieved courtesy of Dr Dieter Fuchs at Fujifilm VisualSonics BV (Science Park 402, Amsterdam, The Netherlands).

Funding

This work was partly supported by ASSEMBLE, the Stazione Zoologica Anton Dohrn and CephRes-ONLUS. GP has been supported by the Flagship Project RITMARE-The Italian Research for the Sea-coordinated by the Italian National Research Council and funded by the Italian Ministry of Education, University and Research within the National Research Program 2011–2013.

Author details

[1]Division of Biomedical Sciences, St George's University of London, Cranmer Terrace, London SW17 0RE, UK. [2]Stazione Zoologica Anton Dohrn, Villa Comunale, 80121 Naples, Italy. [3]King's College London-Centre for Molecular and Cellular Biology of Inflammation, New Hunt's House, Great Maze Pond, London SE1 1UL, UK.

References

1. Norman M. Cephalopods of the world, a world guide. Hakenheim, Germany: ConchBooks; 2003.
2. Budelmann B. Autophagy in octopus. S Afr J Mar Sci. 1998;20(1):101–8.
3. Alupay J, Caldwell R. The costs and benefits of losing an arm: autotomy in the octopus *Abdopus aculeatus*. Integr Comp Biol. 2013;53:E4.
4. Huffard CL. Ethogram of Abdopus aculeatus (d'Orbigny, 1834) (Cephalopoda: Octopodidae): Can behavioural characters inform octopodid taxomony and systematics? J Molluscan Stud. 2007;73(2):185–93.
5. Bush SL. Economy of arm autotomy in the mesopelagic squid *Octopoteuthis deletron*. Mar Ecol Prog Ser. 2012;458:133–40.
6. Florini M, Fiorito G, Hague T, Andrews P. "Monco": A natural model for studying arm usage and regeneration in *Octopus vulgaris*. J Shellfish Res. 2011;30:1002.
7. Voight JR. Movement, Injuries and Growth of Members of a Natural Population of the Pacific Pygmy Octopus, *Octopus digueti*. J Zool. 1992;228:247–63.
8. Bely AE, Nyberg KG. Evolution of animal regeneration: re-emergence of a field. Trends Ecol Evol. 2010;25(3):161–70.
9. Feral J-P. Wound healing after arm amputation in *Sepia officinalis* (Cephalopoda: Sepioidea). J Invertebr Pathol. 1988;52(3):380–8.
10. Adam W. Notes sur les cephalopodes. VII. Sur un cas de regeneration chez Sepioteuthis lessoniana Lesson, 1830. Belguim: Musee Royal d Histoire Naturelle de Belgique; 1937.
11. Murata M, Ishii M, Osako M. On the regeneration of tentacle of the oceanic squid *Ommastrephes bartramii* (Lesueur). Bull Hokkaido Reg Fish Res Lab. 1981;46:1–14.
12. Bush SL, Robison BH, Caldwell RL. Behaving in the dark: locomotor, chromatic, postural, and bioluminescent behaviors of the deep-sea squid *Octopoteuthis deletron* (Young 1972). Biol Bull. 2009;216(1):7–22.
13. Lange MM. On the regeneration and finer structure of the arms of the cephalopods. J Exp Zool. 1920;31(1):1–58.
14. Robison BH, Reisenbichler KR, Hunt JC, Haddock SH. Light production by the arm tips of the deep-sea cephalopod *Vampyroteuthis infernalis*. Biol Bull. 2003;205(2):102–9.
15. Alvarado AS. Metazoan regeneration. Dev Biol. 2000;222(1):231–1.
16. Kier WM, Smith KK. Tongues, tentacles and trunks: the biomechanics of movement in muscular-hydrostats. Zool J Linnean Soc. 1985;83(4):307–24.
17. Young J. The number and sizes of nerve cells in Octopus. Proc Zool Soc London. 1963;140:229–54.
18. Gutfreund Y, Matzner H, Flash T, Hochner B. Patterns of motor activity in the isolated nerve cord of the octopus arm. Biol Bull. 2006;211(3):212–22.
19. Fossati SM, Carella F, De Vico G, Benfenati F, Zullo L. Octopus arm regeneration: Role of acetylcholinesterase during morphological modification. J Exp Mar Biol Ecol. 2013;447:93–9.
20. Polglase JL, Bullock AM, Roberts RJ. Wound Healing and the Hemocyte Response in the Skin of the Lesser Octopus *Eledone cirrhosa* (Mollusca, Cephalopoda). J Zool. 1983;201(Oct):185–204.
21. Rohrbach B, Schmidtberg H. Sepia arms and tentacles: Model systems for studying the regeneration of brachial appendages. Vie et milieu. 2006; 56(2):175–90.
22. Tressler J, Maddox F, Goodwin E, Zhang Z, Tublitz NJ. Arm regeneration in two species of cuttlefish *Sepia officinalis* and *Sepia pharaonis* Invertebr Neurosci. 2014;14(1):37–49.
23. Brockes JP, Kumar A. Comparative aspects of animal regeneration. Annu Rev Cell Dev Biol. 2008;24:525–49.
24. Wong VW, Gurtner GC, Longaker MT. Wound healing: a paradigm for regeneration. Mayo Clin Proc. 2013;88(9):1022–3 .
25. Smith JA, Andrews PL, Hawkins P, Louhimies S, Ponte G, Dickel L. Cephalopod research and EU Directive 2010/63/EU: Requirements, impacts and ethical review. J Exp Mar Biol Ecol. 2013;447:31–45.
26. Fiorito G, Affuso A, Anderson DB, Basil J, Bonnaud L, Botta G, Cole A, D'Angelo L, De Girolamo P, Dennison N. Cephalopods in neuroscience: regulations, research and the 3Rs. Invertebr Neurosci. 2014;14(1):13–36.
27. Canali E, Ponte G, Belcari P, Rocha F, Fiorito G. Evaluating age in *Octopus vulgaris*: estimation, validation and seasonal differences. Mar Ecol Prog Ser. 2011;441:141–9.
28. Fiorito G, Von Planta C, Scotto P. Problem solving ability of *Octopus vulgaris* lamarck (Mollusca, Cephalopoda). Behav Neural Biol. 1990;53(2):217–30.
29. Andrews PL, Darmaillacq A-S, Dennison N, Gleadall IG, Hawkins P, Messenger JB, Osorio D, Smith VJ, Smith JA. The identification and management of pain, suffering and distress in cephalopods, including anaesthesia, analgesia and humane killing. J Exp Mar Biol Ecol. 2013;447:46–64.
30. Boyle P. The UFAW handbook on the care and management of cephalopods in the laboratory. Wheathampstead, Hertfordshi: Universities Federation for Animal Welfare; 1991.
31. Fiorito G, Affuso A, Basil J, Cole A, De Girolamo P, D'Angelo L, Dickel L, Gestal C, Grasso F, Kuba M et al. Guidelines for the Care and Welfare of Cephalopods in Research–A consensus based on an initiative by CephRes, FELASA and the Boyd Group. Lab Anim-Uk. 2015;49:1–90.
32. Amodio P, Andrews P, Salemme M, Ponte G. The Use of Artificial Crabs for Testing Predatory Behavior and Health in the Octopus. ALTEX. 2014;31(4):494–9.

33. Messenger J, Nixon M, Ryan K. Magnesium chloride as an anaesthetic for cephalopods. Comp Biochem Physiol C. 1985;82(1):203–5.
34. Grimaldi AM, Agnisola C, Fiorito G. Using ultrasound to estimate brain size in the cephalopod *Octopus vulgaris* (Cuvier) *in vivo*. Brain Res. 2007;1183:66–73.
35. Andrews PLR, Messenger JB, Tansey EM. Colour Changes in Cephalopods after Neurotransmitter Injection into the Cephalic Aorta. Proc R Soc Ser B-Bio. 1981;213(1190):93.
36. Crook RJ, Lewis T, Hanlon RT, Walters ET. Peripheral injury induces long-term sensitization of defensive responses to visual and tactile stimuli in the squid *Loligo pealeii*, Lesueur 1821. J Exp Biol. 2011;214(19):3173–85.
37. Margheri L, Ponte G, Mazzolai B, Laschi C, Fiorito G. Non-invasive study of *Octopus vulgaris* arm morphology using ultrasound. J Exp Biol. 2011;214(22):3727–31.
38. Wells MJ. Octopus: physiology and behaviour of an advanced invertebrate. London: Chapman and Hall; 1978.
39. Young JZ. The anatomy of the nervous system of Octopus vulgaris. Oxford: Clarendon Press Oxford; 1971.
40. Martin P, Lewis J. Actin cables and epidermal movement in embryonic wound healing. Nature. 1992;360(6400):179–83.
41. Lacy ER. Epithelial restitution in the gastrointestinal tract. J Clin Gastroenterol. 1988;10 Suppl 1:S72–7.
42. Wood W, Jacinto A, Grose R, Woolner S, Gale J, Wilson C, Martin P. Wound healing recapitulates morphogenesis in *Drosophila* embryos. Nat Cell Biol. 2002;4(11):907–12.
43. Wester J, Sixma JJ, Geuze JJ, Van der Veen J. Morphology of the early hemostasis in human skin wounds: influence of acetylsalicylic acid. Lab Investig. 1978;39(3):298–311.
44. Castellanos-Martinez S, Prado-Alvarez M, Lobo-da-Cunha A, Azevedo C, Gestal C. Morphologic, cytometric and functional characterization of the common octopus (*Octopus vulgaris*) hemocytes. Dev Comp Immunol. 2014;44(1):50–8.
45. Wells MJ. Circulation in cephalopods. In: Saleuddin ASM, Wilbur KM, editors. The Mollusca. London: Academic Press, Inc; 1983. p. 239–90.
46. Sandoval-Guzman T, Wang H, Khattak S, Schuez M, Roensch K, Nacu E, Tazaki A, Joven A, Tanaka EM, Simon A. Fundamental differences in dedifferentiation and stem cell recruitment during skeletal muscle regeneration in two salamander species. Cell Stem Cell. 2014;14(2):174–87.
47. Kragl M, Knapp D, Nacu E, Khattak S, Maden M, Epperlein HH, Tanaka EM. Cells keep a memory of their tissue origin during axolotl limb regeneration. Nature. 2009;460(7251):60–5.
48. Gourevitch D, Clark L, Chen P, Seitz A, Samulewicz SJ, Heber-Katz E. Matrix metalloproteinase activity correlates with blastema formation in the regenerating MRL mouse ear hole model. Dev Dynam. 2003;226(2):377–87.
49. Calve S, Odelberg SJ, Simon HG. A transitional extracellular matrix instructs cell behavior during muscle regeneration. Dev Biol. 2010;344(1):259–71.
50. Harty M, Neff AW, King MW, Mescher AL. Regeneration or scarring: An immunologic perspective. Dev Dynam. 2003;226(2):268–79.
51. Bergmann A, Steller H. Apoptosis, Stem Cells, and Tissue Regeneration. Sci Signal. 2010;3(145):re8.
52. Seifert AW, Monaghan JR, Voss SR, Maden M. Skin regeneration in adult axolotls: a blueprint for scar-free healing in vertebrates. PLoS One. 2012;7(4):e32875.

13

On the sighted ancestry of blindness – exceptionally preserved eyes of Mesozoic polychelidan lobsters

Denis Audo[1,2*], Joachim T. Haug[3], Carolin Haug[3], Sylvain Charbonnier[2], Günter Schweigert[4], Carsten H. G. Müller[5] and Steffen Harzsch[6]

Abstract

Background: Modern representatives of Polychelida (Polychelidae) are considered to be entirely blind and have largely reduced eyes, possibly as an adaptation to deep-sea environments. Fossil species of Polychelida, however, appear to have well-developed compound eyes preserved as anterior bulges with distinct sculpturation.

Methods: We documented the shapes and sizes of eyes and ommatidia based upon exceptionally preserved fossil polychelidans from Binton (Hettangian, United-Kingdom), Osteno (Sinemurian, Italy), Posidonia Shale (Toarcian, Germany), La Voulte-sur-Rhône (Callovian, France), and Solnhofen-type plattenkalks (Kimmeridgian-Tithonian, Germany). For purposes of comparison, sizes of the eyes of several other polychelidans without preserved ommatidia were documented. Sizes of ommatidia and eyes were statistically compared against carapace length, taxonomic group, and outcrop.

Results: Nine species possess eyes with square facets; *Rosenfeldia oppeli* (Woodward, 1866), however, displays hexagonal facets. The sizes of eyes and ommatidia are a function of carapace length. No significant differences were discerned between polychelidans from different outcrops; Eryonidae, however, have significantly smaller eyes than other groups.

Discussion: Fossil eyes bearing square facets are similar to the reflective superposition eyes found in many extant decapods. As such, they are the earliest example of superposition eyes. As reflective superposition is considered plesiomorphic for Reptantia, this optic type was probably retained in Polychelida. The two smallest specimens, a *Palaeopentacheles roettenbacheri* (Münster, 1839) and a *Hellerocaris falloti* (Van Straelen, 1923), are interpreted as juveniles. Both possess square-shaped facets, a typical post-larval feature. The eye morphology of these small specimens, which are far smaller than many extant eryoneicus larvae, suggests that Jurassic polychelidans did not develop via giant eryoneicus larvae. In contrast, another species we examined, *Rosenfeldia oppeli* (Woodward, 1866), did not possess square-shaped facets, but rather hexagonal ones, which suggests that this species did not possess reflective superposition eyes. The hexagonal facets may indicate either another type of superposition eye (refractive or parabolic superposition), or an apposition eye. As decapod larvae possess apposition eyes with hexagonal facets, it is most parsimonious to consider eyes of *R. oppeli* as apposition eyes evolved through paedomorphic heterochrony.

(Continued on next page)

* Correspondence: denis.audo@univ-rennes1.fr; denis.audo@edu.mnhn.fr
Denis Audo and Joachim T. Haug share the first authorship
Carsten H. G. Müller and Steffen Harzsch share last authorship
[1]Université de Rennes 1, EA 7316, 263 Avenue du Général Leclerc CS 74205, 35042 Rennes Cedex, France
[2]Centre de Recherche sur la Paléobiodiversité et les Paléoenvironnements (CR2P, UMR 7207), Muséum national d'Histoire naturelle, Sorbonne Universités, UPMC, CNRS, 8 rue Buffon, F-75005 Paris, France
Full list of author information is available at the end of the article

(Continued from previous page)

Conclusion: Polychelidan probably originally had reflective superposition. *R. oppeli*, however, probably gained apposition eyes through paedomorphosis.

Keywords: Polychelida, Solnhofen, La Voulte-sur-Rhône, Osteno, Nusplingen, Posidonia Shale, Heterochrony, Deep-sea adaptations, Superposition eyes

Background

Eyes in the fossil record

The study of eye structures in fossil arthropods has long been limited to that of highly calcified trilobites [1]. Due to their sturdy outer surface, eyes and interior structures, such as substructures of the lenses, are regularly preserved in trilobite fossils [2]. The investigation of exceptionally well-preserved eye structures in less heavily calcified arthropod taxa (e.g., crustaceans) has attracted a great deal of scientific attention in recent years. Although the preservation of such an apparently fragile organ would appear to be exceptional, the preservation of eye structures in sclerotized arthropods is not uncommon. The good preservation of some such fossil eyes has, in certain cases, even allowed the partial reconstruction of putative optical properties [3–6]. Such studies focused on functional aspects of eyes possessed by early representatives of different arthropod lineages to infer possible ancestral features, and thus to shed light on the early evolution of eye structures in sclerotized arthropods. These studies included eyes of presumably early chelicerates [7–11] which, however, have been alternatively interpreted as representatives of the lineage towards Euarthropoda [12, 13] (see also Haug et al. critical review [9]).

These early arthropods include a number of spectacular taxa, such as (1) *Anomalocaris* sp. from Emu Bay Shale [4] which is one of the arthropods with the highest known number of ommatidia in the compound eye, or (2) the early chelicerate *Leanchoilia superlata* Walcott, 1912 which initially was believed to have possessed two pairs of compound eyes [14], but later was suggested to have carried "just" one pair of bilobed compound eyes [10] (as do other species of *Leanchoilia* [15]). In addition, favorable incidents during fossilization of Early Cambrian biota not only led to the preservation of prominent epidermal sense organs (e.g., compound eyes, cuticular sensilla), but also yielded information on the anatomy, as was demonstrated most recently in the early sclerotized arthropod from Chengjiang *Fuxianhuia protensa* Hou, 1987, which displayed a tripartite brain and optic neuropil [16]. Further studies also included early representatives of the mandibulate lineage other than trilobites; i.e., softer (non-calcified) and smaller fossils of early representatives Crustacea *sensu lato* [17] discovered as uncompressed fossils. In these, the cuticle became impregnated by calcium phosphate, in a so-called Orsten-type preservation [18]. Among fossilized arthropods, trilobites and Orsten crustaceans share the advantage of being preserved extraordinarily well, retaining most of their original volume [18].

Besides these early derivatives of different evolutionary lineages, fossilized eye structures of representatives of extant groups have been described in Orsten-type preservation, for example the compound eyes of an ostracod [19], an achelate phyllosoma larva [3] and a possible maxillopod [17]. Eyes preserved uncompressed in three dimensions, i.e. in their original volume, are also reported in many arthropods from the La Voulte-sur-Rhône nodules, notably in thylacocephalans [20–24] and in glypheid lobsters [25].

A further source of exquisitely preserved fossils comes from the plattenkalks. Although the body volume of these fossils was not preserved entirely, as for instance in Orsten's or La Voulte's fossils, at least some specimens may show aspects of their original volume [26]), or even appear almost uncompressed [27]. Specimens with preserved eyes have been described from the famous Solnhofen-type plattenkalks of southern Germany (e.g. an isopod). Additional, although not very well-preserved examples of fossil eyes include those of benthesicymid shrimps [28] and of cirolanid isopods [29, 30] from the Late Cretaceous Sahel Alma Lagerstätte (Lebanon).

Eyes in polychelidan lobsters

Polychelidan lobsters are ascribed to Reptantia, a group comprising mostly benthic crustaceans with elongated pleon, such as spiny, slipper or squat lobsters, crayfishes, hermit crabs, and true crabs. Among Reptantia, Polychelida is probably the sister group to Eureptantia comprising all remaining groups of Reptantia [31]. The predominantly benthic polychelidans have retained some plesiomorphic characters from the ground pattern of natantian shrimps, such as the triangular telson. Polychelida is thus a key group for reconstructing character evolution in Reptantia, and Decapoda as a whole.

Adults of extant species of Polychelida are all thought to be blind and to inhabit deep-sea ecosystems [32]. Their larvae, however, may retain functional eyes [33] which may degenerate over the course of development. In adults, the eye stalks are still present (except in *Willemoesia* Grote, 1873), but the corneae are always reduced;

despite this, live specimens of *Polycheles typhlops* Heller, 1862 still reacted to the intensity of light [34]. However, it is known that in the fossil record, many specimens of polychelidan lobsters possessing eyes are present. Even the oldest occurrence of fossil polychelidan, *Coleia uzume* [35] from Japan and *Tetrachela raiblana* [36] from Italy and Austria, both Carnian (Late Triassic), seem to have had developed eyes as, although the eyes themselves may not have been preserved, their carapace margins are incised by developed ocular incisions [35, 36]. Despite being documented fossilized eyes have not been sufficiently discussed in an evolutionary context [37–40]. Spence Bate [41] was the first to stress the occurrence of eyes in some fossil forms, and assumed their reduction in extant species. More recently, Schweigert & Dietl [42] illustrated an eye-bearing specimen of *Palaeopolycheles longipes* (Fraas, 1855) from the Nusplingen Plattenkalk and discussed its ommatidia. On the basis of the ommatidial facets of *Paleopopolycheles longipes* [24] as reported by Schweigert & Dietl [42], and its phylogenetic position, Ahyong [43] suggested that coleiids may have already shifted to deep water, in which case the stem polychelids may have also evolved in deep habitats.

Palaeoenvironments and associated faunas

Hettangian of Binton (United-Kingdom) The outcrop of Binton (Warwickshire, United-Kingdom) corresponds to the Wilmcote Limestone member of the basal Blue Lias Formation dating from Rhaetian to Hettangian. The Blue Lias Formation was deposited in a shallow epicontinental sea covering England, which was "normally no more than a few tens of meters deep" [44]. The Blue Lias Formation yielded other fossils, including ichthyosaurs.

Sinemurian of Osteno (Italy) The main outcrop is a quarry along the lake Lugano (Ceresio), near Osteno village (Como, Italy). It is a Fossil-Lagerstätte (outcrop with exceptional preservation of remains in connection, with frequent soft-part preservation). It has yielded some terrestrial plants (allochthonous). It more importantly yielded an important exceptionally preserved marine fauna dominated by crustaceans and thylacocephalans, which also included fishes, cephalopods, polychaetes, and acorn worms [45, 46]. The palaeoenvironment is not well-constrained; however, Teruzzi [47] pointed out similarities in the sponge community with those developing at the limit between the modern neritic–pelagic basins. Osteno was therefore probably deposited in a relatively deep palaeoenvironment.

Toarcian of Holzmaden and Gomaringen (Germany) The outcrops of Holzmaden and Gomaringen (Baden-Württemberg, Germany) are parts of the Posidonia Shale Formation. This formation represents a Fossil-Lagerstätte celebrated for its exquisite preservation of an abundant fauna of marine reptiles that included very large specimens of ichthyosaurs, plesiosaurs, and crocodiles. The Posidonia Shale Formation also yielded some terrestrial tetrapods, numerous bony fishes, sharks, crustaceans, crinoids attached to driftwood, ammonites, brachiopods, numerous bivalves and coleoids, including some with soft parts preserved. Some plant fragments (in addition to driftwood) also occur [48]. Deposition of the Posidonia Shale and the nature of its palaeoenvironment are subject of debate. Bottom water and/or sediment seem to have been often anoxic, with short period of oxygenation. The Posidonia Shale was possibly deposited at a depth of around 50–150 m [49]. The anoxia was possibly a result of significant productivity in the overlying water, which may have caused the benthic habitat to be quite dark.

Bajocian-Bathonian of Monte Fallano (Italy) Monte Fallano is a recently discovered Fossil-Lagerstätte. Its flat-bedded, micritic limestones are similar to those of the plattenkalks of southern Germany; it was therefore possibly deposited in a lagoon fringed by reefs in which the (parautochthonous) fauna lived, similarly to the younger southern German plattenkalks. This outcrop yielded terrestrial plants, insects, numerous fishes, including coral-feeding pycnodonts and crustaceans [50].

Callovian of La Voulte-sur-Rhône (France) The Fossil-Lagerstätte of La Voulte-sur-Rhône is celebrated for its rich fauna preserving cephalopods with their soft parts, abundant ophiuroids, numerous thylacocephalans (with huge eyes), numerous crustaceans, sea spiders, coelacanths and sharks. The La Voulte Lagerstätte is thought to correspond to a deep-water environment, based upon geological and palaeontological evidence (absence of orientation of epizoans on sponges in adjacent outcrops, presence of deep-sea cephalopods) summarized in Charbonnier [22, 51]. Polychelidans were probably autochthonous to this outcrop [24].

Kimmeridgian-Tithonian plattenkalks of southern Germany and Cerin (France) The southern Germany plattenkalks are among the most famous Fossil-Lagerstätten. They are often referred as "Solnhofen-type" outcrops. They were deposited in the regressive trend of the end of Jurassic in small lagoons. Cerin belonged to the same palaeogeographic region [52]. These plattenkalks preserved some terrestrial taxa with plants, insects and tetrapods (e.g. *Archaeopteryx*) and a rich marine fauna dominated by fishes and arthropods, and local echinoderms, also including jellyfishes, brachiopods, annelids, and acorn worms [53]. Most of the fauna was probably parautochthonous to allochthonous, since traces of life such as bioturbation, are rare in these plattenkalks. The plattenkalks were probably formed in a quite shallow palaeoenvironment, ranging from 20 to 60 m in depth [54].

Methods

Objectives

We present new data on eye structures of fossil Polychelida. This includes re-description of the specimen of *Palaeopolycheles longipes* previously shown in Schweigert & Dietl [42]. We also illustrate new specimens of Mesozoic Polychelida from Solnhofen-type plattenkalks (Late Jurassic; Figs. 1 and 2), La Voulte-sur-Rhône (Middle Jurassic: Figs. 3, 4 and 5), and Gomaringen, Osteno and Binton (Early Jurassic: Fig. 6) and compare them to other fossil arthropods, whose living relatives possess well-developed eyes (Figs. 7 and 8). Based on these observations, we discuss the evolution of eye types and optical mechanisms within Polychelida and propose new ideas on the visual ecology of fossil polychelidans.

Origin of specimens

In total, fourteen specimens of polychelidan lobsters with distinct eye structures and preserved facets were examined (Tables 1 and 2):

(1) a specimen of *Coleia barrovensis* M'Coy, 1849 from the Hettangian of Binton (Warwickshire, United Kingdom) preserving eyes and ommatidia, first observed by Woods [38] and which represents the oldest fossil examined in this study;
(2) the holotype of *Coleia viallii* Pinna, 1968 from the Sinemurian (Early Jurassic) of Osteno (northern Italy);
(3) a specimen of *Gabaleryon* Audo et al. in press from the Toarcian (Early Jurassic) of Gomaringen

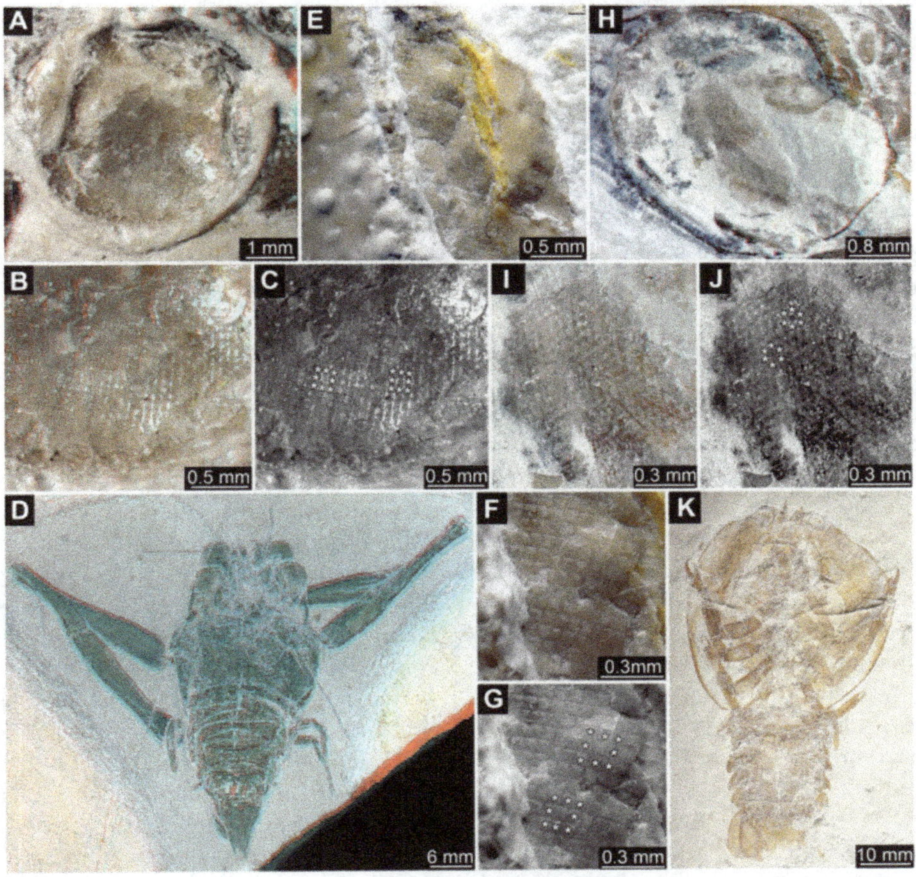

Fig. 1 Late Jurassic Polychelida with preserved compound eye structures. **a-d** *Palaeopolycheles longipes*, SMNS 63724. **e-g**. *Knebelia totoroi*, SMNS 67916. **h-k**. *Rosenfeldia oppeli*, SMNS 66004. **a**. Presumed compound eye with partially preserved cornea. **b**. Detail of the cornea with square facets. **c**. Same detail as in B; note regular alignment of square facets, see asterisks (*) indicating eight facets surrounding a central one. **d**. Overview image of entire specimen. **e-g**. Details of square corneal facets, note asterisk (*) marking some facets in G. **h**. Presumed compound eye with in part preserved cornea. **i-j**. Detail of corneal facets with apparently hexagonal profile, note asterisks (*) marking some facets in J, each facet is surrounded by six neighbored ones. **k**. Overview of entire specimen. **a-b**, **d**, **h-i** are red-cyan stereo images. Please use red-cyan stereo glasses to view

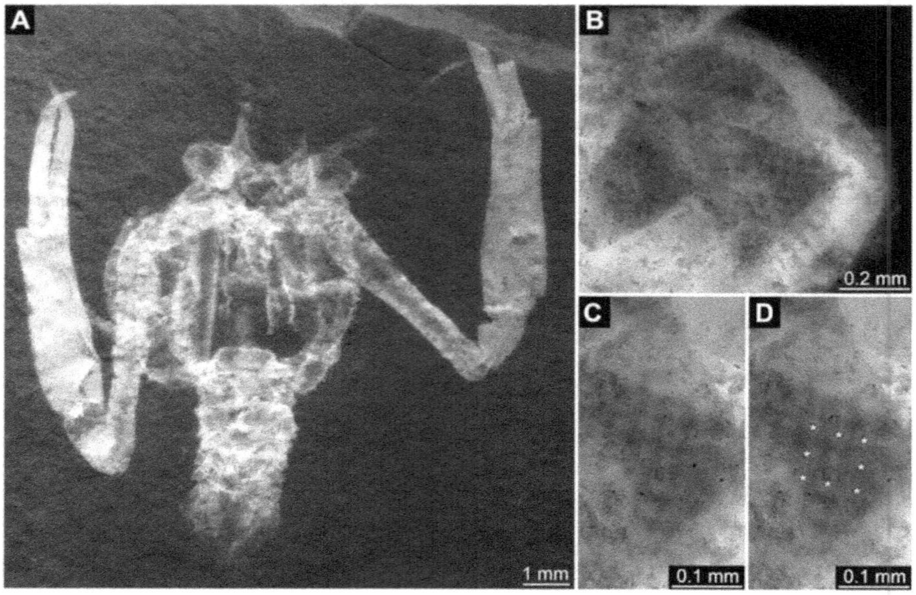

Fig. 2 Late Jurassic Polychelida with preserved compound eyes, continued. **a**. Specimen of *Palaeonpentacheles roettenbacheri* (SMNS 67903). **b**. Close-up on right eye. **c**. Close-up on facetted region. **d**. Same details as C, note asterisks (*) revealing each square facet being surrounded by eight neighbored facets. All images are fluorescence composite images

(southwestern Germany). This specimen is too poorly preserved to be identified to the species level, however, it is the first in which ommatidia are preserved [55];

(4–10) seven specimens from the Callovian (Middle Jurassic) La Voulte-sur-Rhône Lagerstätte (France), among which four specimens of *Proeryon giganteus* [39], two specimens of *Hellerocaris falloti* [39], and one specimen of *Willemoesiocaris ovalis* [39];

(11–14) four specimens from the Late Jurassic Solnhofen-type plattenkalks including one specimen of *Palaeopolycheles longipes* and the holotype of *Knebelia totoroi* Audo et al., 2014b, in which preservation of ommatidia was previously reported by Audo et al. [56], the latter two from Nusplingen (Late Kimmeridgian), a specimen of *Rosenfeldia oppeli* (Woodward, 1866) from Sappenfeld near Eichstätt (Early Tithonian), and a small specimen of *Palaeopentacheles roettenbacheri* (Münster, 1839) from Eichstätt (Early Tithonian) (hence being the youngest occurrences).

All of the above specimens are flattened to a variable degree, including those from La Voulte which retained most of their original volume. To assess the quality of preservation, we compared our sampling of polychelidan fossils with three specimens belonging to different arthropod taxa, such as the isopod *Palaega nusplingensis* Polz, Schweigert & Maisch, 2006 from the Nusplingen Plattenkalk (Germany) or two insects from the Early Cretaceous Santana Formation (Brazil).

For comparison purposes, we also documented 30 additional specimens on which eyes are preserved, but the outline of each ommatidium is not visible (see Table 3). We additionally documented eyes of one isopod and insects that display preservation modes similar to those of fossil polychelidans.

Documentation

Small specimens were documented utilizing composite auto-fluorescence imaging [57–59], macro-fluorescence [58, 60] under polarized light (analyzed; [9, 10, 61–63]) by a single-lens reflex camera or a flatbed scanner. Composite images were processed in CombineZM/ZP, Zerene stacker, Microsoft Image Composite Editor or Photoshop CS3. For some specimens detailed close-ups were documented with SEM (Jeol Neoscope 2 JCM 6000). Details of (isopod and insects) comparative material was documented on a Leica stereomicroscope or Zeiss Axioskop 1 equipped with a Skopetek DCM 510 digital camera producing image stacks. Based on these image stacks, virtual surface reconstructions were calculated and presented as red-blue stereo anaglyphs according to the method recently described by Haug et al. [64].

Calculation of eye parameters

Three values were measured on the high-resolution images:

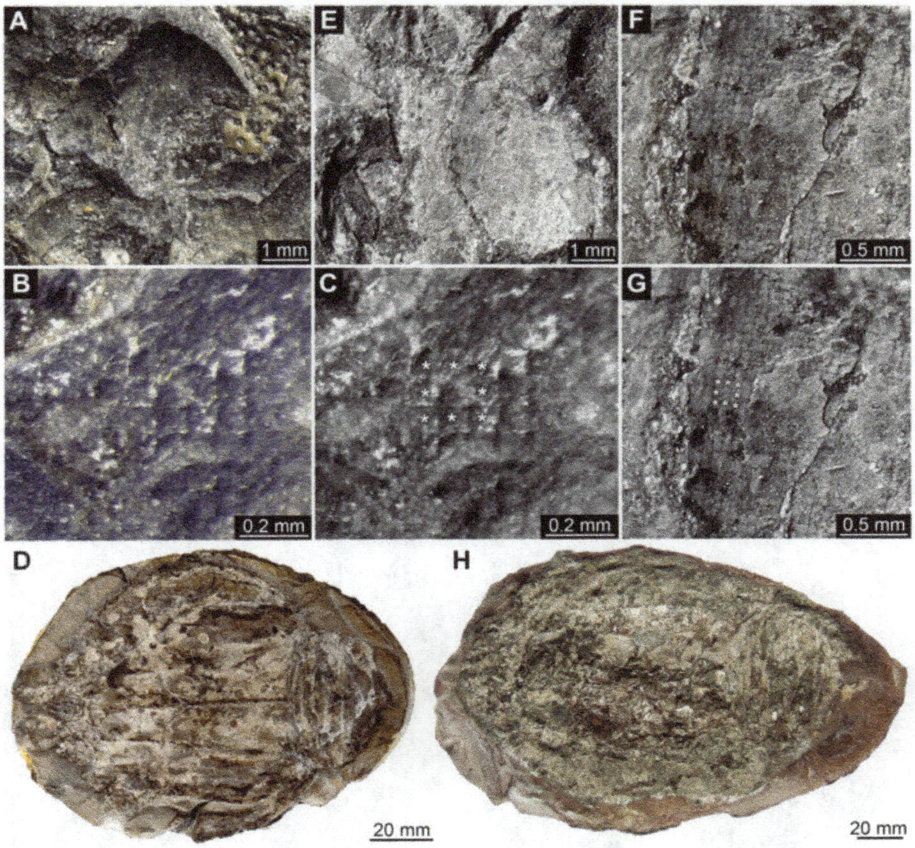

Fig. 3 Middle Jurassic (Callovian) Polychelida (*Proeryon giganteus*) from La Voulte-sur-Rhône (France) displaying preserved eye structures. **a**-**d**. Specimen UJF-ID.11547. **e**-**h**. Specimen UPMC-248. **a**, **e**. Presumed eye structures. **b**-**c**, **f**-**g**. Details of square facets, note that each square facet is surrounded by eight adjacent facets, see asterisks (*) in **c** (same image as **b**) and **g** (same image as **e**). **d**, **h**. Overview of entire specimens

1. Eye diameter d; due to the often incomplete preservation and possible deformation of the eye, we assumed a roughly circular diameter. We measured the highest distance possible (axis) between two opposing margins for each eyes. In the case of a more oval-shaped outline an average of the two axes was used. In the case of incompletely preserved eyes, the greatest measurable length on the eye was used (Tables 1 and 3).
2. Either 2a) the edge length of a square-shaped facet a or 2b) the diameter of a hexagonal facet a, which equals 2× the edge length (Tables 1, 2 and 3). Since ommatidia edge length is variable across eye surface, we measured the edge length of as many adjacent ommatidia as possible and divided this value by the number of ommatidia. Measurements were made along different axes of the eye and at different places; for all specimens but three (Table 2), variations were below the precision of our measurement or ommatidia were of similar sizes across the preserved portion of eye. The three specimens, in which variations in ommatidia edge length were detected, are detailed in Table 2. For these specimens, mean value of ommatidia edge length was compared with other specimens. We note that the preservation of ommatidia varies. Some ommatidia appear convex, others concave or surrounded by walls. These variations, which may be linked to different preservation, may affect slightly the measurements of ommatidia. This probable bias is unfortunately unavoidable when dealing with fossil material, but considering the excellent preservation of measured ommatidia, this bias is probably limited.
3. The carapace total length, measured from the most anterior to the most posterior point of the carapace, parallel to the longitudinal axis (Table 3).

The exact curvature of the eye surface cannot be measured, as all fossils showed a more or less strong (although sometimes only a slight) compaction. To be as cautious as possible, we considered both possible extreme shapes of an eye which may be either flat or display a hemispherical outline. The original morphology most likely ranged somewhere between these

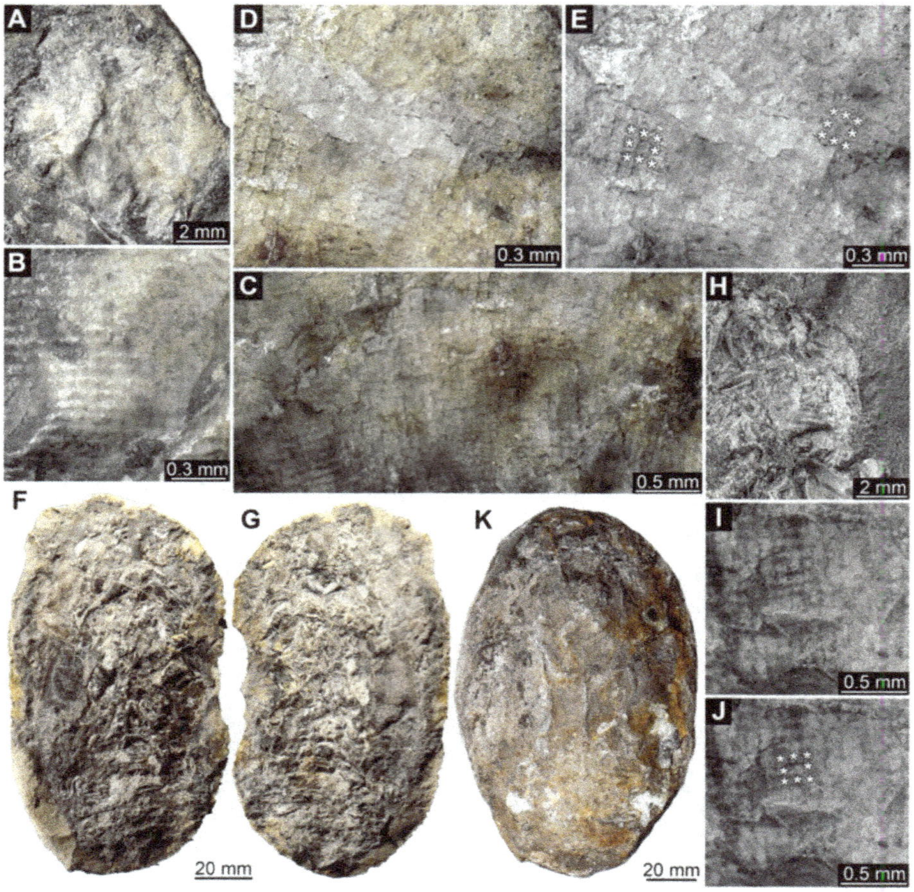

Fig. 4 Middle Jurassic (Callovian) Polychelida (*Proeryon giganteus*) from La Voulte-sur-Rhône (France) with preserved eye structures. **a-g**. Specimen UJF-ID.14020. **h-k**. Specimen FSL 170603. **a**, **h**. Presumed remains of compound eyes. **b-e**, **i-j**. Details of square facets; note that each square facet is surrounded by eight neighbor facets, see asterisks (*) in **e** (same image as **d**) and **j** (same image as **i**). **f**, **g**. Overview of part and counterpart. **k**. Overview of entire specimen

two extremes. In consequence, all further calculations have to be considered estimations, and to refer only to the visible part of the preserved eye, which is the eye most likely formed more than a half sphere, i.e. almost a full sphere; however we have no information on parts below the visible preserved parts.

We then calculated how many ommatidia (n) are present along the longest line across the surface by dividing diameter (assuming a flat eye) or half circumference ($\pi d/2$ assuming a hemispherical eye) divided by (a). Based on this, the average opening angle for each ommatidium was calculated as 180° divided by n (for half sphere; opening angle for flat eyes is 0). The surface of the eye was calculated as $\pi d^2/4$ (flat) or $\pi d^2/2$ (hemispherical). This value was divided by the surface of a single ommatidium (a^2, for square-shaped; $((3\sqrt{3})/2)(a/2)^2$, for hexagonal) to get the total number of ommatidia per visible surface of the eye.

Statistics

To study the possible impact of palaeoenvironments and phylogenetic on eyes parameters, we performed a series of statistical analysis using R (http://www.R-project.org). Measured values were first tested for normality (Shapiro test) for whole samples (eye size, ommatidia) and in subset defined by categories (familial assignment, outcrops); not all subsets were normally distributed, for this reason, non-parametric tests (considered more robust for not normally distributed samples) were used for analysis of the whole sample and a parametric test two compare normally distributed subsets (Coleiidae and Eryonidae ratios—see below; parametric tests are considered more valid when applied to normally distributed samples). We compared the diameter of eyes to the carapace length to search possible link between the two variables (Spearman correlation). Subsequent test were based on the ratio "eye diameter" on "carapace length" and the possible

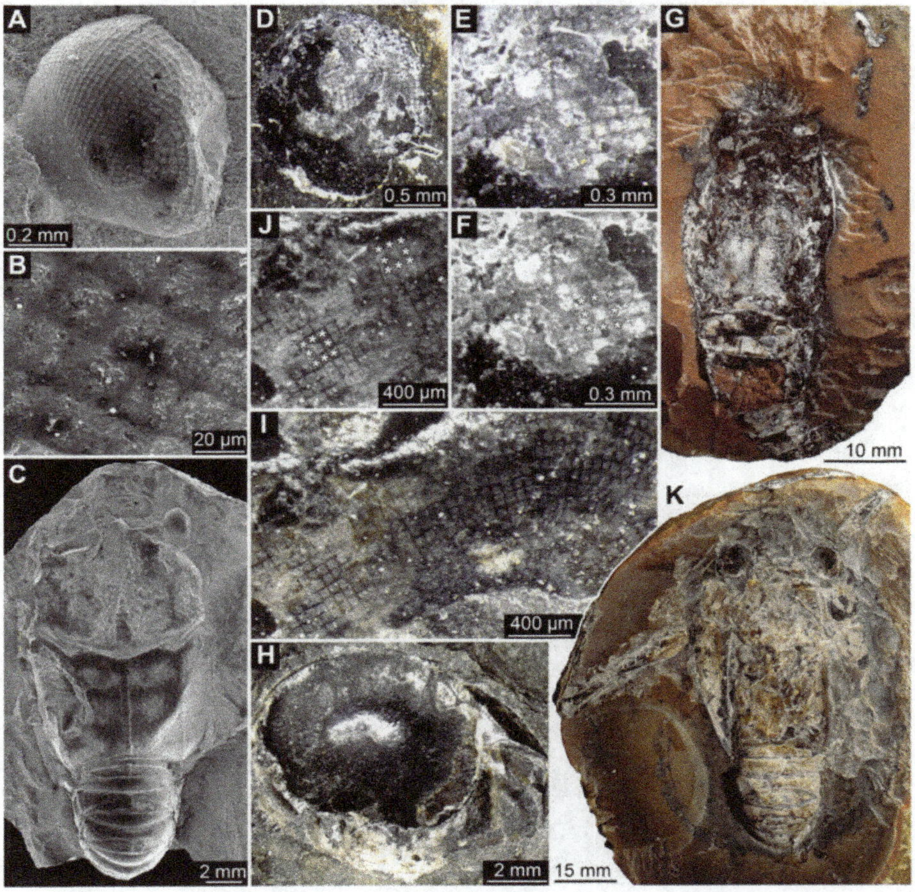

Fig. 5 Middle Jurassic (Callovian) Polychelida from La Voulte-sur-Rhône (France) with preserved compound eye structures, continued: *Hellerocaris falloti* **a-g** and *Willemoesicaris ovalis* **h-k. a-c**. Specimen MNHN.F.A50709, SEM images (Philippe Loubry). **d-g**. Specimen FSL 710108. **h-k**. Specimen MNHN.F.A29521. **a**, **d**, **h**. Overview of compound eye structure. **b**, **e-f**, **i-j**. Details of square facets, note that each square facet is surrounded by eight neighbor facets, see asterisks (*) in **f** (same image as **e**) and **j** (part of **i**).**c**, **g**, **k**. Overview of entire specimens

influence of palaeoenvironments (outcrops) and phylogeny (familial assignment) (Kruskal-Wallis tests). Ratios of "eye diameter" to "carapace length" were compared between Eryonidae and Coleiidae, since number of specimens for other families is not sufficient for statistical analysis. Size of ommatidia was also compared to the diameter of the eyes. We used the square root of ommatidia surface to normalize the size of ommatidia and compare its variation to that of the diameter of the eye (spearman correlation).

Phylogenetic context

Our evolutionary scenario is based on the phylogenies of Scholtz & Richter [31] and Ahyong [43]. Familial assignments follow Audo et al. [24], and Ahyong [43], except in the case of *Rosenfeldia oppeli*, which due to its unusual appearance is not considered as an eryonid.

Results

Eyes and ommatidia

All specimens examined possess anterior pairs of ball- or disk-shaped structures that strongly resemble stalked compound eyes of extant decapod taxa. On 14 specimens, these protuberances display either squared or hexagonal sculpturations. On the remaining specimens, the preservation is not as fine, and these fine sculpturations are not visible. The nature of sculpturations is slightly obscured by compression in flattened specimens. However, objects with fragmentary volume preservation, as for instance in specimens from La Voulte, can be confidently interpreted as fossilized compound eyes. Due to the specific preservation of these eyes, some being partly flattened, incomplete and/or not visible in their entirety, detailed descriptions are difficult. Ommatidia do usually not cover the entire surface. Therefore, the descriptions are used as a basis for simple geometric assumptions

Fig. 6 Early Jurassic Polychelida with preserved compound eye structures. **a**-**d**. *Gabaleryon* sp. specimen SMNS 67631 from the Toarcian of Gomaringen. **e**-**h**. *Coleia viallii*, holotype MSNM i3368 from the Sinemurian of Osteno. **i**-**j**. *Coleia barrovensis*, specimen NHMUK.I6589 (photo © NHMUK) from the Hettangian of Binton. **a**, **e**, **i**. Overview of presumed compound eye structures. **b**-**c**, **f**-**g**. Details of square facets, note that each square facet is surrounded by eight neighbor facets, see asterisks (*) in **c** (same image as **b**) and **g** (same image as **f**). **d**, **h**, **j**. Overview of entire specimens

and generalizations about the entire eye. Measurements and estimations of the properties of these eye structures are summarized in Table 1, see also Table 2.

We find quite a variety of ommatidia sizes from 35–90 μm, their inferred number also varies considerably; depending on high or low estimation there are about 700–17.000 to 1500–34.000 ommatidia per eye.

Differences in eye sizes

Comparison between the diameter of the eye and carapace length shows that both values are linearly linked (Spearmann rho = 0.80; P-value < 10^{-9}: Fig. 9). This suggests, as expected, that larger specimens have larger eyes.

Difference in the proportion of the eyes (diameter of the eye = Eye Ø), relatively to the carapace length (CL) cannot be explained by variations in palaeoenvironments based on our data (Kruskal-Wallis $\chi^2 = 6.95$; P = 0.43). However, phylogenetic variations may help to explain differences in eyes proportion (Kruskal-Wallis $\chi^2 = 11.61$; $P < 0.05$). More precisely, representation of the distribution of values for different families (Fig. 10) shows that: Coleiidae have in proportion larger eyes than Eryonidae and that fossil species more closely related to the modern species of Polychelidae than to fossil species attributed to other families;

- Palaeopentachelidae have in proportion larger eyes than all other taxa (Note that this family is currently composed of a single species: *Palaeopentacheles roettenbacheri*);
- Eryonidae, stem-Polychelidae and *Rosenfeldia oppeli* have eyes of similar proportions.

We compared statistically Eryonidae to Coleiidae (sample-size for other subgroups – here families – is

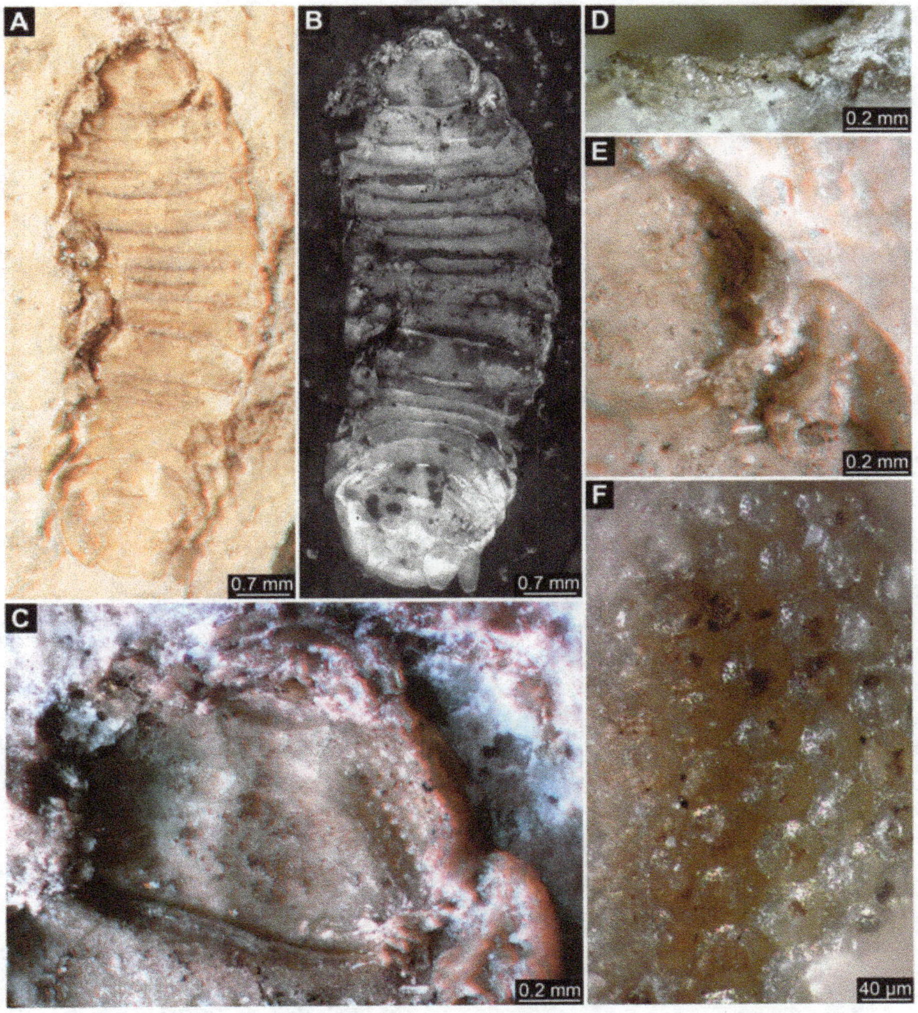

Fig. 7 Comparison with eye-bearing arthropods other than Polychelida: 1. Isopoda. **a**, **b**. Overview of specimen of *Palaega nusplingensis* from the Nusplingen Plattenkalk, SMNS 65512. Note that the specimen is mainly seen from the inside. **c**. Close-up of head. **d**. Close-up on a small part of the compound eye seen on the outside. **e**. Close-up on left side of the head, from a more oblique angle than in **c**. **f**. Close-up of hexagonal facets. **a**, **c** and **e** are red-cyan stereo-images (use red-cyan stereo glasses)

insufficient). Eryonidae eyes are significantly smaller than those of Coleiidae (Welch test: Df = 30.8; $P < 0.01$).

Ommatidial size

Comparison between the square root of ommatidial surface and diameter of the eye shows that both values are linearly linked (Spearmann rho = 0.87; $P < 10^{-4}$: Fig. 11). This implies that large eyes include larger ommatidia. Ommatidia of La Voulte species (Fig. 11) are generally slightly larger than other, unfortunately, our sample-size doesn't allow for detailed comparisons between families or outcrops, so these differences cannot be distinguished from random variations. The small sample size and, as indicated above, variations in the preservation of ommatidia may also have obscured the palaeoenvironmental signal.

Discussion

Optical mechanisms of crustacean eyes and interpretation of fossil polychelidan eyes

Generally, arthropod compound eyes have been classified to function according to two common optical principles: the ancestral apposition type and superposition type which most likely evolved several times in parallel. In apposition eyes, every separate element of the compound eye (ommatidium) acts as independent optical unit. In the different types of superposition the cornea of each ommatidium may spread light onto photoreceptor cells of adjacent ommatidia, thus forming a superimposed image. Apposition eyes are optimized for maximal optical resolution whereas superposition optics increase absolute sensitivity of the eyes [65]. In modern crustaceans, and specifically in decapods, a high diversity of

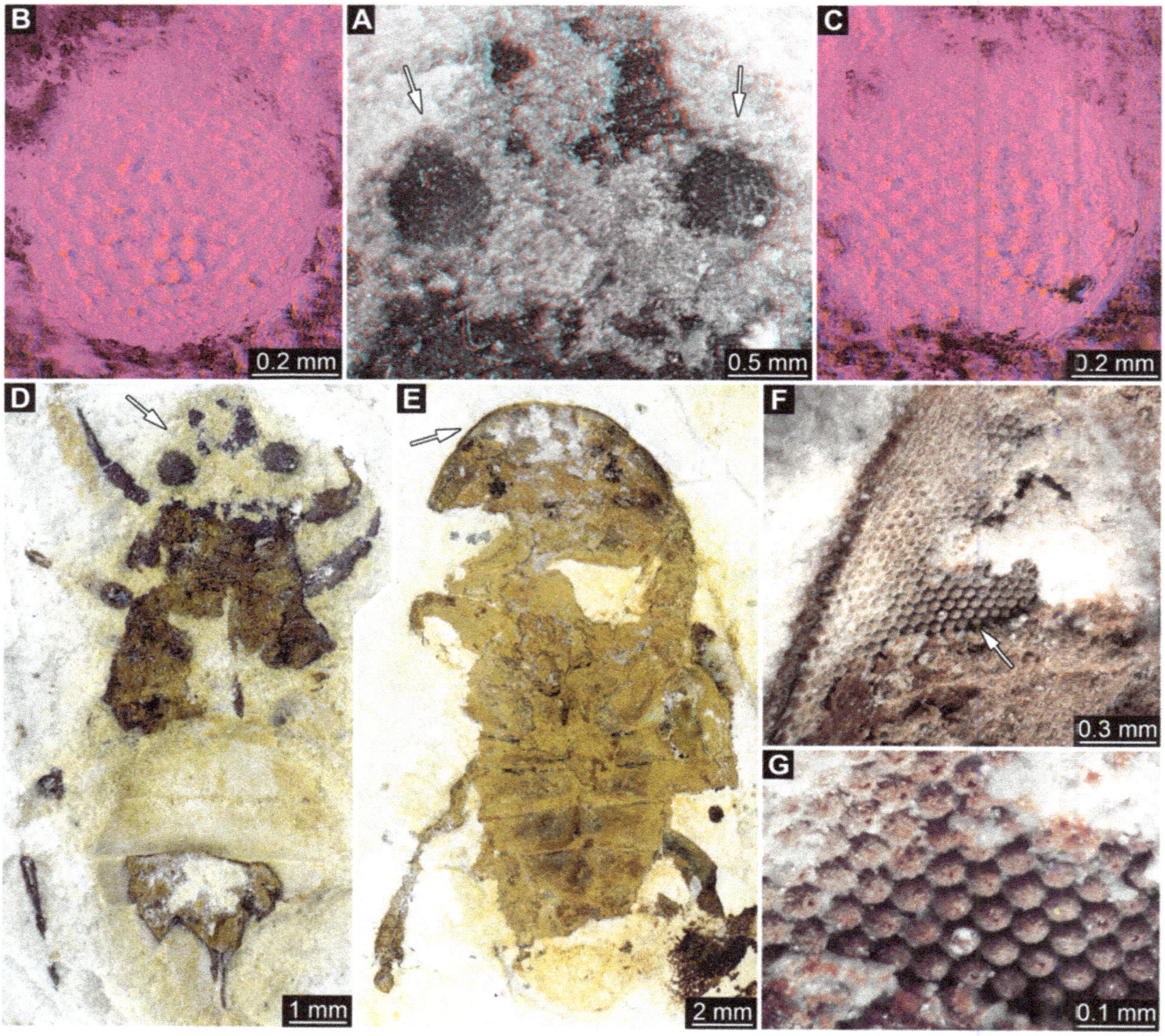

Fig. 8 Comparison with eye-bearing arthropods other than Polychelida: 2. Hexapoda, Pterygota from the Santana Formation, Brazil. **a-d**. Specimen of an undetermined heteropteran species with hexagonal facets. **e-g**. Specimen of an undetermined notonectid species with hexagonal facets. (**a**, **b**, **c**: use red-cyan stereo glasses)

different superposition eye types is found in adults: refracting, reflecting, and parabolic superposition optics [65–70]. Although being quite diverse in adults, larval decapod compound eyes are exclusively of the apposition type [65, 71–74]. The evolutionary transitions of these optical types within Decapoda in general and in Reptantia in particular have been addressed by Gaten [69] and Richter [70]. Gaten [69] did not discuss Polychelida. Richter [70] considered the poylchelid stem species to be blind based on representatives of extant polychelids only. According to our analysis of fossil representatives with preserved eye structures, the view proposed by Richter [70] needs to be revisited.

Interpretation of structures: square ommatidia

We found spherical or subspherical protuberances covered by a distinctive sculpturation of squares or hexagons at the anterior end of the cephalothorax of various Mesozoic polychelids that we interpret as fossilized compound eyes. This interpretation is based on the position, shape and specific surface of these structures in comparison to the compound eyes of extant euarthropods. In all examined species (except *Rosenfeldia oppeli*), compound eyes with square facets are present. This morphology is suggestive of reflective superposition [65], an optical mechanism in which the light received by a given ommatidium is reflected through a "mirror box" onto

Table 1 Measurements and estimations of eyes proprieties in studied samples of polychelidan with preserved ommatidia

			d: Eye diameter (μm)	*a*: Om-matidia size (μm)	Ommatidia shape	Approximation of eye geometry used	*n*: Longest row of ommatidia	Opening angle of ommatidia (degree)	*se*: Surface of eye (μm²)	*so*: surface of one ommatidium (μm²)	Number of ommatidia
		Formula applied →	Measured		Flat	$\frac{d}{a}$	nul, by definition	$\pi \times \left(\frac{d}{2}\right)^2$	Quadrate:	a^2	$\frac{se}{so}$
Age	Species ↓	Specimen ↓			Hemispheric	$\frac{\pi \times d}{2 \times a}$	$\frac{180}{n}$	$2\pi \times \left(\frac{d}{2}\right)^2$	Hexagonal:	$\frac{3\sqrt{3}}{2} \times \left(\frac{a}{2}\right)^2$	
Late Jurassic	*Rosenfeldia oppeli*	SMNS 66004	3,400	53	Hexagonal	Flat	64	0.00	9,079,203	1,824	4,976
						Hemispheric	101	1.79	18,158,406		9,953
	Palaeopentacheles roettenbacheri	SMNS 67903	1,100	35	Quadratic	Flat	31	0.00	950,332	1,225	776
						Hemispheric	49	3.65	1,900,664		1,552
	Palaeopolycheles longipes	SMNS 63724	4,200	50	Quadratic	Flat	84	0	13,854,424	2,500	5,542
						Hemispheric	132	1.36	27,708,847		11,084
	Knebelia totoroi	SMNS 67916	7,200	67.5	Quadratic	Flat	107	0	40,715,041	4,556	8,936
						Hemispheric	168	1.07	81,430,082		17,872
Middle Jurassic	*Proeryon giganteus*	FSL 170603	9,100	90	Quadratic	Flat	101	0	65,038,822	8,100	8,029
						Hemispheric	159	1.13	130,077,644		16,059
		UJF-ID 11547	11,000	88	Quadratic	Flat	125	0	95,033,178	7,744	12,272
						Hemispheric	196	0.92	190,066,356		24,544
		UJF-ID 14020*	11,600	78.8	Quadratic	Flat	147	0	105,683,177	6,209	17,020
						Hemispheric	231	0.78	211,366,354		34,040
		UPMC-248	NA	77	Quadratic	Flat	NA	NA	NA	5,929	NA
						Hemispheric	NA	NA	NA		NA
	Hellerocaris falloti	FSL 710108	1,900	62	Quadratic	Flat	31	0	2,835,287	3,844	738
						Hemispheric	48	3.74	5,670,575		1,475
		MNHN.F.A50709*	850	24.25	Quadratic	Flat	35	0	567,450	588	965
						Hemispheric	55	3.27	1,134,900		1,930
	Willemoesiocaris ovalis	MNHN.F.A29521*	7,000	68.83	Quadratic	Flat	102	0	38,484,510	4,738	8,122
						Hemispheric	160	1.13	76,969,020		16,245
Early Jurassic	*Gabaleryon sp.*	SMNS 67631	1,850	55	Quadratic	Flat	34	0	2,688,025	3,025	889
						Hemispheric	53	3.41	5,376,050		1,777
	Coleia viallii	MSNM.i3368	5600	82	Quadratic	Flat	68	0	24,630,086	6,724	3,663
						Hemispheric	107	1.68	49,260,173		7,326
	Coleia barrovensis	NHMUK.I6589	4100	66	Quadratic	Flat	62	0	13,202,543	4,356	3,031
						Hemispheric	98	1.84	26,405,086		6,062

The size of square ommatidial facets corresponds to the measure of their side length; that of hexagonal ommatidial facets corresponds to the distance between to opposite summits of the hexagon, which is in turn equal to two times the side length. The asterisks after specimens registry numbers indicate that given values are based on multiple ommatidia length measurements, in these cases average values are given. Measured values for these specimens are indicated in Table 2. Distances in μm, surfaces in μm²

Table 2 Measurements and position of ommatidia in specimens having ommatidia of different sizes

Specimen	Ommatidia size Lower	Position on eye	Intermediate	Position on eye	Higher	Position on eye	Mean value
UJF-ID 14020	68	dorsal	76	anterior	92.4	anterior	78.80
MNHN.F.A29521	61	lateral	70	posterior	75.5	posterior	68.83
MNHN.F.A50709	21.5	anterior	27	lateral, dorsal	NA	NA	24.25

Indication is made of the position of measured ommatidia on the surface of the eye. Measurements in µm

the light-guiding structures of neighbouring ommatidia as well. Indeed, the square shape of the ommatidial prism allows two reflections of light to enter the ommatidia against prism walls, and giving rise to an erected image, which allows superposition without changes in the refractive index. Extant representatives of the Palinuridea (also assigned to reptantian decapods) do also possess square facets and reflexive superposition, e.g. representatives of the genera *Jasus* [75] and *Panulirus* [76]. An alternative for this interpretation, although unlikely exists: indeed, some xanthid crabs (e.g. *Trapezia* spp.) also possess square ommatidial facets, but their eyes are of the parabolic superposition type [65, 67]. Because neither the longitudinal profile of the crystalline cone nor its refractive gradients are preserved during fossilization, we cannot entirely exclude the possibility that early polychelidans had parabolic superposition eyes, in analogy to these examples.

The finding of reflective superposition eyes in different representatives of Polychelida closes the gap of knowledge in the early evolution of reptantian decapods [70]. Adult reflecting superposition eyes appear to have been retained from the ground pattern of Decapoda along the entire early evolutionary lineage of Polychelida. Reduced compound eyes are so far known only known in one subgroup: Polychelidae. Therefore, blindness, as found in extant Polychelidae, becomes even more evidently a result of secondary evolution.

Eye preservation

The described specimens here mark the oldest direct evidence of superposition eyes within euarthropods. The presence of reflective superposition eyes is remarkable from the aspect of preservation of eye structures.

In many cases in which compound eyes are fossilized, the ommatidia are either recognizable as 1) dome-shaped swellings on the surface or 2) cavities. In these two cases, apparently inner structures of the ommatidium play a major role in the preservation. In the former case, the swelling may represent an original structural aspect. If we compare it to, for example, SEM images of many extant arthropod compound eyes, we see that similar dome-shaped swellings mark the ommatidia. Alternatively, the ommatidium can be partly collapsed (due to taphonomic processes) but become stabilized by inner structures of the ommatidium. If preserved as a cavity, the outer "walls" of the ommatidium must preserve its stability.

In specimens presented here, such a type of preservation appears to be found in isopods and the insect specimens (Figs. 7 and 8). Among the polychelidan specimens, the single specimen of *Palaeopentacheles roettenbacheri* may possess this type of preservation. Some of the more roughly preserved eyes of some of the polychelidans (Figs. 3a-d; 6a-d) may represent taphonomically collapsed ommatidia. In some specimens from La Voulte, the cornea was probably broken during preparation of the fossil; hence it is likely that what we have observed here is the inner part of the inner lumen of the ommatidia.

In most of the other specimens, the preservation appears to be different and we most likely see the fossilized cornea. In extant decapod crustaceans (e.g. Astacida, Caridea, Galatheoidea, some Brachyura), the square-shaped cornea does reflect the square cross-shape of the crystalline cone in the ommatidia [69, 75, 77–79]). Occasionally, as for instance in some dendrobranchiate crustaceans (prawns), corneal facets may vary considerably in shape, ranging diffusely between square, pentagonal and hexagonal facets, especially to the margins of the eye [80]. This is caused by simple geometrical constraints on a curved surface. The structural variability of the surface is continued further proximally in the ommatidium, hence the crystalline cones display variable cross-profiles. Despite their structural variability, compound eyes of Dendrobranchiata have been interpreted as representing reflective superposition eyes [69, 70]. Strong variations also occur in compound eyes with hexagonal facets, such as the refractive superposition eyes of some mysidacean crustaceans (compare with Richter [80]). This indicates that even the frequent correlation of square facets and reflective superposition may be obscured, at least in some malacostracan taxa.

Hexagonal ommatidia of Rosenfeldia oppeli

In *Rosenfeldia oppeli*, with its hexagonal facets, it is even more difficult to judge from external characters alone which optical type this eye represents. Hexagonal facets occur in different types of eyes with different optical mechanisms. Indeed, eyes with hexagonal facets are known to occur [65, 67, 81]:

Table 3 Measurements of carapace length, eye diameter and ommatidia surface for all studied specimens of polychelidans

Species	Specimen	CL: Carapace Length (μm)	d: Eye diameter (μm)	d/CL	ommatidia surface	Outcrop	Putative palaeodepth	Family
Coleia barrovensis	NHMUK.I6589	28000	4100	0.15	4356	Binton	shallow	Coleiidae
Gabaleryon sp.	SMNS 67631	18300	1850	0.10	3025	Gomaringen	intermediate	Coleiidae
Coleia viallii	MSNM i3368	37000	5600	0.15	6724	Osteno	deep	Coleiidae
Cycleryon elongatus	SNSB-BSPG AS I 939	39700	4700	0.12	NA	Southern Germany plattenkalks	shallow	Eryonidae
Cycleryon elongatus	SNSB-BSPG AS VI 43	39800	4100	0.10	NA	Southern Germany plattenkalks	shallow	Eryonidae
Cycleryon orbiculatus	SMNS 62574	19600	3150	0.16	NA	Southern Germany plattenkalks	shallow	Eryonidae
Cycleryon orbiculatus	SMCU F11408	32500	3900	0.12	NA	Southern Germany plattenkalks	shallow	Eryonidae
Cycleryon orbiculatus	SNSB-BSPG AS VII 762	15800	2200	0.14	NA	Southern Germany plattenkalks	shallow	Eryonidae
Cycleryon propinquus	SMNS (no reg. Number)	62000	4800	0.08	NA	Southern Germany plattenkalks	shallow	Eryonidae
Cycleryon propinquus	MNHN.F.B13436	59000	8300	0.14	NA	Southern Germany plattenkalks	shallow	Eryonidae
Cycleryon propinquus	SNSB-BSPG 1922 I 35	85000	8500	0.10	NA	Southern Germany plattenkalks	shallow	Eryonidae
Cycleryon propinquus	SNSB-BSPG AS VI 42	57000	3000	0.05	NA	Southern Germany plattenkalks	shallow	Eryonidae
Cycleryon propinquus	SNSB-BSPG (no reg. Number)	49000	6500	0.13	NA	Southern Germany plattenkalks	shallow	Eryonidae
Cycleryon propinquus	JME-SOS 6827	55500	5200	0.09	NA	Southern Germany plattenkalks	shallow	Eryonidae
Cycleryon propinquus (female)	SNSB-BSPG AS VI 40	73000	5400	0.07	NA	Southern Germany plattenkalks	shallow	Eryonidae
Cycleryon romani	FSL 170522	112000	4000	0.04	NA	La-Voulte-sur-Rhône	deep	Eryonidae
Hellerocaris falloti	MNHN.F.A50709	6500	850	0.13	588	La-Voulte-sur-Rhône	deep	aff. Polychelidae
Hellerocaris falloti	FSL 710108	24500	1900	0.08	3844	La-Voulte-sur-Rhône	deep	aff. Polychelidae
Hellerocaris falloti	FSL 170598	25800	2350	0.09	NA	La-Voulte-sur-Rhône	deep	aff. Polychelidae
Knebelia bilobata	JME-SOS 6864	40000	4600	0.12	NA	Southern Germany plattenkalks	shallow	Eryonidae
Knebelia bilobata	NHMUK.In 28964	54000	4300	0.08	NA	Southern Germany plattenkalks	shallow	Eryonidae
Knebelia bilobata	SMNS 70044	43500	4000	0.09	NA	Southern Germany plattenkalks	shallow	Eryonidae

Table 3 Measurements of carapace length, eye diameter and ommatidia surface for all studied specimens of polychelidans *(Continued)*

Knebelia totoroi	SMNS 67916	68500	7200	0.11	4556	Southern Germany plattenkalks	shallow	Eryonidae
Palaeopentacheles roettenbacheri	SMNS 67903	4200	1100	0.26	1225	Southern Germany plattenkalks	shallow	Palaeopentachelidae
Palaeopentacheles roettenbacheri	BSPG AS I 989	30000	4900	0.16	NA	Southern Germany plattenkalks	shallow	Palaeopentachelidae
Palaeopentacheles roettenbacheri	BSPG AS I 992	26700	3400	0.13	NA	Southern Germany plattenkalks	shallow	Palaeopentachelidae
Palaeopentacheles roettenbacheri	SNSB-BSPG (no reg. Number)	9250	900	0.10	NA	Southern Germany plattenkalks	shallow	Palaeopentachelidae
Palaeopolycheles longipes	SMNS 63744	29500	4100	0.14	NA	Southern Germany plattenkalks	shallow	Coleiidae
Palaeopolycheles longipes	SMNS 63833	22000	3000	0.14	NA	Southern Germany plattenkalks	shallow	Coleiidae
Palaeopolycheles longipes	SMNS 70203	22000	3300	0.15	NA	Southern Germany plattenkalks	shallow	Coleiidae
Proeryon giganteus	UJF-ID 11547	71000	11000	0.15	7744	La-Voulte-sur-Rhône	deep	Coleiidae
Proeryon giganteus	UJF-ID 14020	79000	11600	0.15	6209	La-Voulte-sur-Rhône	deep	Coleiidae
Proeryon giganteus	FSL 170603	85000	9100	0.11	8100	La-Voulte-sur-Rhône	deep	Coleiidae
Proeryon hartmanni	BSPG 1942 I 20	60000	6100	0.10	NA	Holzmaden	intermediate	Coleiidae
Proeryon hartmanni	SMNS 64019	89000	7700	0.09	NA	Holzmaden	intermediate	Coleiidae
Rosenfeldia oppeli	SMNS 66004	38000	3400	0.09	1824	Southern Germany plattenkalks	shallow	unknown
Soleryon amicalis	MHNL 20271902	72500	2500	0.03	NA	Cerin	shallow	Eryonidae
Tethyseryon campanicus	CSMNF 22000a	21500	2400	0.11	NA	Monte Fallano	shallow	Coleiidae
Tethyseryon campanicus	CSMNF 22000c	8200	1100	0.13	NA	Monte Fallano	shallow	Coleiidae
Tethyseryon campanicus	CSMNF 22000 g	10000	1200	0.12	NA	Monte Fallano	shallow	Coleiidae
Willemoesiocaris ovalis	MNHN F A29521	45000	7000	0.16	4738	La-Voulte-sur-Rhône	deep	Coleiidae
Willemoesiocaris ovalis	FSL 170602	32000	5800	0.18	NA	La-Voulte-sur-Rhône	deep	Coleiidae

Distances in μm, surfaces in μm^2

1. In apposition eyes, including larval apposition eyes, which may eventually be replaced by superposition eyes later in development in some decapod taxa;
2. In refractive superposition eyes;
3. In parabolic superposition eyes.

Our careful examination of the specimen of *Rosenfeldia oppeli* only allowed the description of the external aspect. Therefore, it is not possible to distinguish with confidence to which optical type *R. oppeli* hexagonal ommatidia correspond. However, it is possible to provide

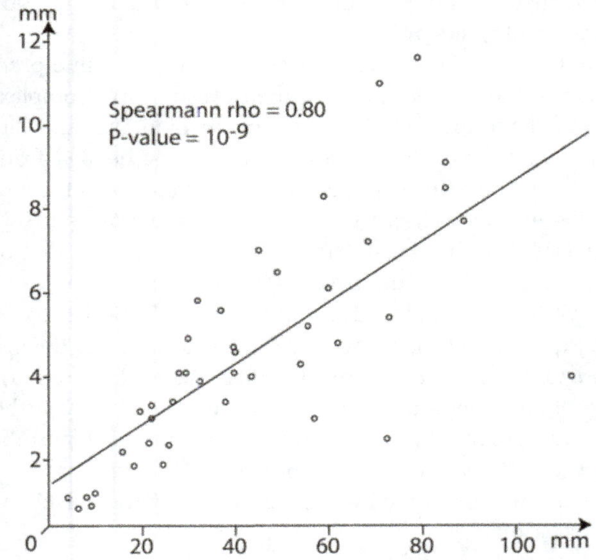

Fig. 9 Scatterplot of the diameter of the eye and the length of carapace with regression line

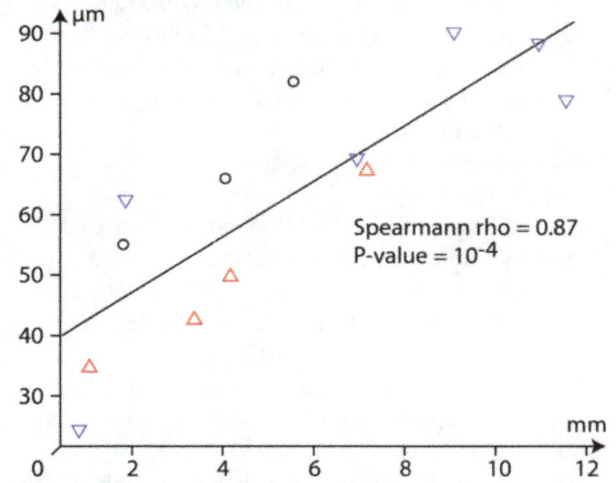

Fig. 11 Scatterplot of the square root of ommatidial surface and the diameter of the eye: blue triangles, specimens from La Voulte (probably deep); red triangles, specimens from plattenkalks (probably shallow); circles, specimens from other outcrops

some background on the probability of each of these possibilities:

1. Apposition optics in adult compound eyes are part of the ground pattern of Malacostraca [70]. Thus, their occurrence is not unexpected in any malacostracan crustacean. However, within Eumalacostraca, refractive or reflective superposition eyes independently evolved several times [70] and therefore cannot be excluded here for *R. oppeli*

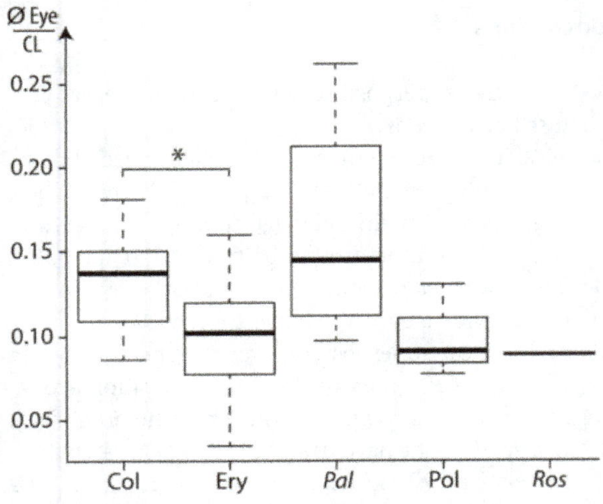

Fig. 10 Boxplot comparing the ratio of the eye diameter on the carapace length (CL) in five different groups of fossil polychelidans: Col = Coleiidae, Ery = Eryonidae, *Pal* = *Palaeopentacheles* (only species of Palaeopentachelidae), Pol = fossil species closely related to Polychelidae, *Ros* = *Rosenfeldia oppeli*. *, Welch test, *P*-value < 0,01. Abbreviations

(see next points). The multiple evolution of different types of superposition in adult crustaceans is linked to alterations of ommatidia adjusted to perform apposition as larvae. However, Nilsson et al. [82] showed that for instance the transformation of apposition into refracting superposition does not require drastic changes of the ommatidial set-up, neither on the cellular nor on the subcellular level. Therefore, the optic apparatus of decapods is likewise prone to convergent modifications of inherited structures, even though these incidents, such as the evolution of refracting superposition, apparently happened less often than theoretically possible [70]. Alternatively, the specimen could represent a larva with apposition eyes because this type is found in all extant malacostracan larvae [65, 71–74]. However, the specimen of *R. oppeli* studied here seems too large (carapace length = 38 mm) to be a larva. Therefore, the occurrence of apposition eyes in this specimen would be best explained by paedomorphosis.

2. Refractive superposition eyes are frequent within Malacostraca [70]. Although the ground pattern of Decapoda is characterized by reflective superposition eyes, several ingroups, such as representatives of the shrimp genus *Gennadas* possess well-functioning refracting superposition eyes with hexagonal facets. The compound eyes of *R. oppeli* may thus represent this type of eye. However, among decapod crustaceans, reflecting superposition, as characterized by square facets and mirror boxes [65, 83, 84], are frequently

encountered. Depending on the phylogenetic concept applied, reflecting superposition optics either evolved in the stem lineage of or very early within Decapoda [70]. Variations of superposition optics, namely the occurrence of refracting superposition in a context of species with reflecting superposition eyes, are not very common in a given decapod taxon but they do occur occasionally. Examples are representatives of the shrimp genus *Gennadas* which possess well-functioning refracting superposition eyes with hexagonal facets, but their closest ancestors had reflecting superposition eyes [68, 85].
3. Parabolic superposition eyes occur in a few groups of decapod crustaceans, such as Xanthidae and Portunidae (e.g. *Macropipus depurator*) [65, 67, 85]. However, if present, their distribution among decapod crustaceans would most likely imply a convergent evolution of such eyes for polychelidan lobsters.

By comparison with the eyes described for other polychelidan lobsters described in the present study, it seems more plausible that the eyes of *R. oppeli* correspond either to apposition eyes derived from the transparent subtype and retained though paedomorphosis or they represent refractive superposition eyes. The occurrence of possibly paedomorphic apposition eyes or refractive superposition eyes of *Rosenfeldia oppeli* is most likely an autapomorphy of this species.

Visual palaeoecology

Eyes of crustaceans possess a wide array of adaptations to different environments and light intensities: the size of ommatidia can vary; superposition eyes allow sensitivity to light while apposition eyes have a higher resolution; in all cases, pigments can modulate the light entering the ommatidia or isolate ommatidia one from another in superposition eyes; reflective layers (proximal tapetum) can also enhance sensitivity by reflecting the light in the eye so it passes twice through the retina [85, 86].

From the fossil perspective, most of the adaptations of eyes are hidden in external view and possibly not preserved in fossil for which only the superficial lenses are available (for instance, in exuviae). However, the shape and size of ommatidia lenses are readily observable.

In the case of *R. oppeli*, if we consider its hexagonal facets to be indicative of apposition eyes, we can draw some conclusions regarding its life-habits: indeed, apposition eyes are by far less sensitive to light than superposition eyes, because the light illuminating the cornea can only come from one facet, and is correlated to its surface. However, these eyes have a higher resolution than that of superposition eyes, because each ommatidium is independent.

In the case of the other species described here, interpretation of their life-habits from their eyes is more complex. All of them possess reflecting superposition eyes and differences between those eyes can only apprehended from our perspective through the size of ommatidia and the size of the eye. Our observations show that the size of eyes is correlated to the size of the carapace (and thus to that of the animal). Differences in visual performance between bright-light-adapted and dim-light-adapted species could be expected to be observed from the proportion of the eye compared to the carapace. However, our analysis did not reveal any significant differences in this ratio between the various groups. Significant differences are however observed between different taxa of polychelidan lobsters.

Multiple factors may explain this apparent lack of correlation:

- Our sample of preserved ommatidia is too small, or variations in the preservation, add noise to a possible palaeobathymetric signal.
- Differences in palaeobathymetry of the various outcrops we sampled may not be sufficient to observe differences in the size of eyes.
- Other mechanisms are probably involved (such as the presence of pigments reducing light input in shallow water species and tapetum reflecting the light through the retina in deep-water species).
- Most extant shallow-water crustaceans tend to hide during the day and search for food at night; for this reason, they may need ommatidia as large as deep water species.

Conclusions

- Reflective superposition eyes can be preserved in fossil arthropods.
- Modern representatives of Polychelida are blind; Jurassic representatives still possessed functional eyes, most of them reflective superposition eyes retained from the ground pattern of Decapoda.
- The examples presented here mark the oldest direct evidence for reflective superposition eyes and of clear superposition eyes in general.
- The eye of *R. oppeli* could be an apposition eye that evolved through paedomorphic heterochrony. *R. oppeli* might have lived in shallower water than its relatives. Alternatively, it could correspond to a refractive or even a parabolic superposition eye.
- Eyes proportions appear linked to phylogeny.

We hope that with the reinvestigation of other fossils we may complete these observations and obtain a clearer picture of how variations in eye design and size is distributed across taxa of polychelidans and if a variation between outcrops can be observed.

Abbreviations

CSMNF, Centro Musei delle Scienze Naturali e Fisiche of the Università degli Studi di Napoli "Federico II" (Napoli, Italy); FSL, Université Claude Bernard Lyon 1 (Villeurbanne, France); JME-SOS, Jura-Museum Eichstätt (Eichstätt, Germany); MNHN.F, Muséum national d'Histoire naturelle, collection de Paléontologie (Paris, France); MSNM, Museo Civico di Storia Naturale di Milano (Milano, Italy); NHMUK, Natural History Museum (London, United Kingdom); SMCU, Sedgwick Museum – Cambridge University (Cambridge, United Kingdom); SMNS, Staatliches Museum für Naturkunde (Stuttgart, Germany); SNSB-BSPG, Bayerische Staatssammlung für Paläontologie und Geologie (Munich, Germany); UJF-ID, Université Joseph Fourier, Institut Dolomieu (Grenoble, France); UPMC, Université Pierre et Marie Curie – Paris 6 (Paris, France)

Acknowledgements

We are indebted to the curators Sergio Bravi (CSMNF, Napoli, Italy), Abel Prieur, Emmanuel Robert (FSL, Villeurbannes, France), Martina Kölbl-Ebert (JME, Eichstätt, Germany), Jean-Michel Pacaud (MNHN, Paris, France), Alessandro Garassino (MSNM, Milano, Italy), Claire Mellish (NHMUK, London, United-Kingdom), Matt Riley (SMCU, Cambridge, United-Kingdom), Martin Nose (SNSB-BSPG, Munich, Germany), Jean-Bernard Caron, Peter Fenton and Janet Waddington (ROM, Toronto, Canada), Fabienne Giraud-Guillot (UJF-ID, Grenoble, France), Isabelle Rouget (UPMC, Paris, France) for their work in the collections, without which this study would not have been possible. We also acknowledge the contribution of Philippe Loubry, who realized the SEM micrograph and Denis Poinsot (Université Rennes 1) who offered insightful advices on statistical analysis. JTH is currently kindly funded by the German Research Foundation (DFG; Ha 6300/3-1). CH is supported by a Bavarian Equal Opportunities Sponsorship at the LMU Munich. CH and JTH are kindly supported by J. Matthias Starck, Munich. We especially thank all private fossil collectors for their long-lasting support and all people involved in providing open source, open access or low cost software. We are glad to acknowledge the precious help of two anonymous referees: their detailed review significantly improved this article.

Funding

JTH is currently kindly funded by the German Research Foundation (DFG; Ha 6300/3-1). CH is supported by a Bavarian Equal Opportunities Sponsorship at the LMU Munich. CH and JTH are kindly supported by J. Matthias Starck, Munich. These funds allowed the documentation of German specimens and supported writing part of this article.

Authors' contributions

JTH and CH initiated the study, implemented the calculation of eyes parameter, provided data and imaged specimens from Germany. DA provided data for and imaged specimens from France, United-Kingdom and Italy and realized the statistical analyses. CHGM and SH provided background and interpretation of eye types evolution and their significance; GS and SC provided access and photographs of specimens housed respectively in Stuttgart and Paris as well as significant improvements to the manuscript. All authors read and approved the final manuscript.

Competing interests

The authors declare that they have no competing interests.

Author details

[1]Université de Rennes 1, EA 7316, 263 Avenue du Général Leclerc CS 74205, 35042 Rennes Cedex, France. [2]Centre de Recherche sur la Paléobiodiversité et les Paléoenvironnements (CR2P, UMR 7207), Muséum national d'Histoire naturelle, Sorbonne Universités, UPMC, CNRS, 8 rue Buffon, F-75005 Paris, France. [3]Department Biologie II und GeoBio-Center, Ludwig-Maximilians-Universität München, Biozentrum der LMU, Großhaderner Str. 2, 82152 Planegg-Martinsried, Germany. [4]Staatliches Museum für Naturkunde Stuttgart, Rosenstein 1, 70191 Stuttgart, Germany. [5]Zoologisches Institut und Museum, Lehrstuhl Allgemeine und Systematische Zoologie, Ernst-Moritz-Arndt-Universität Greifswald, Anklamer Str. 20, 17487 Greifswald, Germany. [6]Zoologisches Institut und Museum, Lehrstuhl Cytologie und Evolutionsbiologie, Ernst-Moritz-Arndt-Universität Greifswald, Soldmannstr. 23, 17489 Greifswald, Germany.

References

1. Clarkson E, Levi-Setti R, Horváth G. The eyes of trilobites: The oldest preserved visual system. Arthropod Struct Dev. 2006;35:247–59. doi:10.1016/j.asd.2006.08.0022.
2. Clarkson ENK. The visual system of trilobites. Palaeontology. 1978;22:1–22.
3. Tanaka G, Smith RJ, Siveter DJ, Parker AR. Three dimensionally preserved decapod larval compound eyes from the Cretaceous Santana Formation of Brazil. Zoological Sciences. 2009;26:846–50. doi:10.2108/zsj.26.846.
4. Paterson JR, Garcia-Bellido DC, Lee MSY, Brock GA, Jago JB, Edgecombe GD. Acute vision in the giant Cambrian predator *Anomalocaris* and the origin of compound eyes. Nature. 2011;480:237–40. doi:10.1038/nature10689.
5. Schoenemann B, Castellani C, Clarkson ENK, Haug JT, Maas A, Haug C, Waloszek D. The sophisticated visual system of a tiny Cambrian crustacean: analysis of a stalked fossil compound eye. Proceedings of the Royal Society, London B. 2012;279:1335–40. doi:10.1098/rspb.2011.1888.
6. Vannier J, Schoenemann B, Gillot T, Charbonnier S, Clarkson E. Exceptional preservation of eye structure in arthropod visual predators from the Middle Jurassic. Nat Commun. 2016;7:10320. doi:10.1038/ncomms10320.
7. Chen J, Waloszek D, Maas A. A new 'great-appendage' arthropod from the Lower Cambrian of China and homology of chelicerate chelicerae and raptorial antero-ventral appendages. Lethaia. 2004;37:3–20. doi:10.1080/00241160410004764.
8. Vannier J, Garcia-Bellido DC, Hu S-X, Chen AL. Arthropod visual predators in the early pelagic ecosystem: evidence from the Burgess Shale and Chengjiang biotas. Proc R Soc B Biol Sci. 2009;276:2567–74. doi:10.1098/rspb.2009.0361.
9. Haug JT, Waloszek D, Maas A, Liu Y, Haug C. Functional morphology, ontogeny and evolution of mantis shrimp-like predators in the Cambrian. Palaeontology. 2012;55:369–99. doi:10.1111/j.1475-4983.2011.01124.x.
10. Haug JT, Briggs DEG, Haug C. Morphology and function in the Cambrian Burgess Shale megacheiran arthropod *Leanchoilia superlata* and the application of a descriptive matrix. BMC Evolutionary Biology. 2012;doi:10.1186/1471-2148-12-162
11. Tanaka G, Hou X, Ma X, Edgecombe GD, Strausfeld NJ. Chelicerate neural ground pattern in a Cambrian great appendage arthropod. Nature. 2013;502:364–7. doi:10.1038/nature12520.
12. Daley AC, Budd GE, Caron JB, Edgecombe GD, Collins D. The Burgess Shale Anomalocaridid Hurdia and Its Significance for Early Euarthropod Evolution. Science. 2009;323:1597–600. doi:10.1126/science.1169514.
13. Kühl G, Briggs DEG, Rust J. A Great-Appendage Arthropod with a Radial Mouth from the Lower Devonian Hunsrück Slate, Germany. Science. 2009;323:771–73. doi:10.1126/science.1166586.
14. Schoenemann B, Clarkson ENK. The eyes of *Leanchoilia*. Lethaia. 2012;45:524–31. doi:10.1111/j.1502-3931.2012.00313.x.
15. Liu Y, Hou X, Bergström J. Chengjiang arthropod *Leanchoilia illecebrosa* (Hou, 1987) reconsidered. GFF. 2007;129:263–72. doi:10.1080/11035890701293263.
16. Ma X, Hou X, Edgecombe GD, Strausfeld NJ. Complex brain and optic lobes in an early Cambrian arthropod. Nature. 2012;490:258–62. doi:10.1038/nature11495.
17. Schoenemann B. The eyes of a tiny 'Orsten' crustacean – A compound eye at receptor level? Vision Res. 2013;76:89–93. doi:10.1016/j.visres.2012.10.013.
18. Maas A, Braum A, Dong X-P, Donoghue PCJ, Müller KJ, Olempska E, Repetski JE, Siveter DJ, Stein M, Waloszek D. The 'Orsten'—More than a Cambrian Konservat-Lagerstätte yielding exceptional preservation. Palaeoworld. 2006;15:266–82. doi:10.1016/j.palwor.2006.10.005.
19. Weitschat W. Myodocopid ostracodes with preserved appendages from the Lower Triassic of Spitsbergen. Paläontol Z. 1983;57:309–23. doi:10.1007/BF02990320.
20. Fröhlich F, Mayrat A, Riou B, Secrétan S. Structures rétiniennes phosphatisées dans l'œil géant de *Dollocaris*, un crustacé fossile. Annales de Paléontologie. 1992;78:193–204.

21. Vannier J, Chen J-Y, Huang D-Y, Charbonnier S, Wang X-Q. The Early Cambrian origin of thylacocephalan arthropods. Acta Palaeontol Pol. 2006;51:201–14.
22. Charbonnier S. Le Lagerstätte de La Voulte : un environnement bathyal au Jurassique. Mémoires du Muséum national d'Histoire naturelle. 2009;199:272.
23. Charbonnier S, Vannier J, Hantzpergue P, Gaillard C. Ecological significance of the arthropod fauna from the Jurassic (Callovian) La Voulte Lagerstätte. Acta Palaeontol Pol. 2010;55:111–32. doi:10.4202/app.2009.0036.
24. Audo D, Schweigert G, Saint Martin JP, Charbonnier S. High biodiversity in Polychelida crustaceans from the Jurassic La Voulte-sur-Rhône Lagerstätte. Geodiversitas. 2014;36:489–525.
25. Charbonnier S, Garassino A, Schweigert G, Simpson M. A worldwide review of fossil and extant glypheid and litogastrid lobsters (Crustacea, Decapoda, Glypheoidea). Mémoires du Muséum national d'Histoire naturelle. 2013;205:304.
26. Haug JT, Waloszek D, Haug C, Maas A. High-level phylogenetic analysis using developmental sequences: The Cambrian †*Martinssonia elongata*, †*Musacaris gerdgeyeri* gen. et sp. nov. and their position in early crustacean evolution. Arthropod Struct Dev. 2010;39:154–73. doi:10.1016/j.asd.2010.01.005.
27. Haug JT, Maas A, Waloszek D. Ontogeny of two Cambrian stem crustaceans, †*Goticaris longispinosa* and †*Cambropachycope clarksoni*. Palaeontographica. 2009;A289:1–43.
28. Audo D, Charbonnier S. Late Cretaceous crest-bearing shrimps from the Sahel Alma Lagerstätte of Lebanon. Acta Palaeontol Pol. 2013;58:335–49. doi:10.4202/app.2011.0056.
29. Feldmann RM. A new cirolanid isopod (Crustacea) from the Cretaceous of Lebanon: dermoliths document the pre-molt condition. J Crustac Biol. 2009; 29:373–8. doi:10.1651/08-3096.1.
30. Feldmann RM, Charbonnier S. *Ibacus cottreaui* Roger, 1946, reassigned to the isopod genus *Cirolana* (Cymothoida: Cirolanidae). J Crustac Biol. 2011;31: 317–9. doi:10.1651/10-3406.1.
31. Scholtz G, Richter S. Phylogenetic systematics of the reptantian Decapoda (Crustacea, Malacostraca). Zool J Linn Soc. 1995;113:289–328. doi:10.1111/j. 1096-3642.1995.tb00936.x.
32. Galil BS. Crustacea Decapoda: Review of the genera and species of the family Polychelidae Wood-Mason, 1874. In: Crosnier A, editor. Résultats des Campagnes MUSORSTOM, Volume 21. Mémoires du Muséum national d'Histoire naturelle, 2000.184. Paris: Publications scientifiques du Muséum; 2000. p. 285–387.
33. Bernard F. Decapoda Eryonidae (*Eryoneicus* et *Willemoesia*). Dana-Report. 1953;37:1–93.
34. Santucci R. Biologia del fondo a"Scampi" nel Mar Ligure. I. *Polycheles typhlops* Heller. R. Comitato talassografico italiano 1933;199:1–48
35. Karasawa H, Takahashi F, Doi E, Ishida H. First notice of the family Coleiidae Van Straelen (Crustacea: Decapoda: Eryonoidea) from the upper Triassic of Japan. Paleontological Research. 2003;7:357–62.
36. Bronn HG. Beiträge zur triasischen Fauna und Flora der bituminösen Schiefer von Raibl. Neues Jahrbuch für Mineralogie, Geognosie, Geologie und Petrefakten-Kunde. 1858;1858:1–32.
37. Oppel A. Ueber jurassische Crustaceen (Decapoda macrura). Palaeontologische Mittheilungen aus dem Museum des koeniglich Bayerischen Staates. 1862;1:1–120.
38. Woods H. A monograph of the fossil macrurous Crustacea of England. London: The Palaeontographical Society; 1925–1931: p. 1–122, pl. 1–27
39. Van Straelen V. Les crustacés décapodes du Callovien de la Voulte-sur-Rhône (Ardèche). C R Hebd Seances Acad Sci. 1923;175:982–3.
40. Garassino A, Schweigert G. The Upper Jurassic Solnhofen decapod crustacean fauna: review of the types from old descriptions (infraorders Astacidea, Thalassinidea, and Palinura). Memorie della Società Italiana di Scienze Naturali e del Museo Civico di Storia Naturale di Milano. 2006; 34(1):1–64.
41. Spence Bate C. Report on the Crustacea Macrura collected by the Challenger during the years 1873–1876. Reports on the Scientific Results of the Voyage of H. M. S. "Challenger" 1888;24:i-xc, 1–942.
42. Schweigert G, Dietl G. Neubeschreibung von '*Eryon longipes* O. Fraas' (Crustacea, Decapoda, Eryonidea) aus dem Nusplinger Plattenkalk (Ober-Kimmeridgium, Schwäbische Alb). Stuttgarter Beiträge zur Naturkunde. Serie B. 1999;274:1–19.
43. Ahyong ST. The Polychelidan Lobster: Phylogeny and Systematics (Polychelida : Polychelidae). In: Martin JW, Crandall KA, Felder DL DL, editors. Decapod Crustacean Phylogenetics. Crustacean Issues 18. Boca Raton, London, New York: CRC Press, Taylor & Francis Group; 2009. p. 369–96.
44. Radley JD. Highlights of Jurassic of Warwickshire. Geol Today. 2005;21:146–52.
45. Pinna G. Gli Erionidei della nuova fauna sinemuriana a crostacei decapodi di Osteno in Lombardia. Atti della Società italiana di Scienze naturali e del Museo civico di Storia naturale in Milano. 1968;107:93–134.
46. Tang CM. Osteno: jurassic preservation to the cellular level. In: Bottjer DJ, Etter W, Hagadorn JW, Tang CM, editors. Exceptional fossil preservation. New York: Columbia University Press; 2002. p. 251–64.
47. Teruzzi G. The genus *Coleia* Broderip, 1835 (Crustacea, Decapoda) in the Sinemurian of Osteno in Lombardy. Atti della Società italiana di Scienze naturali e del Museo civico di Storia naturale in Milano. 1990;131:85–104.
48. Hauff B, Hauff RB. Das Holzmadenbuch. Holzmaden: Hauff-Stiftung; 1981. p. 1–136.
49. Röhl H-J, Schmil-Röhl A, Oschmann W, Frimmel A, Schwark L. The Posidonia Shale (Lower Toarcian) of SW-Germany: an oxygen-depleted ecosystem controlled by sea level and palaeoclimate. Palaeogeogr Palaeoclimatol Palaeoecol. 2001;165:27–52. doi:10.1016/S0031-0182(00)00152-3.
50. Bravi S, Garassino A, Bartiromo A, Audo D, Charbonnier S, Schweigert G, Thévenard F, Longobardi C. Middle Jurassic Monte Fallano Plattenkalk (Campania, southern Italy): first report on terrestrial plants, decapod crustaceans and fishes. Neues Jahrb Geol Palaontol Abh. 2014;272:79–107. doi:10.1127/0077-7749/2014/0398.
51. Charbonnier S, Vannier J, Gaillard C, Bourseau J-P, Hantzpergue P. The La Voulte Lagerstätte (Callovian): evidence for a deep water setting from sponge and crinoid communities. Palaeogeogr Palaeoclimatol Palaeoecol. 2007;250:216–36. doi:10.1016/j.palaeo.2007.03.013.
52. Audo D, Charbonnier S, Schweigert G, Saint Martin J-P. New eryonid crustaceans from the Late Jurassic Lagerstätten of Cerin (France), Canjuers (France), Wattendorf (Germany) and Zandt (Germany). J Syst Palaeontol. 2014;12:459–79. doi:10.1080/14772019.2013.777809.
53. Frickhinger KA. Die fossilien von Solnhofen 2. Korb: Goldschneck. 1999. p. 190.
54. Barthel KW, Swinburne NHM, Conway Morris S. Solnhofen. A study in Mesozoic palaeontology. Cambridge: Cambridge University Press; 1990. p. 236.
55. Audo D, Williams M, Charbonnier S, Schweigert G. *Gabaleryon*, a new genus of widespread early Toarcian polychelidan lobsters. J Syst Palaeontol. in press. doi: 10.1080/14772019.2016.1167786
56. Audo D, Schweigert G, Haug JT, Haug C, Saint Martin J-P, Charbonnier S. Diversity and palaeoecology of the enigmatic genus *Knebelia* (Eucrustacea, Decapoda, Eryonidae) from Upper Jurassic plattenkalks in southern Germany. Palaeontology. 2014;57:397–416. doi:10.1111/pala.12071.
57. Haug JT, Haug C, Ehrlich M. First fossil stomatopod larva (Arthropoda: Crustacea) and a new way of documenting Solnhofen fossils (Upper Jurassic, Southern Germany). Palaeodiversity. 2008;1:103–9.
58. Haug JT, Haug C, Kutschera V, Mayer G, Maas A, Liebau S, Castellani C, Wolfram U, Clarkson ENK, Waloszek D. Autofluorescence imaging, an excellent tool for comparative morphology. J Microsc. 2011;244:259–72.
59. Haug C, Haug JT, Waloszek D, Maas A, Frattigiani R, Liebau S. New methods to document fossils from lithographic limestones of southern Germany and Lebanon. Palaeontol Electron. 2009;12(6T):12.
60. Haug JT, Haug C. Fossilien unter langwelligem Licht: Grün-Orange-Fluoreszenz an makroskopischen Objekten. Archaeopteryx. 2011;29:20–3.
61. Schaarschmidt F. Pflanzenfossilien in ungewöhnlichem Licht. Nat Mus. 1973; 103:247–53.
62. Bengtson S. Teasing fossils out of shales with cameras and computers. Palaeontol Electron. 2000;3(Art. 4):14.
63. Kerp H, Bomfleur B. Photography of plant fossils – new techniques, old tricks. Rev Palaeobot Palynol. 2011;166:117–51. doi:10.1016/j.revpalbo.2011.05.001.
64. Haug JT, Müller CHG, Sombke A. A centipede nymph in Baltic amber and a new approach to document amber fossils. Organisms Diversity & Evolution. 2013;13:425–32. doi:10.1007/s13127-013-0129-3.
65. Nilsson D-E. Optics and evolution in the compound eye. In: Stavenga DG, Hardie RC, editors. Facets of vision. Berlin: Springer; 1989. p. 30–73.
66. Exner S. Die Physiologie der facettirten Augen von Krebsen und Insecten. Leipzig: Franz Deuticke; 1891. p. 198–7.
67. Nilsson D-E. A new type of imaging optics in compound eyes. Nature. 1988; 332:76–8. doi:10.1038/332076a0.
68. Nilsson D-E. Three unexpected cases of refracting superposition eyes in crustaceans. J Comp Physiol A. 1990;167:71–8. doi:10.1007/BF00192407.
69. Gaten E. Optics and phylogeny: is there an insight? The evolution of superposition eyes in the Decapoda (Crustacea). Contrib Zool. 1998;67: 223–36.
70. Richter S. Evolution of optical design in the Malacostraca (Crustacea). P. In: Wiese K, editor. The Crustacean Nervous System. Berlin: Springer; 2002. p. 512–24.

71. Fincham AA. Ontogeny and optics of the eyes of the common prawn *Palaemon* (*Palaemon*) serratus (Pennant, 1777). Zoological Journal of the Linnaean Society. 1984;81:89–111. doi:10.1111/j.1096-3642.1984.tb01173.x.
72. Fincham AA. Ontogeny of anomuran eyes. Symposia of the Zoological Society London. 1988:59:123–55.
73. Nilsson D-E. Evolutionary links between apposition and superposition optics in crustacean eyes. Nature. 1983;302:818–21. doi:10.1038/302818a0.
74. Harzsch S, Dawirs RR. Maturation of the compound eyes and eyestalk ganglia during larval development of the brachyuran crustaceans *Hyas araneus* L. (Decapoda, Majidae) and *Carcinus maenas* L. (Decapoda, Portunidae). Zoology. 1996;99:189–204.
75. Meyer-Rochow VB, Tiang KM. The eye of *Jasus edwardsii* (Crustacea, Decapoda, Palinuridae). Electrophysiology, histology, and behavior. Zoologica. 1984;45:1–58.
76. Meyer-Rochow VB. Larval and adult eye of the Western Rock Lobster (*Panulirus longipes*). Cell Tissue Res. 1975;162:439–57. doi:10.1007/BF00209345.
77. Doughtie DG, Rao RK. Ultrastructure of the eyes of the grass shrimp, *Palaemonetes pugio*. Cell Tissue Research. 1984;238:271–88. doi:10.1007/BF00217299.
78. Gaten E. The ultrastructure of the compound eye of *Munida rugosa* (Crustacea: Anomura) and pigment migration during light and dark adaption. J Morphol. 1990;205:243–53. doi:10.1002/jmor.1052050302.
79. Hafner GS, Tokarski TR. Morphogenesis and pattern formation in the retina of the crayfish *Procambarus clarkii*. Cell Tissue Research. 1998;293:535–50. doi:10.1007/s004410051146.
80. Richter S. The structure of the ommatidia of the Malacostraca – a phylogenetic approach. Verhandlungen des naturwissenschaftlichen Vereins Hamburg (NF). 1999;38:161–204.
81. Cronin TW, Porter ML. Exceptional Variation on a Common Theme: The Evolution of Crustacean Compound Eyes. Evolution: Education and Outreach. 2008;1:463–75. doi:10.1007/s12052-008-0085-0.
82. Nilsson D-E, Hallberg E, Elofsson R. The ontogenetic development of refracting superposition eyes in crustaceans: Transformation of optical design. Tissue Cell. 1986;18:509–19. doi:10.1016/0040-8166(86)90017-0.
83. Vogt K. Zur Optik des Flußkrebsauges. Z Naturforsch. 1975;32:466–8.
84. Vogt K. Die Spiegeloptik des Flusskrebsauges. J Comp Physiol A. 1980;135:1–19. doi:10.1007/BF00660177.
85. Land MF, Nilsson DE. Animal eyes. Oxford: Oxford University Press; 2012. p. 1–271.
86. Meyer-Rochow VB. The crustacean eye: dark light adaptation, polarization sensitivity, flicker fusion frequency, and photoreceptor damage. Zoolog Sci. 2001;18:1175–97. doi:10.2108/zsj.18.1175.

14

Mode of reduction in the number of pharyngeal segments within the sarcopterygians

Victoria Shone[1], Silvan Oulion[2], Didier Casane[2], Patrick Laurenti[2] and Anthony Graham[1*]

Abstract

Background: Pharyngeal segmentation is a defining feature of vertebrate embryos and is apparent as a series of bulges found on the lateral surface of the embryonic head, the pharyngeal arches. The ancestral condition for gnathostomes is to have seven pharyngeal segments: jaw, hyoid, and five posterior branchial arches. However, within the sarcopterygians, the pharyngeal region has undergone extensive remodelling that resulted in a reduction in the number of pharyngeal segments, such that amniotes have only five pharyngeal arches. The aim of this study is to probe the developmental basis of this loss of pharyngeal segments.

Results: We have therefore compared the development of the pharyngeal arches in an amniote, the chick, which has five segments, with those of a chondrichthyan, the catshark, which has seven segments. We have analysed the early phase of pharyngeal segmentation and we find that in both the most anterior segments form first with the posterior segments being added sequentially. We also documented the patterns of innervation of the pharynx in several vertebrates and note that the three most anterior segments receive distinct innervation: the first arch being innervated by the Vth nerve, the second by the VIIth and the third by the IXth. Finally, we have analysed Hox gene expression, and show that the anterior limit of *Hoxa2* aligns with the second pouch and arch in both chick and catshark, while *Hoxa3* is transiently associated with the third arch and pouch. Surprisingly, we have found that *Hoxb1* expression is spatially and temporally dynamic and that it is always associated with the last most recently formed pouch and that this domains moves caudally as additional pouches are generated.

Conclusion: We propose that the first three pharyngeal segments are homologous, as is the posterior limit of the pharynx, and that the loss of segments occurred between these two points. We suggest that this loss results from a curtailment of the posterior expansion of the pharyngeal endoderm in amniotes at relatively earlier time point, and thus the generation of fewer segments.

Keywords: Pharyngeal, Segmentation, Sarcopterygians, Tetrapods, Hoxb1, Endoderm

Background

Pharyngeal segmentation is a characteristic of all vertebrates. During development it is first seen in the appearance of a series of bulges on the lateral surface of the head, the pharyngeal arches [1]. These primordia consist of epithelia with a mesenchymal filling. Externally, the epithelium is derived from the ectoderm and internally from the endoderm, while the mesenchyme consists of a core of mesoderm surrounded by neural crest cells. Between the arches the ectoderm and the endoderm contact each other, and it is this apposition between these tissues that defines the anterior and posterior margins of each of the pharyngeal arches [1].

A general feature of gnathostomes is to have seven pharyngeal segments; jaw, hyoid, and five posterior branchial (gill-bearing) arches [2]. This arrangement represents the plesiomorphic state as this situation is observed in many chondrichthyans and all actinopterygians, and this arrangement is also seen in sarcopterygians, such as the coelacanth *Latimeria* and the

* Correspondence: anthony.graham@kcl.ac.uk
[1]Department of Developmental Neurobiology, Kings College London, London SE1 1UL, UK
Full list of author information is available at the end of the article

lungfishes, the sister group to the tetrapods [3–6]. However, extant tetrapods show a reduced number of pharyngeal segments. Thus, larval amphibians have six pharyngeal segments, while amniote embryos have only five. Moreover, there has also been a loss of pharyngeal segmentation in the adult form as a result of the remodelling of the pharynx, which occurs during metamorphosis in amphibians and embryogenesis in amniotes. The reduction in the number of pharyngeal segments and the loss of explicit segmentation in adult tetrapods clearly reflects the shift from respiration in water via gills to air breathing using lungs.

These alterations to the pharyngeal region that occurred with the evolution of the tetrapods are obviously underpinned by changes to the developmental programme. The loss of overt pharyngeal segmentation is due to the overgrowth of the second arch to cover the more posterior arches, followed by the fusion of the caudal edge of the second arch to the subjacent tissue which results in the internalisation of the posterior arches [7]. However, it is less clear how the reduction in the number of pharyngeal segments has been achieved, although this is likely to involve changes to the early organisation of the pharyngeal endoderm.

The segmentation of the endoderm, which results in the formation of the pharyngeal pouches, is central to the development of the pharyngeal arches [8]. The pharyngeal pouches form at distinct positions along the anteroposterior axis. The two most anterior pouches form first with the more posterior pouches forming subsequently and sequentially [9, 10]. The pouches grow to contact the overlying ectoderm, which invaginates to meet them, generating the pharyngeal clefts, and thus neural crest cells and mesoderm migrate into these preformed units. Significantly, in mutants in which the endoderm fails to segment, the pharyngeal arches fail to form [11, 12]. Consequently, a reduction in the number of pharyngeal segments in sarcopterygians must have involved alterations to the development of the pharyngeal pouches.

To begin to address the route through which the number of pharyngeal segments has been reduced, we have compared the development of the pharyngeal arches in the chick (*Gallus gallus*), an amniote, which has five segments, to those in the catshark (*Scyliorhinus canicula*), a chondrichthyan, which has seven segments. We have analysed the formation of pharyngeal pouches, patterns of innervation of the pharyngeal arches, and the expression of *Hox* genes, which are markers of axial identity in the pharynx, and in particular how these relate to the endodermal pouches. Our data support the view that the three most anterior pharyngeal segments are homologous between different vertebrate classes, and that the reduction in the number of segments must have been achieved by a loss of those lying more caudally. Notably, we find that in both chick and catshark embryos *Hoxb1* expression marks the posterior limit of the pharynx. However, this expression pattern is spatially and temporally dynamic, with *Hoxb1* labelling the posterior limit of the pharynx at early and late stages irrespective of the number of segments that have been generated. Thus we conclude, that the posterior limit of the pharynx is likely to be homologous across the gnathostomes and that the reduction in the number of pharyngeal segments that accompanied the evolution of the tetrapods will have involved a heterochronic shift such that the posterior expansion of the pharynx was curtailed at a relatively earlier time point and thus less segments were generated.

Materials and methods

Embryo collection

Fertile hen's eggs were incubated at 38 °C to the required stages (HH st) [13] and the embryos were fixed in 4 % PFA. *Scyliorhinus canicula* embryos were taken from the egg cases, anaesthetized (MS222), staged [14] and fixed in 4%PFA. Lamprey embryos were kindly donated by Dr Sebastian Shimeld, University of Oxford.

Immunohistochemistry

Previously fixed embryos were washed three times 30 min in PBS/1 % TritonX-100 (PBSTx) before being washed in a blocking solution of 10 % goat serum in PBSTx twice for one hour at room temperature. The relevant primary antibody was diluted in the blocking solution with 0.02 % sodium azide and the embryos incubated at 4 °C for 1–2 weeks. Embryos were then rinsed in blocking solution and washed three times for one hour in blocking solution before adding the secondary antibody diluted in blocking solution with 0.02 % sodium azide. This was incubated at 4 °C for 1–2 weeks. The primary antibodies used were rabbit anti-laminin at 1:100 (Sigma); mouse anti- NFM 1:10000 (Zymed). Secondary antibody was Alexa 488-conjugated goat anti-mouse IgG, and Alexa 568 goat anti-rabbit IgG, both used at 1:1000 (Molecular Probes). For sectioning, embryos were washed into PBS, embedded in gelatin, fixed, and vibratomed at 50 μm slices.

In situ hybridisation

Previously fixed embryos were washed twice in PBST (PBS + 0.1 % Tween-20), dehydrated through a methanol series in PBST, bleached with 6 % H_2O_2 in methanol for 1 h and then rehydrated. Embryos were then treated with 10 μg/ml Proteinase K in PBST for 20 min, post-fixed with 4 % PFA/0.1 % gluteraldehyde and then washed twice in PBST. The embryos were incubated at 70 °C in hybridisation buffer (50 % formamide, 1.3X SSC, 5 mM EDTA, 50 μg/ml tRNA, 100 μg/ml heparin,

0.2 % Tween-20, 0.5 % CHAPS) for 1 h, followed by overnight at 70 °C in digoxigenin-labelled riboprobes. Embryos were washed four times with hybridisation buffer at 70 °C, then three times 30 min with MABT (100 mM maleic acid, 150 mM NaCl, 1 % Tween-20; pH7.5), and blocked with 2 % BBR (Roche Diagnostics)/ 20 % goat serum/MABT and incubated overnight in anti-DIG-AP antibody (Roche) diluted 1:2000 in the same block. Embryos were washed extensively with MABT and the alkaline phosphatase activity detected using NBT and BCIP in NTMT. The reaction was stopped by washing in MABT and fixing in 4 % PFA.

Results

Sequential generation of pharyngeal segments

We used anti-laminin immunostaining to conduct a general analysis of the formation of the pharyngeal pouches in both chick and catshark embryos. The pharyngeal segments form first anteriorly and then subsequently posteriorly, in both species (Fig. 1). The first two pharyngeal pouches in chick form at stage 13 [15], and by stage 14 (Fig. 1a) these have fully formed and are in intimate contact with the ectoderm. At this stage it is also clear that the third pharyngeal pouch is forming and has contacted the ectoderm, defining the posterior limit of the third arch. By stage 19 of chick development, the fourth, and last pouch, has formed and four distinct pharyngeal segments are apparent (Fig. 1b). The final segment, the sixth arch will soon form between the fourth pouch and the posterior limit of the pharynx; even though this is numerically the fifth arch, this is termed the sixth due to the long held, but erroneous, belief that a transient fifth arch formed between this segment and the fourth [16]. By this stage, the first pouch has now receded from the ectoderm and mesenchyme lies between these tissues, while the second pouch has broken through and created an external opening. The posterior pouches, 3 and 4, remain in contact with the ectoderm. Thus the anterior and posterior pouches interface with the ectoderm in different ways and are morphologically distinct. In catshark, the anterior pharyngeal segments are also delineated first and by stage 19 the three most anterior pouches have formed and are in contact with the ectoderm (Fig. 1c). By stage 22 of catshark development, pouches 4 and 5 have formed, yielding five distinct arches. The anterior pouches have also matured and these have pushed through and created external pharyngeal openings, with only pouch 5 not having broken through, as the ectoderm and endoderm remain opposed to each other at this position (Fig. 1d). In contrast to the chick, however, the relationships between the pouch endoderm and the ectoderm are relatively consistent across the pouches and these share a similar morphology.

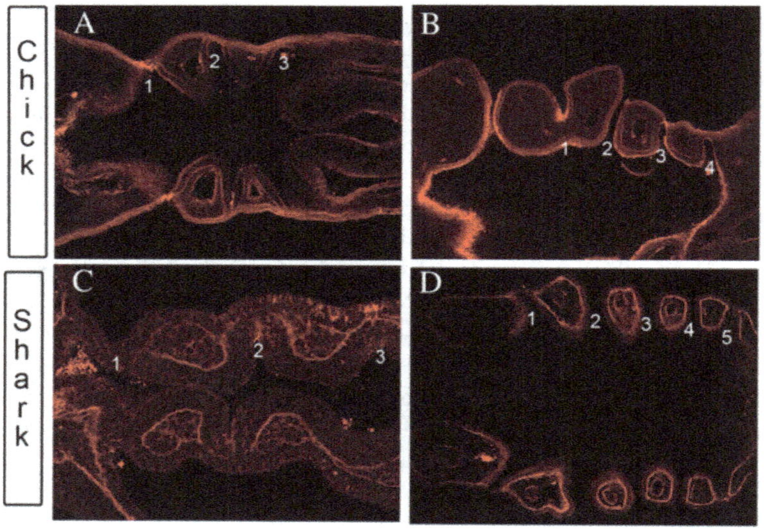

Fig. 1 Overview of pharyngeal segmentation in Chick (*Gallus gallus*) and catshark (*Scyliorhinus canicula*) embryos. Longitudinal sections through the pharyngeal region of whole mount anti-laminin immunostained chick (**a, b**) and catshark (**c, d**) embryos during the period of segmentation. **a** Section through a stage 14 chick embryo, at which point the first three pharyngeal pouches have formed. The contact between the pouches and the endoderm can be clearly seen. **b** Section through a stage 19 embryo, by which time all four pharyngeal pouches have formed. It is now quite clear that the different pouches have assumed different morphologies. The first pouch has receded from the ectoderm and mesenchyme lies between, while the second pouch has broken through to create an external opening. The interface between pouch 3 and the ectoderm is thinning at this point while pouch 4 is still abutting the ectoderm. **c** Section through a stage 19 catshark embryo, at which point the first three pouches have formed and are in contact with the ectoderm. **d** Section through a stage 22 catshark embryo which has formed 5 pouches. At this stage pouches 1 to 4 have broken through externally while the fifth pouch is just beginning to do so

Analysis of pharyngeal innervation further identifies the most posterior segments as being reduced in number

An examination of the patterns of cranial nerve innervation of the pharynx in different vertebrate embryos can prove useful in focussing on which specific arches are likely to have been lost. We have therefore documented the patterns of innervation of the pharyngeal segments in two amniotes, chick and mouse, which have five segments, in a chondrichthyan, the catshark, which has seven segments, and in the lamprey which has nine. Wholemount immunofluorescence with an anti-neurofilament antibody is shown in Fig. 2. It is readily apparent that clear and distinct innervation arches 1, 2 and 3 can be seen across all species; i.e., innervation of the first segment by the Vth nerve, the second by the VIIth nerve, and the third by the IXth nerve. It is also clear that the posterior limit of the pharynx is skirted by the XIIth nerve in all. Finally, the variable number of posterior segments are innervated by a variable number of branches of the Xth nerve. This suggests that the anterior three segments are homologous across the vertebrates, and that there has been a loss of segments posterior of the third.

***HOX* gene expression boundaries and their relationship to the pharyngeal arches and pouches in chick embryos**

To ascertain which pharyngeal segments/pouches have been lost from extant tetrapods, we need to be able to align the reduced number of posterior pouches of an amniote, in this case the chick, with those of the catshark, which represents the more basal condition. With regard to this issue, *Hox* genes are useful as they can be used as markers of axial identity. Although the expression patterns of *Hox* genes in the pharyngeal arches have been well documented in amniotes, these studies have largely reported expression within the mesenchyme of the arches and they have often not explicitly analysed the limits of expression of these genes within the pharyngeal endoderm [17, 18].

We therefore analysed the expression patterns of *Hox1-5* gene paralogues, as these are known to be expressed in the pharyngeal region [17, 18]. As anticipated, the different paralogues labelled different arches (Fig. 3a–g). We then sectioned these embryos to determine the expression domain of these genes within the different populations that constitute the arches, and in particular in the endoderm, at stages when all four pharyngeal pouches had formed, stage 20. We note that the first pharyngeal pouch is not associated with the expression of any *Hox* genes. However, we found that *Hoxa2* was expressed in the second arch mesenchyme and the endoderm posteriorly from the second pouch (Fig. 3h) while *Hoxa3* was expressed in the third arch mesenchyme and the third pouch endoderm at stage 17

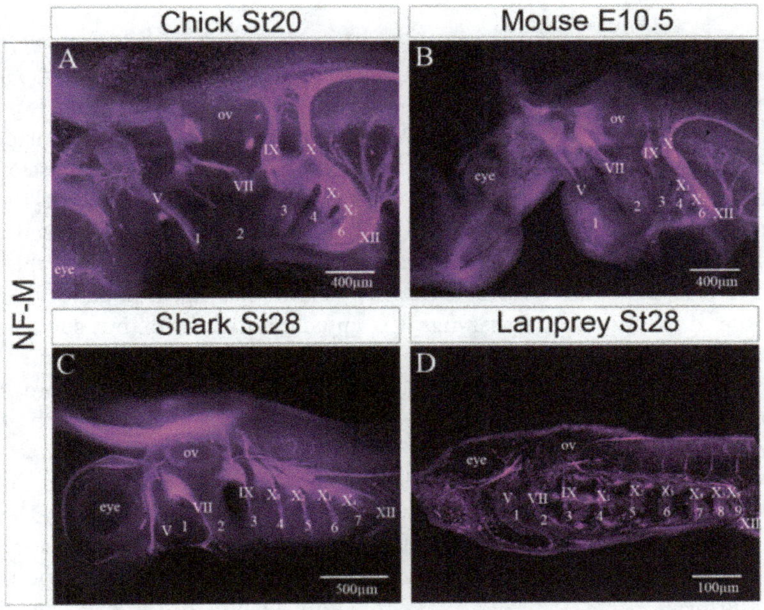

Fig. 2 Comparative analysis of pharyngeal innervation across the verterbates. Patterns of innervation of the developing pharyngeal arches in different vertebrates at the point at which the full complement of pharyngeal segments have been generated. The embryos were wholemount immunostained with and anti-neurofilament antibody and the either bisected and photographed or photographed whole. **a** Chick **b** Mouse **c** Catshark **d** Lamprey. The cranial nerves are labelled with roman numerals. V – trigeminal, VII – facial, IX – glossopharyngeal, X – vagus, XII – hypoglossal. The pharyngeal arch are numbered with Arabic numerals. The position of the eye is indicated as is the position of the otic vesicle (OV). It is clear that in all embryos there is distinct innervation of the first arch by the Vth nerves, of the second arch by the VIIth nerve and of the third by the IXth nerve. In all embryos the XIIth nerve skirts the caudal edge of the pharynx. It is also apparent that there are a variable number of branches of the Xth nerve innervating the variable number of posterior segments

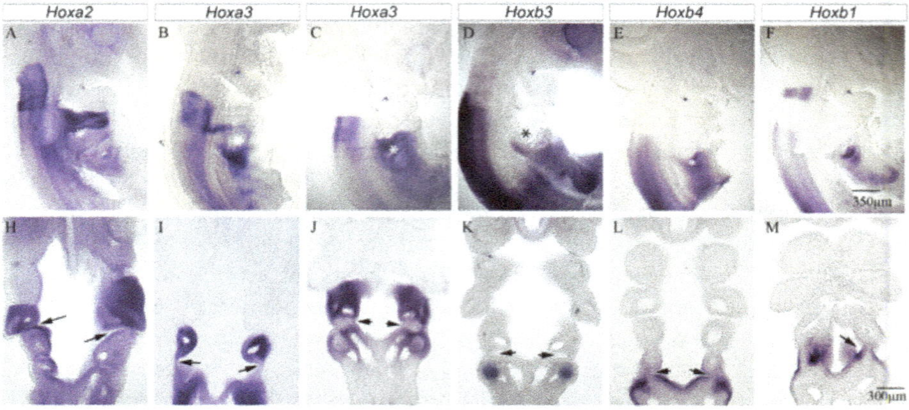

Fig. 3 Hox gene expression boundaries and their relationship to the pharyngeal arches and pouches in the chick embryo. Patterns of expression of *Hox* genes in the developing pharynx at stage 20 (**a**, **b**, **d-g**, **h**, **i**, **k-n**) when all of the pharyngeal segments have formed. This figure additionally shows the expression of *Hoxa3* at stage 17 (**b**, **i**). The expression patterns of these genes are shown in bisected embryos (**a** – **g**) and in section (**h-n**). The expression pattern of *Hoxa2* is shown in (**a**, **h**) and this gene can be seen to be expressed in the mesenchyme of the second and more posterior arches and in the endoderm of the second pouch and posterior of that. The position of the second pouch is marked by an asterisk in (**a**) and an arrow in (**h**). The expression of *Hoxa3* is shown at stage 17 (**b**, **i**) and at stage 20 (**c**, **j**). In these panels the position of the third pouch is marked by an asterisk in (**b**, **c**) and and arrow in (**i**, **j**). The expression pattern of *Hoxa3* in the mesenchyme does not change between these stages and can be seen to be expressed in the third arch and more posterior. However, while *Hoxa3* can be seen to be expressed in the third pharyngeal pouch at stage 17 (**i**), expression is lost from here at stage 20 (**j**). The expression of *Hoxb3* is shown in (**d**, **k**) and this gene can be seen to be expressed in the mesenchyme of the fourth arch but its expression is noticeably absent from the pharyngeal endoderm (**k**). The position of the third pouch is marked by an asterisk in (**d**) and by an arrow in (**k**). The expression of *Hoxb4* is shown in (**e**, **l**) and this gene can be seen to be expressed is not expressed in the mesenchyme of any of the arches, nor in any of the pharyngeal pouches but is expressed in the most posterior pharyngeal endoderm and the ectoderm lying posterior of the pharynx. The position of the fourth pouch is marked by an asterisk in (**e**) and by an arrow in (**l**). The expression of *Hoxb5* is shown in (**f**, **m**) and it can be seen that this gene is not expressed in the pharyngeal region. *Hoxb1* expression is shown in (**g**, **n**) and at stage 20 it is clear that the expression of this gene is restricted to the last formed, fourth, pharyngeal pouch, marked by an asterisk in (**g**) and an arrow in (**n**)

(Fig. 3i). However, although expression of *Hoxa3* persists in the third arch mesenchyme the expression in the third pouch was subsequently lost (Fig. 3j). This is in keeping with what has been observed in the mouse [19]. Contrastingly, we found that *Hoxb3* expression was absent from the endoderm, but that this gene was expressed in the mesenchymal cells that will populate the fourth arch (Fig. 3k). We found *Hoxb4* expression was not associated with any of the pharyngeal pouches but was expressed in the most posterior pharyngeal endoderm and the ectoderm lying posterior of the pharynx (Fig. 3l). We also found that *Hoxb5* was not expressed in the developing pharynx (Fig. 3f, m). However, we did find that the expression of *Hoxb1* was localised to the fourth pouch, the last formed, and at this stage the expression in the mesoderm of the second arch was no longer evident (Fig. 3g, n). This would suggest that *Hoxb1* may be useful as a specific marker of the fourth and/or last pouch formed.

Hox gene expression boundaries and their relationship to the pharyngeal pouches in catshark embryos

The overall relationship between the pharyngeal arches and *Hox* gene expression in catshark (*Scyliorhinus canicula*) has been described previously and it is broadly similar to the situation observed in amniotes, with *Hox* genes of paralogous groups (PG) 1–4 being expressed in the pharyngeal region, and *Hox* gene of paralogous groups (PG) 5–8 not being expressed [20]. There were, however, also some important detailed differences, such as the fact that *Hoxd3* expression had shifted posteriorly and was associated with the fourth arch, and that this segment additionally expressed *Hoxd1, d2* and *d4*. The precise relationship between the expression of *Hox* genes and the pharyngeal pouches has not been scrutinised in any chondrichthyan, but such analysis is vital if we wish to align the pharyngeal pouches across the gnathostomes. The key questions are: Does *Hoxa2* expression align with the second pouch in both amniotes and a chondrichthyan, and does *Hoxb1* expression which highlights the fourth and last pouch of amniotes align with the fourth pouch in a chondrichthyan, or is this gene expressed by multiple posterior pouches, or indeed only in the sixth and last formed pouch?

Our analysis of the chick highlighted the fact that the different pouches were primarily marked by the expression of *Hoxa2, Hoxa3* and *Hoxb1* and we therefore sectioned wholemount in situ hybridisations of catshark embryos processed for these genes. We found that *Hoxa2* expression was present in the mesenchyme of the second and more posterior arches and the endoderm posterior of the second pouch between stages 20 to 24

(Fig. 4a, d). Contrastingly, *Hoxa3* expression was found in the mesenchyme of arch 3 at stages 21 to 24, but that expression in the endoderm of the third pouch was weak at stage 21 and absent at later stages (Fig. 4b, e). Finally, the expression of *Hoxb1* was found in the third pouch at stage 20, and at stage 24 in the fifth pouch (Fig. 4c, f). Thus *Hoxb1* expression is associated with different pharyngeal pouches at different developmental stages.

Transient Hoxb1 expression marks the caudal limit of the pharynx

The association between the expression of *Hoxb1* with different pouches at different stages of development in *Scyliorhinus* was surprising, and we therefore set out to determine if this was a general feature of pharyngeal development in both chick and catshark. The first two pharyngeal pouches form in the chick at stage 13 and we found that even at this early time point *Hoxb1* expression was already established in the most caudal pouch, pouch 2 (Fig. 5a). By stage 15 the third pouch had formed and we noted that *Hoxb1* expression was now associated with that newly formed pouch (Fig. 5b). Finally, we found that by the time all four pouches had formed, *Hoxb1* expression had shifted to this last pouch (Fig. 5c, d). Intriguingly, a more detailed examination of the expression profile of this gene in catshark revealed a similar situation. At stage 21 when four pouches had formed *Hoxb1* expression was associated with the newly formed fourth pouch (Fig. 5e) while at stage 22 when the fifth pouch was forming, *Hoxb1* expression had shifted to this pouch (Fig. 5f). Finally, *Hoxb1* expression shifted to the lastly formed sixth pouch from stage 23 onwards Fig. 5g, h).

Discussion

In the present study, we have attempted to address the route through which the reduction in the number of pharyngeal segments that occurred within the sarcopterygians, and which is evident in extant tetrapod embryos. To do this, we compared the development of the pharyngeal segments in an amniote, the chick, with that in a chondrichthyan, the catshark. We show that in both the overall trend in the formation of the pharyngeal pouches is similar, with the most anterior forming first and then the posterior segments forming sequentially. We further analysed the patterns of innervation and the expression profiles of *Hox* genes in the forming arches, and in particular how these relate to the pharyngeal pouches, which dictate the number of segments formed. Our results show support for the first three anterior

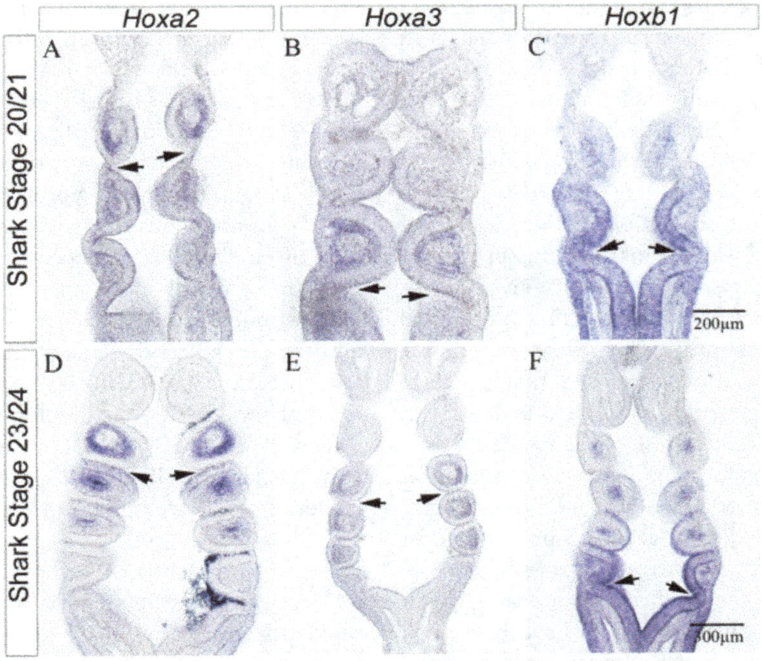

Fig. 4 Hox gene expression boundaries and their relationship to the pharyngeal pouches in *Scyliorhinus canicula*. Expression patterns of *Hoxa2*, *Hoxa3* and *Hoxb1* in early (**a**-**c**), stage 20/21which have formed 3 to 4 pharyngeal arches, and later (**d**-**f**) embryos, stage 23/24 which have formed 5 to 6 pharyngeal arches. At both early and late stages *Hoxa2* is expressed in the mesenchyme of the second and more posterior arches and in the endoderm caudal of the second pouch, marked by an arrow (**a**, **c**). *Hoxa3* shows expression in the mesenchyme of the third and more posterior arches at both early and late stages (**b**, **e**) but only weak (**b**) to no expression in the pharyngeal endoderm. The position of the third pouch (**b**, **e**) is indicated by the arrow. *Hoxb1* shows prominent expression in fourth pharyngeal pouch at early stages, marked by an arrow (**c**) and in the fifth pouch at later stages, marked by an arrow (**f**)

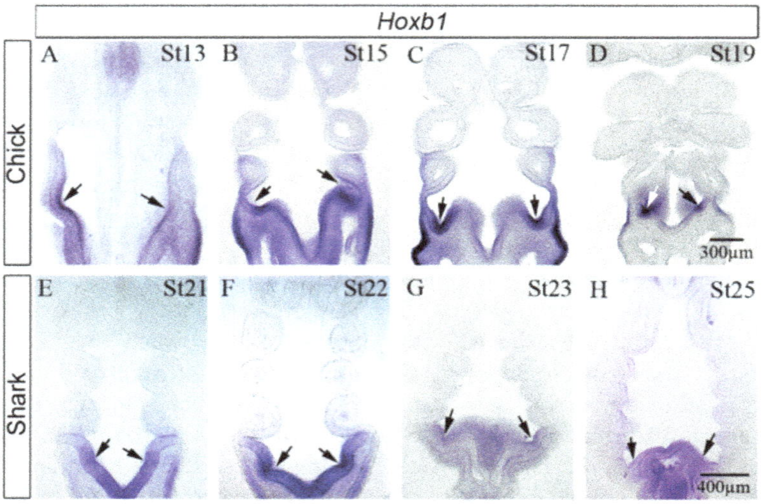

Fig. 5 Hoxb1 expression is spatially and temporally associated with the last formed pharyngeal pouch in both chick and catshark. Spatially and temporally dynamic expression of *Hoxb1* is associated with the last formed pharyngeal pouch in both chick (**a-d**) and catshark (**e-h**) embryos. At stage 13 in the chick *Hoxb1* expression is associated with the newly formed second pouch (**a**), while at stage 15 the expression of this gene is found in the newly formed third pouch (**b**). From stage 17 onwards, *Hoxb1* expression is associated with the fourth pouch (**c**, **d**). In catshark embryos a similar spatially and temporally dynamic association between the expression of *Hoxb1* and the last formed pouch is also seen, Thus while at stage 21, the expression of this gene is found at the fourth pouch (**e**), by stage 22 it is located to the fifth pouch (**f**) and by stage 23 and later with the sixth pouch (**g**, **h**). In all panels the position of the last formed pouch is highlighted by an arrow

segments being homologous across the gnathostomes. Each of these had a distinct innervation, and both the mesenchyme of these arches and the corresponding pouches expressed the same repertoire of *Hox* genes in both chick and catshark. To gain insights into how the posterior pouches align between chick and catshark, we analysed the expression of *Hoxb1*, which is expressed in the most posterior pouch in amniotes. Somewhat unexpectedly, we found that *Hoxb1* exhibits a dynamic expression profile during the formation of the pharyngeal segments, with the expression domain of this gene progressively moving caudally such that at any given stages it is expressed in the last formed pouch. This leads us to suggest that the posterior limit of the pharynx is homologous between chick and catshark, and that a reduction in the number of pharyngeal segments has been achieved by the earlier termination of the caudal expansion of the pharyngeal endoderm. Thus the caudal limit of the pharynx is established at a relatively more anterior position and therefore fewer pharyngeal pouches form and correspondingly fewer pharyngeal segments.

From our analysis we can begin to understand which of the posterior, or branchial, segments have been lost. Our results clearly indicate that the most anterior of the branchial series, the third pharyngeal segment is homologous across the gnathostomes and is retained in amniotes. In lamprey, catshark, chick and mouse, this segment is distinctly innervated by the IXth nerve. Furthermore, we show that the expression of *Hoxa3*, in chick and catshark in both the mesenchyme and the corresponding pouch endoderm is also a conserved feature of the third segment; although, the expression of this gene is lost as development progresses. However, it is less clear which of the more posterior arches have been retained and which have been lost. In part this is due to the fact that the expression patterns of *Hox* PG 3 and 4 do not simply align with distinct arches or pouches in either chick or catshark embryos. Finally, our study suggests that the posterior limit of the pharynx may be homologous, in that it is defined by the expression of *Hoxb1* in both species (Fig. 6). Indeed, we find that the anterior limit of Hoxb1 expression initially marks pouch 2, i.e. the anterior limit of the third arch and thus the anterior limit of the branchial apparatus in anamniotes. This is consistent with the view that the third pharyngeal arch is homologous across the gnathostomes. Additionally, we show here that in both chick and catshark the limit of expression of this gene shifts posteriorly as development progresses and thus it labels sequentially the caudal most brachial pouch. Taken together our observations indicate that the four most anterior arches were retained in amniotes but that the two segments that form subsequently were lost between the fourth arch and the caudal limit of the pharynx.

While our work gives an insight into the possible mechanism that underpinning the reduction in arch number that occurred within the sarcopterygians, the morphological data form fossils also provides a very useful complement that give us insights into when the reduction in the number of pharyngeal segments started

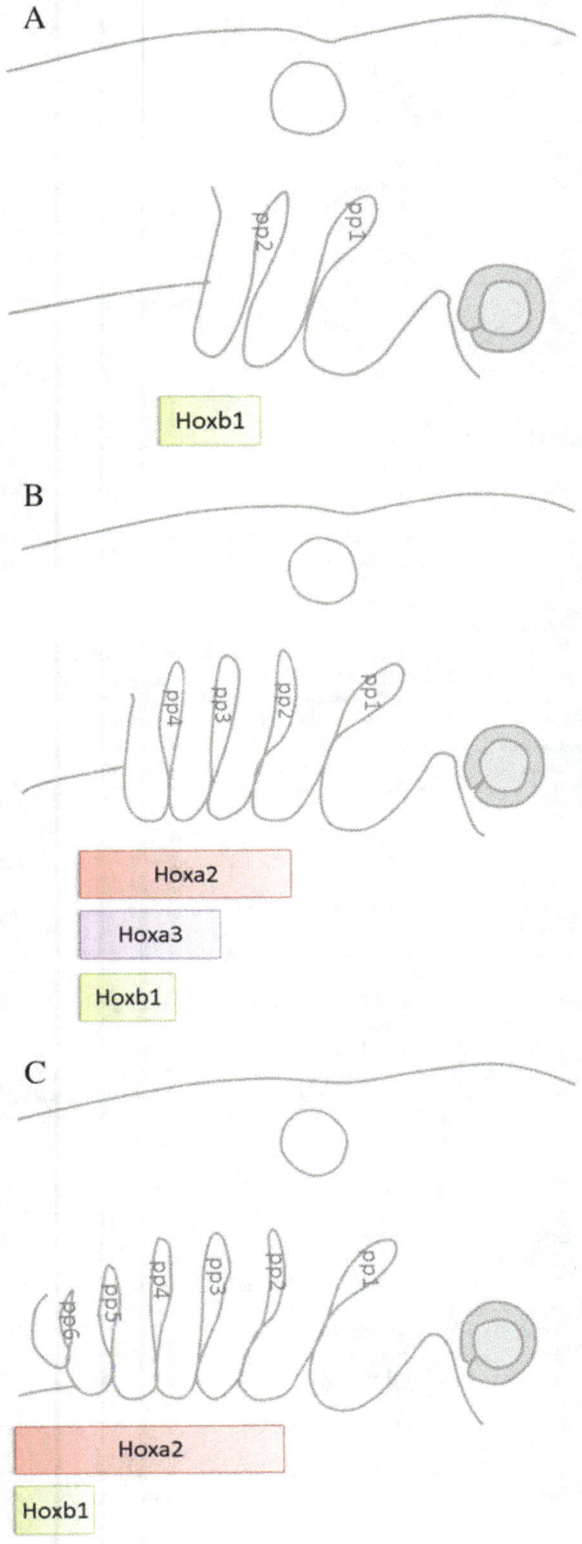

Fig. 6 A conserved pattern of Hox gene expression aligns with the pharyngeal pouches. Schematic representation of *Hox* gene expression in the pharyngeal pouches of a generalised gnathostome throughout pharyngeal development. **a** Initial pharyngeal pouch formation begins with the simultaneous budding off the pharyngeal endoderm of the first and second pharyngeal pouches. *Hoxb1* aligns with the second, most posterior pharyngeal pouch. **b** At slightly older stages, four pharyngeal pouches present. *Hoxa2* is now expressed with an anterior limit in the second pouch, *Hoxa3* with an anterior limit at the third pouch. *Hoxb1* no longer has an anterior limit at the second pouch but is now present in the most posterior pouch, which at this stage is the fourth. **c** Once all pouches have developed, the anterior limit of *Hoxa2* expression still aligns with the second pouch while *Hoxa3* expression is no longer seen in the endoderm. *Hoxb1* expression no longer aligns with the fourth pouch but rather with the most posterior pouch, which at this stage is the sixth pouch. This represents a conserved dynamic expression pattern of *Hoxb1* in the most posterior pouch present of gnathostomes at any given time point during pharyngeal pouch development irrespective of the number of pouches that have formed at that time. ;(pp = pharyngeal pouch)

[21]. Thus while extant and fossil coelacanths have five branchial arches, it has been reported that *Gogonasus* and *Eusthenopteron*, which are tetrapodomorph fish, have lost the fifth arch. Thus it has been suggested that the absence of the fifth branchial arch is a derived feature of advanced sarcopterygians.

Importantly, other studies suggest that the mode of arch reduction that we have identified, in which deletion occurs between a *Hox1*-defined posterior limit of the pharynx and more anterior segments, is likely to extend beyond events within the sarcopterygians. An analysis of Hox gene expression in the lamprey, *Lethenteron japonicum*, found that the *Hox1* gene, *LjHox1w*, in this this species also marks the caudal limit of the pharynx, although it is unclear whether *LfHox1w* displays a similar temporally and spatially dynamic pattern of expression [22]. Perhaps even more significantly, an analysis of pharyngeal gill slit formation in a hemichordate, *Saccoglossus kowalevskii*, which concludes that this process is homologous to the formation of the pharyngeal segments of vertebrates, also reported spatially and temporally dynamic expression of *Hox1* [23]. In that study, the authors noted that at *Hox1* expression was restricted to the posterior boundary of the pharyngeal endoderm at the three gill-slit stage and that it was still restricted to the posterior boundary of the pharyngeal endoderm at the four gill-slit stage. Collectively, our work and those of others therefore suggest that a possible role for *Hox1* genes in defining the caudal limit of the pharynx extends beyond the vertebrates, and is likely to have evolved much earlier with the emergence of the deuterostomes.

This is also significant in that it could help us understand how the number of pharyngeal segments has been modified across the deuterostomes. Thus while enteropneust hemichordates and cephalochordates have numerous pharyngeal gill slits [24, 25], which extend significantly along the length of the body, those in vertebrates are fewer in number and are focussed just caudal of the mouth. Thus, the number of pharyngeal segments could be decreased by terminating the posterior expansion of the pharynx prematurely and so decreasing the distance between the *Hox1* expression domain and the last formed anterior segment. Correspondingly, an increase in the number of pharyngeal segments, Thus the number of pharyngeal segments could be increased, such as is seen in the extinct jawless vertebrate *Endeiolepis* which had up to thirty pairs of gill slits [26], by allowing the posterior of the pharynx to extend for a relatively longer time and thus increase the region between the *Hox1* expression limit and the anterior.

Conclusions

Our results support the view that the three most anterior pharyngeal segments are conserved across the vertebrates and that the caudal limit of the pharynx is also conserved. Thus, within the sarcopterygians, the segments that were lost were those form the "branchial"/posterior region which lie between the third arch and the caudal limit. Furthermore, our results would suggest the reduction in the number of pharyngeal segments was achieved as a result of the premature termination of the posterior extension of the pharyngeal endoderm which in turn would result in the generation of fewer segments. Finally, we suggest that such a mechanism may also account for the variability in the number of pharyngeal segments seen across the vertebrates and other deuterostomes.

Competing interests

The authors declare that they have no competing interests.

Authors' contributions

VS, PL and AG devised the study. VS, SO and PL performed the lab work. SO, PL and DC supplied embryos. VS, SO, DC, PL and AG wrote and approved the paper.

Acknowledgements

We would like to thank the Anatomical Society for support for part of this work and Dr Seb Shimeld for lamprey embryos. We would like to thank Clemens Kieckers for comments on the manuscript.

Author details

[1]Department of Developmental Neurobiology, Kings College London, London SE1 1UL, UK. [2]Laboratoire Évolution, génomes, comportement, écologie, CNRS université Paris-Sud UMR 9191, IRD UMR 247, Avenue de la Terrasse, bâtiment 13, boîte postale 1, 91198 Gif-sur-Yvette, France.

References

1. Graham A, Richardson J. Developmental and evolutionary origins of the pharyngeal apparatus. Evodevo. 2012;3:24.
2. Kardong KV. Vertebrates : comparative anatomy, function, evolution. 6th ed. New York: McGraw-Hill; 2012.
3. Hughes G. Respiration of Latimeria-chalumnae. Zool J Linn Soc. 1976;59:195.
4. Millot J, Anthony J. De lexistence chez LATIMERIA-chalumnae Smith (Coelacanthidae) dun organe regulateur du courant sanguin supra-branchial. Comptes Rendus Hebdomadaires Des Seances De L Academie Des Sciences. 1958;246:1600–2.
5. Amemiya C, Alfoldi J, Lee A, Fan S, Philippe H, MacCallum I, Braasch I, Manousaki T, Schneider I, Rohner N, et al. The African coelacanth genome provides insights into tetrapod evolution. Nature. 2013;496:311–6.
6. Bartsch P. Development of the cranium of Neoceratodus forsteri, with a discussion of the suspensorium and the opercular apparatus in Dipnoi. Zoomorphology. 1994;114:1–31.
7. Richardson J, Shono T, Okabe M, Graham A. The presence of an embryonic opercular flap in amniotes. Proceedings of the Royal Society B: Biological Sciences. 2012;279:224–9.
8. Graham A, Okabe M, Quinlan R. The role of the endoderm in the development and evolution of the pharyngeal arches. J Anat. 2005;207:479–87.
9. Crump J, Swartz M, Kimmel C. An integrin-dependent role of pouch endoderm in hyoid cartilage development. PloS Biology. 2004;2:1432–45.
10. Shone V, Graham A. Endodermal/ectodermal interfaces during pharyngeal segmentation in vertebrates. J Anat. 2014;225:479–91.
11. Lindsay E, Vitelli F, Su H, Morishima M, Huynh T, Pramparo T, Jurecic V, Ogunrinu G, Sutherland H, Scambler P, et al. Tbx haploinsufficiency in the DiGeorge syndrome region causes aortic arch defects in mice. Nature. 2001; 410:97–101.
12. Piotrowski T, Nusslein-Volhard C. The endoderm plays an important role in patterning the segmented pharyngeal region in zebrafish (Danio rerio). Dev Biol. 2000;225:339–56.
13. HAMBURGER V, HAMILTON H. A series of normal stages in the development of the chick embryo. J Morphol. 1951;88:49.
14. Ballard W, Mellinger J, Lechenault H. A series of normal stages for development of *Scyliorhinus canicula* the lesser spotteddogfish (Chondrichthyes: Scyliorhinidae). J Exp Zool. 1993;267:318–36.
15. Veitch E, Begbie J, Schilling TF, Smith MM, Graham A. Pharyngeal arch patterning in the absence of neural crest. Curr Biol. 1999;9:1481–4.
16. Lewis WH. Anatomy of the human body, by Henry Gray. 20th ed., thoroughly rev. and re-edited by Warren H. Lewis. Philadelphia: Lea & Febiger; 1918.
17. Hunt P, Gulisano M, Cook M, Sham M, Faiella A, Wilkinson D, Boncinelli E, Krumlauf R. A distinct Hox code for the branchial region of the vertebrate head. Nature. 1991;353:861–4.
18. Couly G, Grapin-Botton A, Coltey P, Ruhin B, Le Douarin N. Determination of the identity of the derivatives of the cephalic neural crest: incompatibility between Hox gene expression and lower jaw development. Development. 1998;125:3445–59.
19. MANLEY N, CAPECCHI M. The role of HOXA-3 in mouse thymus and thyroid development. Development. 1995;121:1989–2003.
20. Oulion S, Borday-Birraux V, Debiais-Thibaud M, Mazan S, Laurenti P, Casane D. Evolution of repeated structures along the body axis of jawed vertebrates, insights from the Scyliorhinus canicula Hox code. Evol Dev. 2011; 13:247–59.
21. Carvalho M, Bockmann F, de Carvalho M. Homology of the Fifth Epibranchial and Accessory Elements of the Ceratobranchials among Gnathostomes: Insights from the Development of Ostariophysans. PloS One. 2013;8.
22. Takio Y, Kuraku S, Murakami Y, Pasqualetti M, Rijli F, Narita Y, Kuratani S, Kusakabe R. Hox gene expression patterns in Lethenteron japonicum embryos - Insights into the evolution of the vertebrate Hox code. Dev Biol. 2007;308:606–20.
23. Gillis J, Fritzenwanker J, Lowe C. A stem-deuterostome origin of the vertebrate pharyngeal transcriptional network. Proceedings of the Royal Society B-Biological Sciences. 2012;279:237–46.
24. Rychel A, Smith S, Shimamoto H, Swalla B. Evolution and development of the chordates: Collagen and pharyngeal cartilage. Mol Biol Evol. 2006;23:541–9.
25. Gonzalez P, Cameron C. The gill slits and pre-oral ciliary organ of Protoglossus (Hemichordata: Enteropneusta) are filter-feeding structures. Biol J Linn Soc. 2009;98:898–906.
26. Janvier P, Desbiens S, Willett J, Arsenault M. Lamprey-like gills in a gnathostome-related Devonian jawless vertebrate. Nature. 2006;440:1183–5.

15

Two-headed mutants of the lamprey, a basal vertebrate

Daichi G. Suzuki[1,2]

Abstract

Background: This is the first report of two-headed (bicephaly) lamprey twins. Although lampreys sit at a crucial phylogenetic position, there are only a few reports on their teratology and developmental abnormalities.

Results: Two-headed mutants were obtained by artificial fertilization in the laboratory as spontaneous occurrences. All mutants were derived from single fertilizations using single male and female gametes, suggestive of a genetic background. The anterio-posterior position of the axonal bifurcation and symmetricity varied in each mutant. Other malformations were coincidently observed, including pericardial edema, yolk sac edema and axial bending. Asymmetrical (lateral - branched) mutants displayed more severe abnormalities in the cranial nerves than symmetrical mutants.

Conclusion: Two-headed mutants of the lamprey are described. These mutants have similar malformations to dorsal blastopore lip-transplanted lamprey embryos, suggesting that they could be generated by a disorder in head-organizing activity.

Keywords: Lamprey, Teratology, Two-headed twin, Axial duplicity, Bicephaly

Background

Malformed mutants, or so-called "monsters", have attracted the attention of morphologists since the inception of the discipline. For example, Étienne Geoffroy Saint-Hilaire and his colleagues, including his son Isidore, described many developmental anomalies and sought to explain the mechanisms underlying their production [1, 2]. As described in their works, conjoined twins, referred to as "axial duplicity" or "double monsters" in Bateson [3], also represent striking examples of such anomalies.

More recently, in the field of evolutionary developmental biology, researchers have focused more on malformations, as these can represent developmental constraint, variability, and evolvability [4–6]. In fact, axial bifurcation in the caudal fin of the twin-tail goldfish is caused by ventralization during early embryonic development, which is in turn due to a mutation in a *chordin* gene that may have occurred during domestication [7]. Developmental malformations in lampreys, which belong to a basal group of vertebrates (cyclostomes), may similarly provide new insights into the early evolution of vertebrates. However, only a few reports on lamprey teratology and developmental abnormalities have been published to date (e.g., [8–10]).

Here I report two-headed conjoined (bicephaly) mutants in the Arctic lamprey, *Lethenteron camtschaticum*, which were unexpectedly obtained following artificial fertilization in the laboratory. These mutants have similar malformations to dorsal blastopore lip-transplanted embryos [11], suggesting that they may be generated by a disorder in head-organizing activity.

Methods

Adult lampreys (*Lethenteron camtschaticum*) were collected from the Shiribeshi-Toshibetsu River, Hokkaido, Japan in 2012. In the next spawning season (May to June 2013), the animals were anesthetized in ethyl,3-aminobenzoate methanesulfonate (MS-222). Mature eggs and sperm were squeezed from adults and fertilized in vitro. Embryos were cultured at 16 °C, fixed in 4% paraformaldehyde in 0.1 M phosphate-buffered saline (PBS) overnight, dehydrated in a graded methanol series, and stored in 100% methanol at −20 °C.

Whole-mount immunofluorescence with anti-acetylated tubulin (Sigma, T6793) antibodies was performed according to the protocol described by Kuratani et al. [12] with some minor modifications. Briefly, fixed embryos stored in methanol were washed in TBST containing 5% dimethylsulfoxide (TSTd). The embryos were then

Correspondence: daichi1207@gmail.com
[1]Graduate School of Life and Environmental Sciences, University of Tsukuba, 1-1-1 Tennodai, Tsukuba, Ibaraki 305-8572, Japan
[2]Nobel Institute for Neurophysiology, Department of Neuroscience, Karolinska Institutet, SE-17177, Stockholm, Sweden

blocked with 5% non-fat dry milk in TSTd (TSTM). They were incubated with the primary antibody (diluted 1:1000 in TSTM) for 2–4 days at room temperature. After washing with TSTd, samples were incubated with a secondary antibody (Invitrogen, Alexa fluor 555, A21424) diluted 1:200 in TSTM. Then embryos are washed by TSTd several times, treated RNase and counter-stained by YOYO-1. After a final wash in TSTd, embryos were dehydrated and clarified in a 1:2 mixture of benzyl alcohol and benzyl benzoate (BABB) and then examined using a confocal laser microscope (LSM 510, Zeiss).

Paraffin sections were cut at a thickness of 8 μm and stained with hematoxylin and eosin, according to a standard protocol.

Results and discussion

I obtained two-headed embryos from one female lamprey by artificial fertilization (at least 25 embryos). As *L. camtschaticum* females spawn large numbers of eggs (tens of thousands [13]) and dead embryos were immediately removed to keep the batch clean, I could not specify how many mutants were generated and calculate the malformation rate precisely. However, it is clear that the occurrence of two-headed mutants is very rare. The antero-posterior position of the axonal bifurcation and symmetricity varied among mutants (e.g., Fig. 1a–d). Some embryos had exo-embryonic yolk clusters (Fig. 1d), as is often the case with normal lamprey embryos. Over

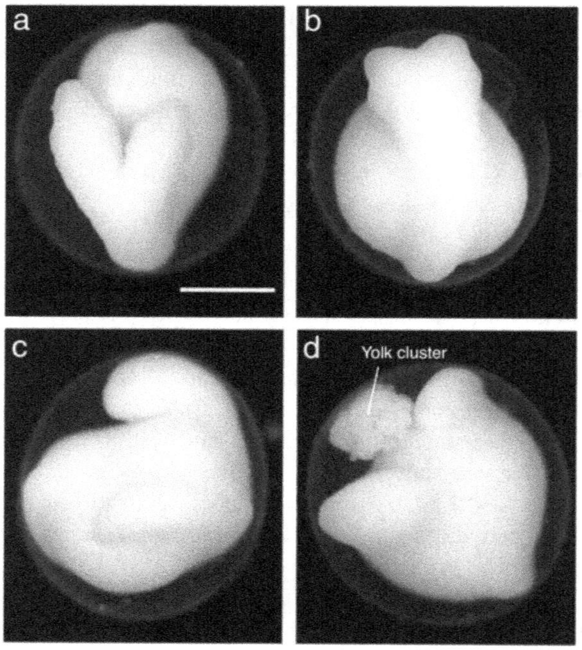

Fig. 1 Two-headed lamprey embryos. **a-d** Pre-hatching embryos at the same magnification. The antero-posterior position of the axonal bifurcation and symmetricity varied among mutants. The exo-embryonic yolk cluster can be seen in (**d**). Scale bar: 500 μm

the 10 years that lamprey embryos have been collected in this laboratory, two-headed mutants have only been obtained from one female, even though sperm from the same male was used to fertilize eggs from other females and the same fertilization/incubation conditions have always been used. Although some environmental effect to the female individual could cause maternal effects and the two-headed mutation, migrant adults collected upstream were exposed to the same condition during their handling and maintenance process, without exhibiting similar mutations. Therefore, I infer that the malformation may be attributable not to an artificial or environmental effect, but to some natural (spontaneous) occurrence in the genetic background, similar to those observed in other vertebrates, including humans [3, 14–16]. This is the first known report of two-headed lamprey twins, although complete twin embryos have been reported previously [10]. It would be difficult to observe these lamprey mutants in the wild, as they die quickly and, even if they manage to survive, tend to hide in the sandy bottom.

During incubation, some two-headed embryos (particularly severely malformed mutants) died before hatching. I obtained 21 mutants (identity numbers are provided as #01–#21, see also Additional file 1) after hatching (e.g., Fig. 2a–h), which were nearly the same size as normal prolarvae (Fig. 2i). There were two general types of malformation: lateral and symmetrical bifurcation. Laterally bifurcated mutants possessed secondary head, which was smaller than primary head, while symmetrically bifurcated mutants showed no such significant difference between the two heads. Depending on the antero-posterior position of the axonal bifurcation, they showed various types of defects, for example, a shared mouth (Fig. 2c) and double hearts (Fig. 2d). While normal prolarvae usually show rhythmic neck movements, most of the mutants moved non-rhythmically, suggesting that twin heads "compete" with each other for motor control. In addition, other malformations were coincidently observed, including pericardial edema (Fig. 2a, d, 24% of obtained mutants), yolk sac edema (Fig. 2b, e–h, 48%) and axial bending (Fig. 2e–g, 57%). This suggests that the causal factors of these malformations are linked and that one malformation tends to involve others (for the sample list with summary of malformations in each mutants, see Additional file 1).

To survey the morphological effects of the malformation on peripheral nerves, I performed immunofluorescence using an anti-acetylated tubulin antibody (for the normal development of the lamprey peripheral nerve, see [12]). The lateral-branched twins had a malformed cranial nerve in the secondary head, which was more severe than that seen in symmetrically bifurcated twins (Fig. 3a, b). Some lacked mouth openings (Fig. 3a), and

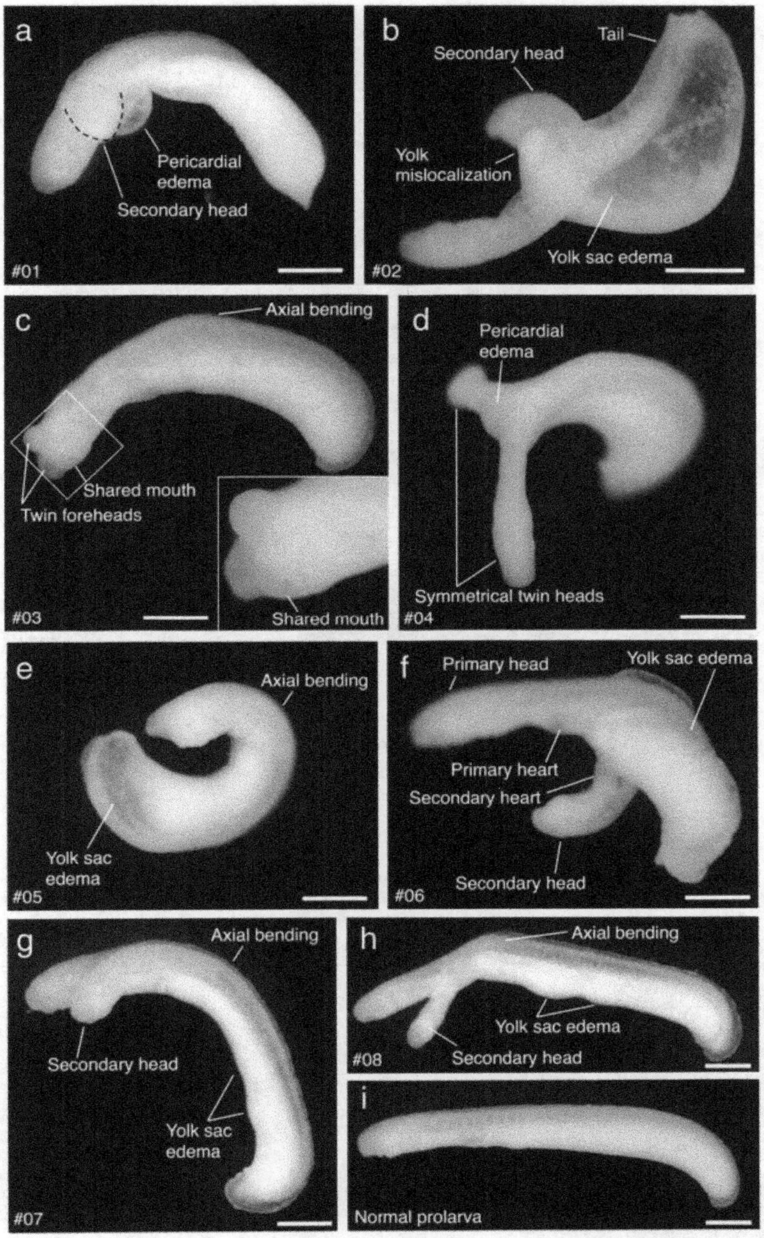

Fig. 2 Two-headed lamprey prolarvae. **a-h** Two-headed mutants after hatching. Identity numbers for each mutants are shown as #01–08. Mutants show various types of defects. For a summary of malformations, see Additional file 1. **i** A normal prolarva (st. 26). Scale bars: 500 μm

cranial nerve fibers were entangled (e.g., ophthalmicus profundus nerve [V_1] in Fig. 3a) or fused in the ventral midline (e.g., maxillomandibular [$V_{2,3}$] and facial [VII] nerves in Fig. 3b). This may be due to a restricted cell source due to the asymmetrical bifurcation, resulting in cranial malformation and axon misguidance. In addition, peripheral nerves in the primary axis were also disordered because of the bifurcation (e.g., vagus nerve [X] in Fig. 3b). Also, as shown in Fig. 3a, secondary pronephric ducts were observed immediately adjacent to the heart, consistent with normal heart development in the lamprey [17]. As the nephric system is derived from intermediate mesoderm, which arises between the somite and lateral plate mesoderm of the trunk [18], this observation supports the idea that the heart is formed in the rostral-most trunk region in the lamprey.

In the symmetrical two-headed twins, cranial nerves were entangled in the conjoined regions (e.g., putative facial nerve fibers [VII?] in Fig. 3c). In these regions, molecular cues for axon guidance may be provided from

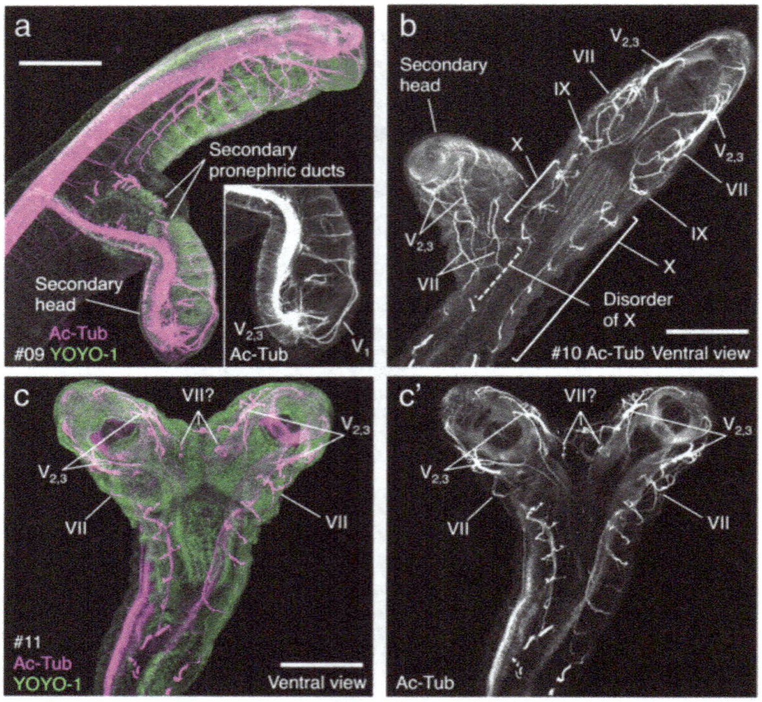

Fig. 3 Immunofluorescence analysis. Identity numbers for each mutants are shown as #09–11. **a**, **b** Lateral bifurcated mutants. **c** A symmetrically bifurcated mutant. Nerve fibers and pronephric ducts are visualized by anti-acetylated tubulin antibody in magenta and cell nuclei are labeled by YOYO-1 in green in (**a**) and (**c**). In (**b**), (**c**′) and the inset in (**a**), only the signal of anti-acetylated tubulin antibody is shown. Scale bars: 200 μm

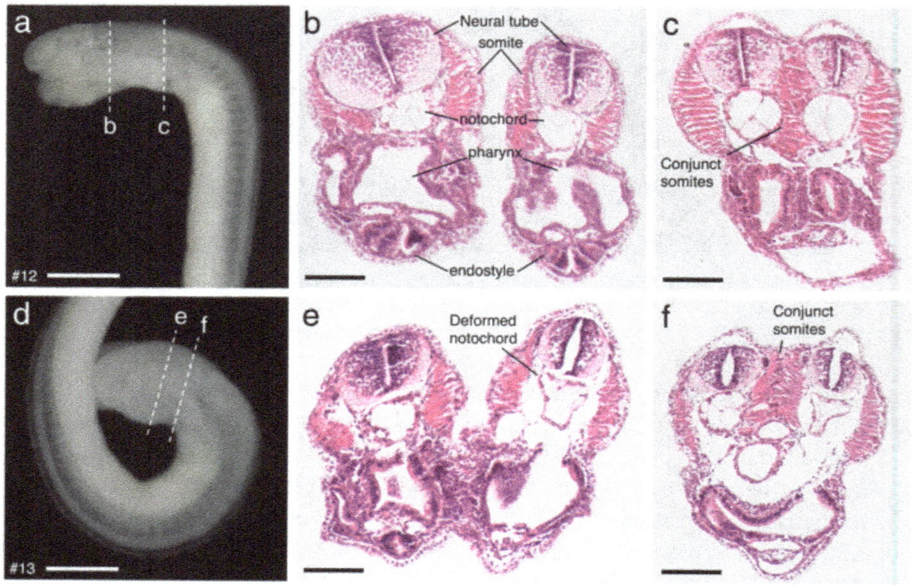

Fig. 4 Histological analysis for symmetrically bifurcated mutants by hematoxylin-eosin (HE) staining. Identity numbers for each mutants are shown as #12 and #13. **a**, **d** Lateral view of the mutants. Sectioned levels are shown. **b**, **e** Transversal sections of two-headed region. **c**, **f** Transversal sections of conjunct region. Scale bars: 500 μm for **a**, **d**, 100 μm for (**b**, **c**, **e**, **f**)

both sides of the bifurcated axes, resulting in the disruption of cranial innervation. I performed histological analysis to survey other morphological effects of the malformation (Fig. 4). In symmetrical two-headed twins, most organs are symmetrically present in both primary and secondary axes (see Fig. 4b), although some organs (e.g., the notochord in Fig. 4e) were morphologically deformed. In the bifurcating region, muscle fibers from primary and secondary axes meshed with each other in the same somite (Fig 4c, f), indicating that the bifurcation could occur *within* the somite (i.e., not *between* somites).

The mutants described in this study are similar to the experimentally induced conjoined lamprey twins, produced through dorsal blastopore lip transplantation, reported by Yamada [11]. The monumental work of Spemann and Mangold [19] demonstrated that transplantation of the dorsal blastopore lip in amphibians could induce conjoined twin embryos and this region acts as the head organizer. Although the developmental mechanisms for the head-organizer in lampreys are not known in detail, it is possible that the two-headed mutants described in this study were generated by some disruption of the head-organizing activity. In *Xenopus*, Siamois (Sia) and Twin (Twn) induce expression of head-organizer-specific genes, such as *Goosecoid* (*Gsc*), and double knockdown of Sia and Twn results in two-headed malformations [20]. However, Sia and Twn orthologs are absent in non-amphibian vertebrates, suggesting that there is species-specific diversity in the developmental mechanisms of head-organizer activity [20]. To investigate the degree to which these mechanisms are conserved or diversified in vertebrates, and to reveal the evolutionary origin of the vertebrate head, further comparative studies on the vertebrate head-organizer are needed.

Conclusions

I report two-headed mutants in a lamprey, obtained from artificial fertilization using single male and female gametes, suggestive of a genetic background. These mutants have similar malformations to dorsal blastopore lip-transplanted lamprey embryos, suggesting that they may be generated by a defect in head-organizing activity.

Abbreviations

V_1: Ophthalmicus profundus nerve; $V_{2,3}$: Maxillomandibular nerve; VII: Facial nerve; X: Vagus nerve; XI: Glossopharyngeal nerve

Acknowledgements

I thank Prof. Hiroshi Wada (University of Tsukuba) for his critical reading of the manuscript and constructive comments.

Funding

This study was supported by Japan Society for the Promotion of Science (JSPS); Grant number: 13J00621.

Competing interests

The author declares that he/she has no competing interests.

References

1. Appel TA. The Cuvier-Geoffroy debate: French biology in the decades before Darwin. New York/Oxford: Oxford Univ Press; 1987.
2. Geoffroy Saint-Hilaire I. Traité de tératologie. Paris: JB Baillière; 1836.
3. Bateson W. Materials of the study of variation: treated with special regard to discontinuity in the origin of species. London: Macmillan; 1894.
4. Alberch P. The logic of monsters: evidence for internal constraints in development and evolution. Geobios. 1989;12:21–57.
5. Guinard G. Introduction to evolutionary teratology, with an application to the forelimbs of Tyrannosauridae and Carnotaurinae (Dinosauria: Theropoda). Evol Biol. 2015;42:20–41.
6. Diogo R, Smith CM, Ziermann M. Evolutionary developmental pathology and anthropology: A new field linking development, comparative anatomy, human evolution, morphological variations and defects, and medicine. Dev Dyn. 2015;244:1357–74.
7. Abe G, Lee SH, Chang M, Liu SC, Tsai HY, Ota KG. The origin of the bifurcated axial skeletal system in the twin-tail goldfish. Nat Commun. 2014;5:3360.
8. Korschelt E. Regeneration und Transplantation. Berlin: Gebrüder Bornträger; 1927/1931.
9. Piavis GW. Embryological stages in the sea lamprey and effects of temperature on development. Fishery Bull. 1961;61:111–43.
10. Hanson LH. Observations on twinning in the sea lamprey (*Petromyzon marinus*) in the laboratory. J Great Lakes Res. 1985;11:549–51.
11. Yamada T. Induktion der sekundären Embryonalanlage im Neunaugenkeim. Okajimas Folia Anat Jpn. 1938;17:369–88.
12. Kuratani S, Ueki T, Aizawa S, Hirano S. Peripheral development of cranial nerves in a cyclostome, *Lampetra japonica*: morphological distribution of nerve branches and the vertebrate body plan. J Comp Neurol. 1997;384:483–500.
13. Hardisty MW. Gonadogenesis, sex differentiation and gametogenesis. In: Hardisty MW, Potter IC, editors. The biology of lampreys, vol. 1. London: Academic; 1971. p. 295–359.
14. Laale HW, Lerner W. Teratology and early fish development. Amer Zool. 1981;21:517–33.
15. Rothschild BM, Schultze HP, Pellegrini R. Herpetological osteopathology: Annotated bibliography of amphibians and reptiles. New York: Springer; 2012.
16. Kaufman MH. The embryology of conjoined twins. Childs Nerv Syst. 2004;20:508–25.
17. Higashiyama H, Hirasawa T, Oisi Y, Sugahara F, Hyodo S, Kanai Y, Kuratani S. On the vagal cardiac nerves, with special reference to the early evolution of the head–trunk interface. J Morphol. 2016;277:1146–58.
18. Ditrich H. The origin of vertebrates: A hypothesis based on kidney development. Zool J Linn Soc. 2007;150:435–41.
19. Spemann H, Mangold H. Induction of embryonic primordial by implantation of organizers from a different species. Roux's Arch Entw Mech. 1924;100: 599–638. reprinted and translated in Int J Dev Biol. 2001;45:13–38.
20. Bae S, Rei CD, Kessler DS. Siamois and Twin are redundant and essential in formation of the Spemann organizer. Dev Biol. 2011;352:367–81.

Permissions

All chapters in this book were first published in ZL, by BioMed Central; hereby published with permission under the Creative Commons Attribution License or equivalent. Every chapter published in this book has been scrutinized by our experts. Their significance has been extensively debated. The topics covered herein carry significant findings which will fuel the growth of the discipline. They may even be implemented as practical applications or may be referred to as a beginning point for another development.

The contributors of this book come from diverse backgrounds, making this book a truly international effort. This book will bring forth new frontiers with its revolutionizing research information and detailed analysis of the nascent developments around the world.

We would like to thank all the contributing authors for lending their expertise to make the book truly unique. They have played a crucial role in the development of this book. Without their invaluable contributions this book wouldn't have been possible. They have made vital efforts to compile up to date information on the varied aspects of this subject to make this book a valuable addition to the collection of many professionals and students.

This book was conceptualized with the vision of imparting up-to-date information and advanced data in this field. To ensure the same, a matchless editorial board was set up. Every individual on the board went through rigorous rounds of assessment to prove their worth. After which they invested a large part of their time researching and compiling the most relevant data for our readers.

The editorial board has been involved in producing this book since its inception. They have spent rigorous hours researching and exploring the diverse topics which have resulted in the successful publishing of this book. They have passed on their knowledge of decades through this book. To expedite this challenging task, the publisher supported the team at every step. A small team of assistant editors was also appointed to further simplify the editing procedure and attain best results for the readers.

Apart from the editorial board, the designing team has also invested a significant amount of their time in understanding the subject and creating the most relevant covers. They scrutinized every image to scout for the most suitable representation of the subject and create an appropriate cover for the book.

The publishing team has been an ardent support to the editorial, designing and production team. Their endless efforts to recruit the best for this project, has resulted in the accomplishment of this book. They are a veteran in the field of academics and their pool of knowledge is as vast as their experience in printing. Their expertise and guidance has proved useful at every step. Their uncompromising quality standards have made this book an exceptional effort. Their encouragement from time to time has been an inspiration for everyone.

The publisher and the editorial board hope that this book will prove to be a valuable piece of knowledge for researchers, students, practitioners and scholars across the globe.

List of Contributors

Ekin Tilic and Thomas Bartolomaeus
Institute of Evolutionary Biology and Ecology, Rheinische Friedrich Wilhelms Universität Bonn, An der Immenburg 1, 53121 Bonn, Germany

Thomas Schwaha and Andreas Wanninger
Department of Integrative Zoology, University of Vienna, Althanstraße 14, Vienna 1090, Austria

Masato Hirose
Atmosphere and Ocean Research Institute, University of Tokyo, 5-1-5 Kashiwanoha, Kashiwa-shi, Chiba 277-8564, Japan

Daniela Minelli and Gabriele Larocca Conte
Dipartimento di Scienze Biologiche, Geologiche e Ambientali, Alma Mater Studiorum, Università di Bologna, Via Zamboni 67, Bologna 40126, Italy

Federico Fanti
Dipartimento di Scienze Biologiche, Geologiche e Ambientali, Alma Mater Studiorum, Università di Bologna, Via Zamboni 67, Bologna 40126, Italy
Museo Geologico Giovanni Capellini, Alma Mater Studiorum, Università di Bologna, Via Zamboni 63, Bologna 40126, Italy

Tetsuto Miyashita
Department of Biological Sciences, University of Alberta, Edmonton, Alberta T6G 2E9, Canada

Nicole R. Rudolf, Carolin Haug and Joachim T. Haug
Ludwig-Maximilians-Universität München, Fakultät für Biologie, Biozentrum, Großhaderner Str. 2, 82152 Planegg-Martinsried, Germany

Tatsuya Hirasawa and Shigeru Kuratani
Evolutionary Morphology Laboratory, RIKEN, 2-2-3 Minatojima-minami, Chuo-ku, Kobe, Hyogo 650-0047, Japan

Yasuhiro Oisi
Development and Function of Inhibitory Neural Circuits, Max Planck Florida Institute for Neuroscience, Jupiter, FL 33458, USA

Michael Griesser
Department of Anthropology, University of Zurich, Winterthurerstrasse 190, 8057 Zurich, Switzerland
Department of Evolutionary Studies of Biosystems, SOKENDAI (The Graduate University for Advanced Studies), Kanagawa, Japan

Toshitaka N. Suzuki
Department of Evolutionary Studies of Biosystems, SOKENDAI (The Graduate University for Advanced Studies), Kanagawa, Japan

Wanlada Klangnurak
Integrated Bioscience Section, Graduate School of Science and Technology, National University Corporation Shizuoka University, Ohya 836, Suruga-ku, Shizuoka 422-8529, Japan

Toshinobu Tokumotot
Integrated Bioscience Section, Graduate School of Science and Technology, National University Corporation Shizuoka University, Ohya 836, Suruga-ku, Shizuoka 422-8529, Japan
Department of Biology, Faculty of Science, National University Corporation Shizuoka University, Shizuoka 422-8529, Japan

Nicolas Bekkouche and Katrine Worsaae
Marine Biological Section, Department of Biology, University of Copenhagen, Universitetsparken 4, 2100 Copenhagen Ø, Denmark

Hiroshi Kakiuchida
Structural Materials Research Institute, National Institute of Advanced Industrial Science and Technology (AIST), Moriyama, Nagoya, Aichi 463-6560, Japan

Daisuke Sakai
Department of Electrical and Electronic Engineering, Kitami Institute of Technology, Koen-cho, Kitami, Hokkaido 090-8507, Japan

Jun Nishikawa
Department of Marine Biology, School of Marine Science and Technology, Tokai University, Orido, Shimizu, Shizuoka 424-8610, Japan

Euichi Hirose
Faculty of Science, University of the Ryukyus, Nishihara, Okinawa 903-0213, Japan

Daichi G. Suzuki, Miho Yoshimura and Hiroshi Wada
Graduate School of Life and Environmental Sciences, University of Tsukuba, 1-1-1 Tennodai, Tsukuba, Ibaraki 305-8572, Japan

Shigeru Kuratani
Laboratory for Evolutionary Morphology, RIKEN, Kobe 650-0047, Japan

Yuma Fukumoto
Laboratory for Evolutionary Morphology, RIKEN, Kobe 650-0047, Japan
Department of Cytology and Histology, Okayama University Graduate School of Medicine, Dentistry and Pharmaceutical Sciences, 2-5-1 Shikata-cho, Okayama 700-8558, Japan
Sumitomo Besshi Hospital, 3-1 Oji-cho, Niihama, Ehime 792-8543, Japan

Jun Kosaka
Department of Cytology and Histology, Okayama University Graduate School of Medicine, Dentistry and Pharmaceutical Sciences, 2-5-1 Shikata-cho, Okayama 700-8558, Japan
Center for Medical Science, International University of Health and Welfare, 2600-1 Kitakanemaru, Ohtawara, Tochigi 324-8501, Japan

Yuji Yamazaki
Graduate School of Science and Engineering for Research, University of Toyama, 3190 Gofuku, Toyama 930-8555, Japan

Takeshi Takeuchi, Fuki Gyoja, Kanako Hisata and Noriyuki Satoh
Marine Genomics Unit, Okinawa Institute of Science and Technology Graduate University, Onna, Okinawa 904-0495, Japan

Takeshi Kawashima
Marine Genomics Unit, Okinawa Institute of Science and Technology Graduate University, Onna, Okinawa 904-0495, Japan
Graduate School of Life and Environmental Science, University of Tsukuba, Ibaraki 305-8572, Japan

Ryo Koyanagi, Miyuki Kanda, Manabu Fujie, Hiroki Goto and Shinichi Yamasaki
DNA Sequencing Section, Okinawa Institute of Science and Technology Graduate University, Onna, Okinawa 904-0495, Japan

Kiyohito Nagai
Pearl Research Institute, Mikimoto CO. LTD, Shima, Mie 517-0403, Japan

Yoshiaki Morino
Graduate School of Life and Environmental Science, University of Tsukuba, Ibaraki 305-8572, Japan

Hiroshi Miyamoto
Department of Genetic Engineering, Faculty of Biology-Oriented Science and Technology, Kinki University, 930 Nishimitani, Kinokawa, Wakayama 649-6493, Japan

Kazuyoshi Endo
Department of Earth and Planetary Science, Graduate School of Science, The University of Tokyo, Bunkyo-ku, Tokyo 113-0033, Japan

Hirotoshi Endo
Atmosphere and Ocean Research Institute, The University of Tokyo, Kashiwa, Chiba 277-8564, Japan

Hiromichi Nagasawa
Department of Applied Biological Chemistry, Graduate School of Agricultural and Life Sciences, The University of Tokyo, Bunkyo-ku, Tokyo 113-8657, Japan
College of Life Sciences, Zhejiang University, Hangzhou, Zhejiang 310058, People's Republic of China

Shigeharu Kinoshita and Shuichi Asakawa
Department of Aquatic Bioscience, Graduate School of Agricultural and Life Sciences, The University of Tokyo, Bunkyo-ku, Tokyo 113-8657, Japan

Shugo Watabe
Department of Aquatic Bioscience, Graduate School of Agricultural and Life Sciences, The University of Tokyo, Bunkyo-ku, Tokyo 113-8657, Japan
Kitasato University School of Marine Bioscience, Sagamihara, Kanagawa 252-0373, Japan

Molly Osborne and Paul L.R. Andrews
Division of Biomedical Sciences, St George's University of London, Cranmer Terrace, London SW17 0RE, UK

Tanya J. Shaw
Division of Biomedical Sciences, St George's University of London, Cranmer Terrace, London SW17 0RE, UK
King's College London-Centre for Molecular and Cellular Biology of Inflammation, New Hunt's House, Great Maze Pond, London SE1 1UL, UK

Giovanna Ponte and Graziano Fiorito
Stazione Zoologica Anton Dohrn, Villa Comunale, 80121 Naples, Italy

Denis Audo
Université de Rennes 1, EA 7316, 263 Avenue du Général Leclerc CS 74205, 35042 Rennes Cedex, France
Centre de Recherche sur la Paléobiodiversité et les Paléoenvironnements (CR2P, UMR 7207), Muséum national d'Histoire naturelle, Sorbonne Universités, UPMC, CNRS, 8 rue Buffon, F-75005 Paris, France

Sylvain Charbonnier
Centre de Recherche sur la Paléobiodiversité et les Paléoenvironnements (CR2P, UMR 7207), Muséum national d'Histoire naturelle, Sorbonne Universités, UPMC, CNRS, 8 rue Buffon, F-75005 Paris, France

Joachim T. Haug and Carolin Haug
Department Biologie II und GeoBio-Center, Ludwig-Maximilians-Universität München, Biozentrum der LMU, Großhaderner Str. 2, 82152 Planegg-Martinsried, Germany

Günter Schweigert
Staatliches Museum für Naturkunde Stuttgart, Rosenstein 1, 70191 Stuttgart, Germany

Carsten H. G. Müller
Zoologisches Institut und Museum, Lehrstuhl Allgemeine und Systematische Zoologie, Ernst-Moritz-Arndt-Universität Greifswald, Anklamer Str. 20, 17487 Greifswald, Germany

Steffen Harzsch
Zoologisches Institut und Museum, Lehrstuhl Cytologie und Evolutionsbiologie, Ernst-Moritz-Arndt Universität Greifswald, Soldmannstr. 23, 17489 Greifswald, Germany

Victoria Shone and Anthony Graham
Department of Developmental Neurobiology, Kings College London, London SE1 1UL, UK

Silvan Oulion, Didier Casane and Patrick Laurenti
Laboratoire Évolution, génomes, comportement, écologie, CNRS université Paris-Sud UMR 9191, IRD UMR 247, Avenue de la Terrasse, bâtiment 13, boîte postale 1, 91198 Gif-sur-Yvette, France

Daichi G. Suzuki
Graduate School of Life and Environmental Sciences, University of Tsukuba, 1-1-1 Tennodai, Tsukuba, Ibaraki 305-8572, Japan
Nobel Institute for Neurophysiology, Department of Neuroscience, Karolinska Institutet, SE-17177, Stockholm, Sweden

Index

CPSIA information can be obtained
at www.ICGtesting.com
Printed in the USA
BVHW061304240519
549249BV00003B/169/P

9 781641 160599